Methods of Testing Protein Functionality

Methods of Testing Protein Functionality

Edited by

G.M. HALL
Department of Chemical Engineering
Loughborough University of Technology

BLACKIE ACADEMIC & PROFESSIONAL
An Imprint of Chapman & Hall
London · Weinheim · New York · Tokyo · Melbourne · Madras

Published by
Blackie Academic & Professional, an imprint of Chapman & Hall,
2–6 Boundary Row, London SE1 8HN

Chapman & Hall, 2–6 Boundary Row, London SE1 8HN, UK

Chapman & Hall GmbH, Pappelallee 3, 69469 Weinheim, Germany

Chapman & Hall USA, Fourth Floor, 115 Fifth Avenue, New York NY 10003, USA

Chapman & Hall Japan, ITP-Japan, Kyowa Building, 3F, 2-2-1 Hirakawacho, Chiyoda-ku, Tokyo 102, Japan

DA Book (Aust.) Pty Ltd, 648 Whitehorse Road, Mitcham 3132, Victoria, Australia

Chapman & Hall India, R. Seshadri, 32 Second Main Road, CIT East, Madras 600 035, India

---

First edition 1996

Softcover reprint of the hardcover 1st edition 1996
Typeset in 10/12pt Times by Cambrian Typesetters, Frimley, Surrey

ISBN-13:978-1-4612-8517-5 e-ISBN-13:978-1-4613-1219-2
DOI:10.1007/978-1-4613-1219-2

A catalogue record for this book is available from the British Library
Library of Congress Catalog Card Number: 95–80000

∞ Printed on permanent acid-free text paper, manufactured in accordance with ANSI/NISO Z39.48-1992 (Permanence of Paper).

Preface

Protein functionality is big business in the food industry, both academic and commercial. This reflects the fascination of protein chemistry and the commercial value of proteins in giving texture and substance to foods. My first encounter with the subject came about through an interest in assessing the extent of thermal damage in spray-dried proteins. A change in functional properties seemed a useful guide to the thermal damage. I found the literature replete with methods (and theory) and so did what everybody else seemed to do – I devised my own methods which were slight variations on the general theme. However, being an analyst at heart I thought there must be a systematic approach to the various tests available.

This book is an attempt to assess the practice of protein functionality testing, keeping theory to a minimum, followed by a suggested 'standard' method or protocol to follow. Each chapter is devoted to a single functional property and the reader will see that attempts to 'standardise' them have had varied success, perhaps reflecting the complexity of the mechanisms at work. As mentioned above theory has been kept to a minimum but hopefully it supports the practical methodology described. I hope that the book stimulates debate on the subject; it is certainly not the last word!

G.H.

Contributors

Dr G.E. Arteaga	Centro de Investigacion en Alimentation y Desarrollo, Hermosillo, Sonora AP 1735, Mexico
Dr. S. Barbut	Department of Animal and Poultry Science, University of Guelph, Ontario N1G 2W1, Canada
Dr D.C. Clark	DMV International, NCB-Iaan 80, PO Box 13, 5460 BA Veghel, The Netherlands
Dr G.M. Hall	Food Engineering and Biotechnology Group, Department of Chemical Engineering, Loughborough University of Technology, Loughborough, Leicestershire LE11 3TU, UK
Dr S.E. Hill	Department of Applied Biochemistry and Food Science, University of Nottingham School of Agriculture and Food Science, Sutton Bonington Campus, Leicestershire LE12 5RD, UK
Professor E. Li-Chan	Department of Food Science, University of British Columbia, 6650 N W Marine Drive, Vancouver, BC V6T 1Z4, Canada
Dr Y. Matsumura	Research Institute for Food Science, Kyoto University, Uji, Kyoto 611, Japan
T. Mori	Research Institute for Food Science, Kyoto University, Uji, Kyoto 611, Japan
Professor C.V. Morr	Department of Food Science, Ohio State University, Columbus, Ohio 43210, USA
Professor S. Nakai	Department of Food Science, University of British Columbia, 6650 N W Marine Drive, Vancouver, BC V6T 1Z4, Canada
T.W. Schenz	Ross Products Division, Abbot Laboratories, 625 Cleveland Ave, Columbus, Ohio 43215, USA
Dr F. Vojdani	Department of Food Science and Technology, University of California, Davis, CA 95616–8515, USA

Dr P.J. Wilde Food Biophysics Department, Institute of Food Research, Norwich Laboratory, Norwich Research Park, Norwich NR4 7UA, UK

Contents

1 Basic concepts

G.M. HALL

1.1 Introduction

The subject of this book is protein functionality and the methods which can be used to measure the range of properties covered by the term. However, the area of functionality (or functional properties) of proteins in food systems has, it seems, always been one of confusion and inconsistency. Pour-El (1981) defined functionality as "any property of a food or food ingredient, except its nutritional ones, that influences its utilisation." This definition has, for many years, served very well as a general description of the subject but today it may require modification. The reason for this is that foods claiming nutritional or health benefits are being described as 'functional' foods. Other terms used are 'probiotics', 'pharmafoods' or 'nutraceuticals'. This whole area is itself controversial, with claims and counter-claims on behalf of these products, so that to use the term 'functional' seems to pile confusion on confusion.

As far as this book is concerned we will limit ourselves to the old definition, which includes the performance of proteins in food systems. We will also omit those 'functional' properties which are in the domain of the senses (taste and flavour) or of biochemistry, such as enzyme activity. Both these areas, the senses and biochemistry, present problems of their own in determination and interpretation of results. If this is accepted then confusion still occurs because of the multitude of methods employed in the measurement of each of the functional properties. This situation exists because protein functionality is often assessed for a particular purpose in a specific food system and the tests are designed to determine this purpose. Moreover, the test (or assay) may involve a model system or the preparation of the food product itself and assessment of its quality. Whilst these approaches are practical and suited to individual needs they do not represent a basis of comparison of different proteins or development of standard methods for different proteins. The object of the succeeding chapters of this book is to discuss the range of methods employed to measure specific protein functional properties and to suggest those which might form the basis of a standard method. It is the reader's job to decide if this approach is realistic.

The rest of this chapter will indicate some general principles which should be borne in mind when measuring any of the properties.

1.2 Sources of functional proteins

The importance of functional proteins to the food industry can be gauged by the number of specialist ingredients traded and transported either as food protein groups or as individual proteins. For example, the range of proteins found in milk can be bought as whole dried milk or they can be fractionated and sold as casein(s) or whey protein(s). Egg white and wheat gluten can be produced as individual proteins also, indicating the degree of sophistication in the functional protein supply industry. Table 1.1 gives a list of the main functional proteins, their sources and applications.

The search for new functional proteins is increasing so, for example, the range of proteins found in various legumes and seeds in Africa has received some attention (Oshodi and Ekperigin, 1989). If economically viable products can be derived from these crops, their value as cash crops will increase. This will act as a stimulus to research into their agronomy and will generate welcome money for their producers.

The range of functional properties of a given protein is usually required to be very wide but few proteins achieve this breadth of properties. Milk proteins, being a mixture, might be considered the nearest to a 'broad-range', functional product (Giese, 1994). Proteins with limited functionality can find specific applications which require this property and hence find commercial usage, for example gelatin. Certain proteins are limited in application because of associated flavours; thus, whilst milk and egg proteins are used in a wide variety of foods, proteins derived from fish should be restricted to fish-based products.

1.3 Factors affecting protein functional properties

These factors have been divided into three areas of influence: those which are intrinsic to the individual protein, processing treatments and, finally, factors affecting the application of the protein (Kinsella, 1982).

1.3.1 Intrinsic factors

The conformation of proteins in nature is divided into four structural levels which reflect their *biological function*, not necessarily their food functional use. However, the physical and chemical properties of the proteins in their food applications may reflect their biological nature, so an understanding of the latter may indicate possibilities for the former.

The first (or primary) structural level of proteins is the amino acid sequence of the linear chain. Amino acids can be classified into one of four groups (non-polar; neutral–polar; positively charged and negatively charged) depending on the nature of the side chains. The proportion of

Table 1.1 Major functional proteins, their sources and applications

Source	Protein	Applications
Cereals	Wheat gluten, maize zein	Bakery, breakfast foods; whipping agents
Egg	Egg white (albumin), egg yolk lipoprotein, whole egg	Wide application as emulsifying, binding and foaming agents, gelation
Fish	Muscle, collagen (gelatin)	Gelling; surimi products
Meat	Muscle, collagen (gelatin), porcine/ bovine blood	Gelling, emulsification, water holding
Milk	Whole/skim milk powder, caseins, whey proteins	Wide application as emulsifying agents; adhesion, viscosity (smoothness)
Oilseed	Soybean, groundnut, sesame as flours, concentrates, isolates	Bakery, soy milk, meat (texture)

each class of amino acid and their distribution along the chain will determine properties such as solubility, surface hydrophobicity and the ability to stabilise foams and emulsions. Proteins with extreme amino acid compositions (such as collagen with a large proportion of glycine and hydroxyproline) might have limited functional properties, as mentioned earlier, but still find a niche usage. Amino acid composition is easily determined, the sequence rather less so, but the overall amino acid composition does not uniquely identify possible functional use.

Despite this fact, attempts have been made to relate amino acid composition to functionality, particularly through hydrophobicity. Bigelow (1967) assigned average hydrophobicity values to a range of proteins based on amino acid composition and the hydrophobicity of each amino acid. Hydrophobic interactions are important in the tertiary level of protein structure (see below) and as hydrophobic amino acids make up 30–50% of total amino acids, typically, variations may have a profound effect on structure and hence functionality. This approach has some use but falls down over issues such as values assigned to some amino acids, and the use of average, total or surface hydrophobicity (see Chapter 8) as the true reflection of protein behaviour.

The long linear chain of amino acids can in theory take up a number of folded secondary structures. The favoured (or native) structure is determined by rotation of the amino acids about the single-bonded backbone of the protein until the side chains reach a thermodynamically stable distribution. The size and nature of the side chains will determine whether they are exposed or hidden from the environment, which is usually aqueous in nature. Polar groups will be exposed and non-polar groups hidden in the internal structure of the protein.

Tertiary structure is determined by a three-dimensional twisting of the linear chain already organised into primary and secondary structures. This tertiary structure is stabilised by hydrogen bonds, hydrophobic interaction, electrostatic interactions and even covalent bonding through sulphur–sulphur bridges. Quaternary structure describes association of two or more protein chains into an oligomeric unit stabilised by much the same forces as the tertiary structure.

These complex levels of structure give proteins in nature a great variety of forms and functions, and retention or loss of this structure will determine food functional applications.

1.3.2 Processing treatments

Information on the native state of the functional protein is important but the processing of the raw materials to produce the commercial product may alter this state. Some process operations may be unavoidable in extracting the protein in sufficiently pure state but the effects of the processing must be recognised. Examples of the processes applied include changes in pH and temperature to precipitate proteins; solvent extraction to remove lipids; addition of acids/alkalis/salts; enzyme treatments; drying and size reduction to produce a powdered form which is easy to handle and transport. For example, soybeans can be processed into a wide variety of products depending on the level of lipid present, purity of proteins and physical size (see Figure 1.1) and each of these products has a different application depending on their residual functionality after the treatments.

Some treatments may be deliberately applied to change the nature of the protein, such as chemical and enzyme modification of solubility and molecular weight. For example, fish proteins have been succinylated to improve their solubility and hence other functional properties (Groninger and Miller, 1975). Enzyme treatment, particularly hydrolysis, has been used to modify proteins with little functionality. The breakdown of large molecular weight proteins into polypeptides has increased solubility and again this affects other properties. The enzyme treatment must be carefully controlled to give a product with the required molecular weight and a number of variables, such as pH, temperature, enzyme type and residence time have been identified (Adler-Nissen, 1986). Enzyme hydrolysates can be fractionated into useful molecular weight ranges by ultrafiltration (Vega and Brennan, 1987).

1.3.3 Application conditions

As mentioned earlier, the vast number of functional property measurement systems in the literature has given rise to a lack of conformity and standardisation. This is particularly due to the variety of conditions under

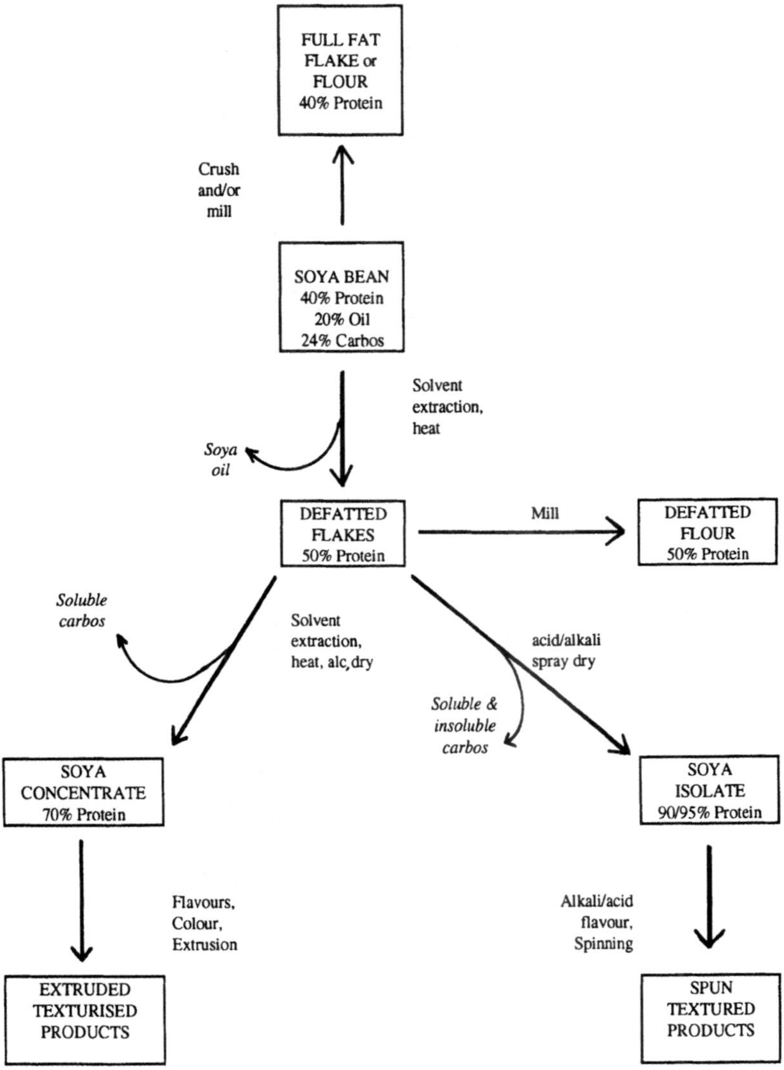

Figure 1.1 Processing treatments giving a range of functional products from the soybean.

which the measurements take place. The ionic conditions, pH, protein concentration, temperature and mixing methods (and time) are often varied considerably. Attempts to repeat a reported method using similar equipment are often doomed to failure. One objective of this book is to assess the methods used and the variables which are pertinent, and to indicate those features which might be built into a standard method. Table 1.2 shows the variables commonly associated with functionality testing,

Table 1.2 Important application variables for functional protein assessment

Functional property	pH	Temperature[a]	Time	Mixing method[b]	Centrifugation speed (g)	Other variables
Solubility	+	+	+	+	+	Centrifugation and mixing times varied
Emulsion capacity and stability	+	–	+	+	–	Stability measured over a wide variety of times
Water/oil adsorption or binding	+	–	+	+	+	Centrifugation and mixing times varied; type of oil for oil adsorption
Viscosity	–	–	+	–	–	Several viscometers available
Gelation	+	+	+	–	–	Time of heating varied, gel strength measured by penetrometers or inversion methods; additives often used

[a]Room temperature used if no other is specified.
[b]Mixing of the functional protein with water is always important, but here refers to the means of mixing the phases, e.g. addition of oil in emulsification or introduction of air/gas for whippability/foaming.

although it is possible to find papers which do not justify the combinations used or even indicate that they have been considered. For example, pH is such an important influence on protein behaviour that it should *always* be indicated in the method description. There is some debate over whether protein functionality should be tested in model systems or those which mimic food production processes (sometimes called utility tests). The latter have the advantage of determining the potential use of the protein in the food application, but the methods are the least standardised and give no indication of the properties of the protein which are significant. Model systems offer the possibility of standardisation and correlation to known properties of the protein. Furthermore, model systems performed quickly and in replicate can weed out those least suitable proteins and reduce the number of utility tests performed.

1.3.3.1 Ionic conditions. The most important consideration here is usually pH. Operation at or near the isoelectric pH will emphasise the low solubility of the protein at this point and have a profound effect on other properties through low solubility. Measurement at other pH values might be done to establish a range of activity, although it can be argued that at high solubility the activity of the test proteins might be difficult to resolve. The pH values selected could be limited to those which would be found in the proposal utility system. Other ionic species might affect specific functional properties, such as sodium, potassium and calcium ions in protein gel formation. The presence of these ions might be deliberate or adventitious (a further justification for thorough analysis of the test material). Given the importance of these effects, a statement on the pH and ionic strength/composition of the test system is essential for all procedures.

1.3.3.2 Temperature. Again, this is an important parameter affecting protein structure and behaviour which needs rigorous control. Many functionality tests are done at room temperature (or rather, are claimed to be done at room temperature) but mixing and dissolution may generate heat which should be acknowledged as a possible source of error. Careful control of temperature is accepted when measuring viscosity and gelation but should be extended into all test systems.

1.3.3.3 Mixing considerations. This area of testing process appears to be one of the most difficult to master and standardise. For the functional properties described in this book, dissolution of the protein powder in water (as the usual solvent) is an essential first step. However, as it is seen as a precursor to the actual test, standardisation of the mixing process is almost unknown. The conditions of mixing are of obvious importance in solubility testing as this fundamental property is measured directly after mixing with water. Morr *et al.* (1985) described a collaborative trial aimed at developing a standard food protein solubility procedure. The mixing procedure involved forming a paste between the powdered protein and 0.1 M sodium chloride solution in a beaker, adding more salt solution to 40 ml and stirring this dispersion for 1 h using a magnetic stirrer bar (2.5 cm smooth, plastic-coated) at a rate *that just failed to form a vortex.* The pH was adjusted initially to the desired value and checked and corrected throughout the stirring period. An open-meshed pad under the beaker was used to prevent heating of the solution during stirring.

This precise methodology was developed during the course of the collaboration because the biggest source of error between laboratories was due to slight differences in analytical practice. If this problem occurs in a tightly controlled collaborative trial the variation between laboratories repeating published methods is not surprising and neither is their decision to develop their own version of protein dissolution.

Emulsification and foaming properties require mixing of an aqueous solution of the protein with a non-aqueous (oil) phase or a gaseous (air) phase, respectively. The mixing methods for these tests are many and varied (see the relevant chapters), again reflecting the diverse applications being tested. Water and fat holding properties are tested by contacting the appropriate medium with the protein, and the means of contact will affect the ability of the protein to retain water or fat under stress conditions.

1.3.3.4 Time considerations. This factor is important in two areas of functionality testing. Firstly, during mixing procedures (as above) and secondly when assessing stability for foams and emulsions. The duration of mixing is important where the test protein may be exposed to strong shear forces and when heat may be generated. A minimum exposure, sufficient to achieve solution, should be aimed for or a longer exposure at low shear (see Morr *et al.* (1985) in section 1.3.3.3). Stability testing should be assessed over a range of times consistent with the protein used in order to get a good comparison. Proteins giving very stable systems might need assessment using stress conditions (heat, pressure or centrifugal forces).

1.3.3.5 Centrifugation. Centrifugation is a widely used means of stressing model functional protein systems in a controlled (time, *g* force) and variable manner. It is useful where a solid–liquid separation indicates the extent of functional behaviour. However, this flexibility of use is open to abuse if insufficient thought is given to the time–force combination used, which should be balanced to give discrimination between proteins. A band of values omitting extremes of performance should be found for the test used, to prevent assessment of unusable proteins.

1.3.3.6 Protein concentration. One of the attractions of functional proteins is their ability to have an impact on a food system even when added in quite small amounts. Given their high price, it is also commercially important to add them at the minimum effective amount. Testing a wide range of concentrations is an unattractive proposition, hence some knowledge (guess?) of the appropriate range must be made usually based on past experience. Certain tests, such as viscosity, depend for reasonable results on a range of concentrations suitable for the measuring system used. Actual protein concentrations may not reflect the amount of material used, depending on other components, so the compositional analysis of the product must be consulted.

1.4 Conclusions

The scope of this book is limited to those functional properties most prized by the food manufacturer for processing, storage, preparation and formulation of products, old and new. The term functionality is expanding to include organoleptic properties (colour, flavour and odour) the nutritionally 'functional' foods and, even, antioxidant behaviour. The properties of emulsification, foaming, gelation, etc., which are so valued, must retain their position and one way to achieve this would be if their determination and assessment could be formalised and given wide applicability.

As suggested earlier, certain information concerning the food protein must be known. This includes a full composition analysis and processing history. However, knowledge of the nature of the protein being tested is of limited use if the tests applied cannot be related to protein structure, hydrophobicity, etc. Until this connection can be made discriminatory tests in model systems, comparing one protein with another, might be an intermediate position between the multitude of utility tests and the sparse fundamentally related tests. Model systems themselves are not perfect and devising a test which is *a fair test for a range of proteins* is not easy. Many authors have called for such tests but the tests are conspicuous by their absence. Similarly, the use of a reference protein, against which others can be assessed, has been proposed. This would require a standard which has broad functionality, and only egg white and milk proteins could fit the bill to any extent. Given a standard protein the next problem would be development of a *standard protocol* which could be followed without ambiguity by any laboratory with the appropriate equipment. From the earlier description of the variations in method application (section 1.3.3) this goal is not easily attained. However, the problems are not greatly different to those faced by analytical chemists in analyte extraction and detection, and these chemists seem to have overcome the difficulties.

Above all, model systems must be discriminatory such that the conditions applied do not result in extremes of performance by a variety of proteins known to exhibit the property assessed. If, with appropriate conditions, extremes are seen they should be used immediately to exclude such proteins from further testing. Such then is the background to the succeeding chapters of this book, each of which deals with a specific property. Theory is kept to a minimum but the present situation with regard to test methods is described and from this position protocols are suggested for a widely applicable method. As the reader will find, this objective has been rather easier to propose in some cases than in others. If the goal of universal methods is not attainable then, at the very least, some of the practical pitfalls and solutions will have been aired.

References

Adler-Nissen, J. (1986) *Enzymic Hydrolysis of Foods*, Elsevier Applied Science, London.

Bigelow, C.C. (1967) On the average hydrophobicity of proteins and the relation between it and protein structure. *J. Theoret. Biol.*, **16**, 187–211.

Giese, J. (1994) Proteins as Ingredients: Type, Functions, Applications. *Food Technol.*, **48**, 50–60.

Groninger, H.S. and Miller, R. (1975) Preparation and aeration properties of an enzyme modified succinylated fish protein. *J. Food Sci.*, **40**, 427–439.

Kinsella, J.E. (1982) Relationships between structure and functional properties of Food Proteins. In *Food Proteins*, Fox, P.F. and London, J.J. (Eds), Elsevier Applied Science, London, pp. 51–103.

Morr, C.V., German, B., Kinsella, J.E., Regenstein, J.M., Van Buren, J.P., Kilara, A., Lewis, B.A. and Mangino, M.E. (1985) A collaborative study to develop a standardised food protein solubility procedure. *J. Food Sci.*, **50**, 1715–1718.

Oshodi, A.A. and Ekperigin, M.M. (1989) Functional properties of pigeon pea (*Cajanus cajan*) flour. *Food Chem.*, **74**, 187–191.

Pour-El, A. (1981) Protein Functionality: Classification, Definition and Methodology. In *Protein Functionality in Foods*, Cherry, J.P. (Ed.), American Chemical Society Symposium Series 147, Washington D.C., pp. 1–20.

Vega, R.E. and Brennan, J.G. (1987) Fractionation of Fish Protein by Ultrafiltration. In *Separations for Biotechnology*, Verral, M.S. and Hudson, M.J. (Eds), Ellis Harwood, Chichester, pp. 373–382.

2 Solubility

F. VOJDANI

2.1 Introduction

Proteins, as macromolecules, perform important roles in functionality in foods and pharmaceuticals, as well as in biological systems. Therefore, the growing demand for proteins as important ingredients in formulated food or in pharmaceutical and industrial mixtures has created a necessity for proteins with specific and consistent functional properties. Proteins exhibit many functional properties governed by their physicochemical activities in a bulk liquid phase.

Among the functional properties of proteins, solubility is of primary importance due to its significant influence on the other functional properties of proteins (Halling, 1981). Many researchers use the protein solubility as a second category of test under the thermodynamic index of protein structure (Kilara and Sharkasi, 1986). In general, proteins used for functionality are required to have high solubility, in order to provide good emulsion, foam, gelation and whipping properties.

Many researchers have tried to increase protein solubility by different techniques for improved functional properties. Some protein concentrates are either naturally insoluble or rendered insoluble by processing treatments which denature proteins. Other protein concentrates tend to be insoluble at acidic pH values near the isoelectric point (pI), and require higher pH conditions for their solubilization. Supplementation of an acidic soft drink or citrus juice with protein, for example, requires solubilization of the protein at acidic pH values. Such applications require solubility testing as a primary test. However, a lack of standard protein solubility tests, as well as other protein functionalities, makes it difficult to correlate the results from different studies (Morr and Ha, 1993). Collaborative research was conducted by Morr *et al.* (1985) to develop a standardized solubility procedure. The results obtained between laboratories revealed major discrepancies. However, the precision of the results within a laboratory was quite high. Many laboratories use their own developed solubility test, based on the specific information they need to obtain. The purpose of this chapter is to recommend reproducible solubility testing methodology, giving simple, rapid, reliable, reproducible and accurate results.

Proteins, as for other macromolecules, have two important properties that are responsible for various functional properties. These are solubility and hydrodynamic properties. Solubility is the result of the surface active properties of proteins, as well as foaming, emulsification, fat and flavour binding properties. Hydrodynamic properties of proteins influence viscosity and gelation. The surface active properties of proteins are the result of amino acid composition, amino acid distribution, molecular flexibility and, to a lesser extent, the shape and size. In contrast, the hydrodynamic properties are affected largely by the shape and size of the protein and are relatively independent of amino acid composition and distribution (Morr and Ha, 1993). Solubility of proteins relates to surface hydrophobic (protein–protein) and hydrophilic (protein–solvent) interactions with water. The relative proportion of surface hydrophobic and hydrophilic groups dictates the degree of solvation by water. Therefore, the solubility of a protein under a given set of solution conditions at equilibrium between the hydrophilic and hydrophobic interactions can be expressed as (Damodaran, 1989):

$$\text{Protein–Solvent} \leftrightarrows \text{Protein–Protein} + \text{Solvent–Solvent}$$

Solubility of proteins can be improved by increasing the number of hydrophilic groups present. This could be accomplished by deamidation, or by covalent attachment of hydrophilic amino acids residues, glyco groups or phosphate groups (Chobert *et al.*, 1987).

Protein conformation is dictated by the environment for solvation. In solution, amino acid side chains are involved in different interactions, such as van der Waals forces (or London dispersion forces), hydrogen bonding, and electrostatic and hydrophobic interactions. In general, to minimize thermodynamically the free energy of stabilization, the hydrophilic amino acid residues are oriented to the outside of the molecule and the majority of hydrophobic amino acid residues are buried in the interior of the molecule. In some native proteins, while almost all the hydrophilic residues are located at the surface, not all the hydrophobic residues are buried. This, creates hydrophobic patches on the surface of the protein; about 40–50% (34–42% by side chain and 5–7% by main chain contributions) of the surfaces of myoglobin, lysozyme and ribonuclease S are non-polar patches (Lee and Richards, 1971). Thus, the proportion and distribution of surface hydrophilic and hydrophobic patches are the main factors in determining the degree of solubility of protein, rather than total hydrophobicity and charge density based on amino acid composition.

The surface characteristics of a protein are greatly affected by environmental conditions around the protein, which influence the inter-relations among physical forces. Such environmental factors are pH, ionic strength, temperature, solvent component and other food components. Physical, chemical and thermal treatments during processing, method of isolation, interaction with other food components, and mechanical treatments prior to solubility testing are other parameters which influence the degree of protein solubility.

Several terms used to express protein solubility of food protein are summarized in Table 2.1. Some of the terms are used interchangeably.

2.1.1 Solubility definition and theory

Protein solubility is, thermodynamically, the protein concentration in the solvent in a single- or two-phase system (protein solution in liquid–liquid phases or in liquid–solid phases) at the equilibrium state. It is a function of many factors, such as the amino acid and non-amino acid composition of proteins, native or denatured state and environmental factors (i.e. pH, temperature, pressure, nature and concentration of salts, and concentration of protein). In general, proteins go into solution until a maximum concentration is reached, after which the soluble concentration remains constant and a solid phase appears; as further protein is added the two phases remain at equilibrium under a given experimental condition. Degree of solubility of a protein is mathematically the amount of protein present in liquid phase in relation to the total amount of protein in liquid and solid phase at equilibrium. Technically, however, protein solubility is an operational parameter determined by retention of protein in the supernatant after centrifugation of the solution for a given time and centrifugal force.

In the early 1900s, initial recognition of some of the remarkable behavioral features of protein solutions led to the idea that these were not true solutions at all. However, it is now agreed that proteins, and higher

Table 2.1 Common terms used to express protein solubility values

Terms	Abbreviations[a]
Protein dispersibility index	PDI
Water dispersible protein	WDP
Degree of turbidity	–
Percent insolubility	–
Dispersibility	–
Maximum protein solubility	–
Nitrogen solubility index	NSI
Solubility index	SI
Water soluble nitrogen	WSN
Water soluble protein	WSP
Percent protein solubility	%PS
Percent nitrogen solubility	%NS

[a]Dash indicates no abbreviation suggested.

polymers in general, form true solutions with small freezing-point depression and slow diffusion (Van Holde, 1985).

Since proteins are highly complex polyelectrolytes with both positive and negative charges, the molecule in aqueous solution is surrounded by an ionic layer of counter ions (the Helmholtz double layer). This second layer of counter ions decreases the electrostatic free energy of the protein, resulting in a decrease in its activity and an increase in its solubility, provided that the chemical potential of the protein in the solid phase remain constant (Arakawa and Timasheff, 1985). For compact proteins, based on the Debye–Hückel electrostatic theory for small ionic molecules, the solubility is described by:

$$\ln (S/S_0) = Z^2\varepsilon^2 N\varkappa/[2DRT(1 + \varkappa a)] \tag{2.1}$$

where S_0 is the solubility of a substance at zero electrolyte concentration, S is the solubility of a substance in a solution of given electrolyte concentration, Z is the net charge of the protein, $\varkappa$ is the electronic charge, N is Avogadro's number, D is the dielectric constant of the medium, R is the universal gas constant, T is Kelvin temperature, a is the sum of the radius of the protein and the average thickness of the supporting electrolyte ionic layer in the solution, and $\varkappa$ is given by:

$$\varkappa = (8\pi N\varepsilon^2/1000\ DkT)^{1/2}(\mu)^{1/2} \tag{2.2}$$

where k is the Boltzmann constant and μ is ionic strength. According to equation (2.1), protein solubility increased with the square root of the ionic strength. At the isoelectric point ($Z = 0$), the protein solubility should be increased by addition of salt (Z^2 is not zero). As shown in Figure 2.1, at low salt concentration all salts have a salting-in effect on proteins. The logarithm of the protein solubility is proportional to $(\mu)^{1/2}$, in the limit as $\mu \rightarrow 0$. Equation (2.1) also suggests that protein solubility should be at a minimum at the isoelectric point at low ionic strength and this is usually observed (Arakawa and Timasheff, 1985).

Again, from the Debye–Hückel limiting theory, the activity coefficient (γ) for a given dipolar ion in dilute salt solution is related to ionic strength by the simpler expression of Kilara and Sharkasi (1986):

$$-\log \gamma = \log (S/S_0) = K\mu \tag{2.3}$$

where K, a constant, is a function of the dielectric constant of the medium and, the dipole moment of the dipolar ion, and γ is the activity coefficient of the substance in the solution (Kirkwood, 1943). From equation (2.3), increasing the ionic concentration stabilizes the charges on the protein thereby increasing its solubility.

Therefore, assuming a solution independent of activity of the solid state, and treating a protein as a dipolar ion:

$$\log (S/S_0) = (K/RT)\mu \tag{2.4}$$

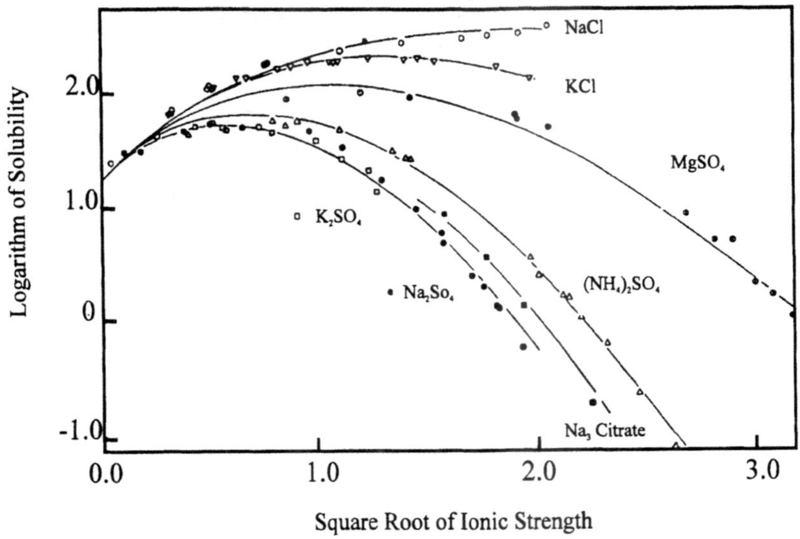

Figure 2.1 The solubility of carboxyhemoglobin in various electrolytes at 25°C. Reproduced with permission from Green, A.A. (1932), Studies in the physical chemistry of the proteins. *J. Biol. Chem.*, **95**, 47–66.

In this case, the logarithm of the protein solubility is proportional to the first power of μ, rather than its square root. Figure 2.2 shows that the solubility of β-lactoglobulin increases in this manner when measured near its isoelectric point at low NaCl concentration. However, at high salt concentrations, the solubility behavior of protein decreases; this is discussed in the next section.

2.1.2 Factors that influence protein solubility

Protein solubility is affected by many environmental factors in addition to the intrinsic properties of the protein itself. These factors are discussed in detail in the following sections.

2.1.2.1 pH. It is well known that the solubility of a protein depends on the pH of the solution. A protein in an aqueous system has a zero net charge at its isoelectric point (pI) and no migration occurs if a DC electric field is applied. At the pI, a protein usually has the least solubility. Protein–protein interaction increases because the electrostatic forces of the molecules are at a minimum and less water interacts with the protein molecules. This is a favorable condition for protein molecules to approach each other and aggregate, and possibly precipitate.

The ionic properties of proteins affect the solubility properties. Many proteins are completely insoluble at their pI, while some proteins are still

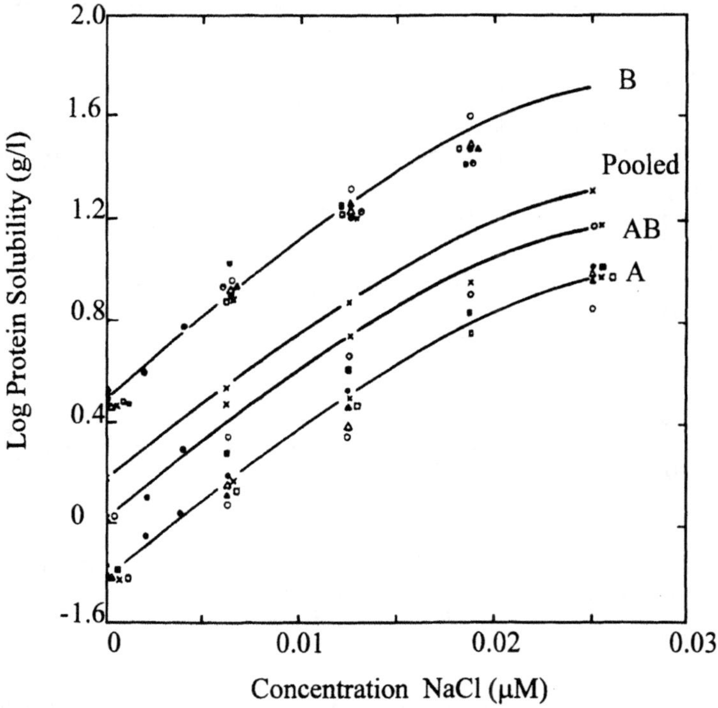

Figure 2.2 Solubilities of β-lactoglobulins as a function of ionic strength (μ) at pH 5.2. Reproduced with permission from Treece, J.M., Sheinson, R.S. and McMeekin, T.L. (1964), The solubilities of β-lactoglobulins A, B, and AB. *Arch. Biochem. Biophys.*, **108**, 99–108.

soluble. For example β-lactoglobulin is almost 90% soluble at its pI in aqueous solution (1 mg/ml in water, pH 5.1). Casein and α-lactalbumin are 2.4% (pH 4.6) and 68% (pH 5.0) soluble respectively at a concentration of 1 mg/ml.

At pH values above and below the pI, where a protein has a net negative or positive charge, more water interacts with the protein charges. Net charges and charge repulsion contribute to greater protein solubility and the protein may stay in the solution. At extreme acidic or basic pH values, the protein may unfold, exposing more hydrophobic groups. Unfolding (denaturation) at low or high pH values occurs because of a decrease in electrostatic bonds. Circular dichroism (CD) has shown that β-lactoglobulin is more unfolded at higher pH than at the pI. For example β-lactoglobulin had 35% random structure at pH 5, compared to 47% at pH 10 (Vojdani, 1994). Sometimes, at high or low pH values, due to a decrease in electrostatic bonds, the protein undergoes denaturation followed by aggregation and precipitation. Some proteins that are unfolded at extreme pH values, under conditions of low ionic strength, may refold and stabilize when the concentration of salts is increased (Hamaguchi, 1992).

The rate of precipitation is enhanced when the bulk densities of the aggregate differ greatly from that of the solvent, and when the sizes of the newly formed aggregates are large (Cheftel *et al.*, 1985).

Usually the pH–protein solubility curve is U-shaped for pure proteins. The minimum solubility occurs at the pI of the protein (Figure 2.3). However, a different pH–protein solubility pattern may be observed for a mixture of proteins, such as whey protein concentrate, caseinate, or in food systems (Figure 2.4). Precipitation of a protein does not necessarily occur at the pI because of protein concentration and the solubility product constant. Therefore, some other perturbation of the solvent properties might be necessary to cause protein precipitation. These perturbations could be addition of a salt such as NaCl, or a water miscible organic solvent such as ethanol or acetone.

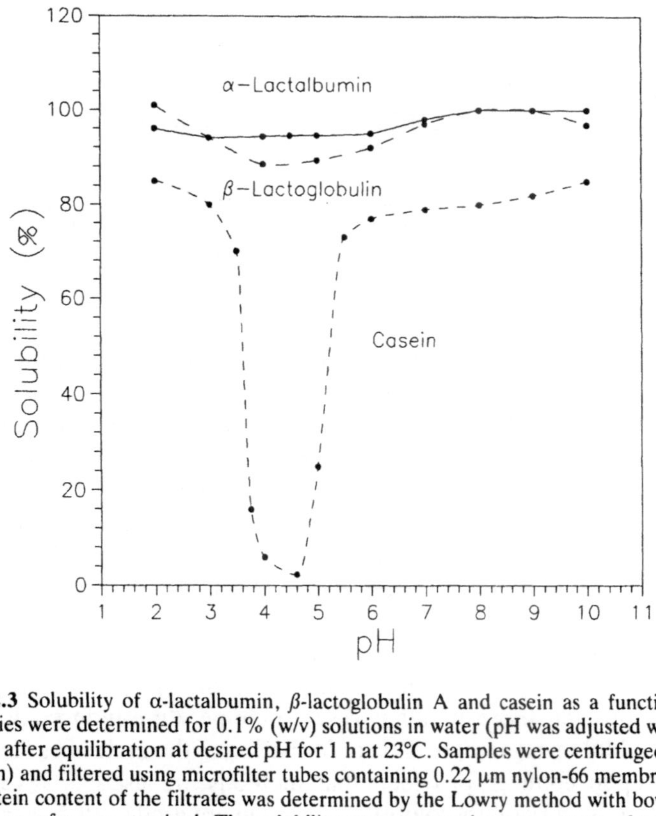

Figure 2.3 Solubility of α-lactalbumin, β-lactoglobulin A and casein as a function of pH. Solubilities were determined for 0.1% (w/v) solutions in water (pH was adjusted with HCl or NaOH), after equilibration at desired pH for 1 h at 23°C. Samples were centrifuged (18 500 *g* × 15 min) and filtered using microfilter tubes containing 0.22 μm nylon-66 membrane filters. The protein content of the filtrates was determined by the Lowry method with bovine serum albumin as reference standard. The solubility was expressed as percentage of total protein concentration.

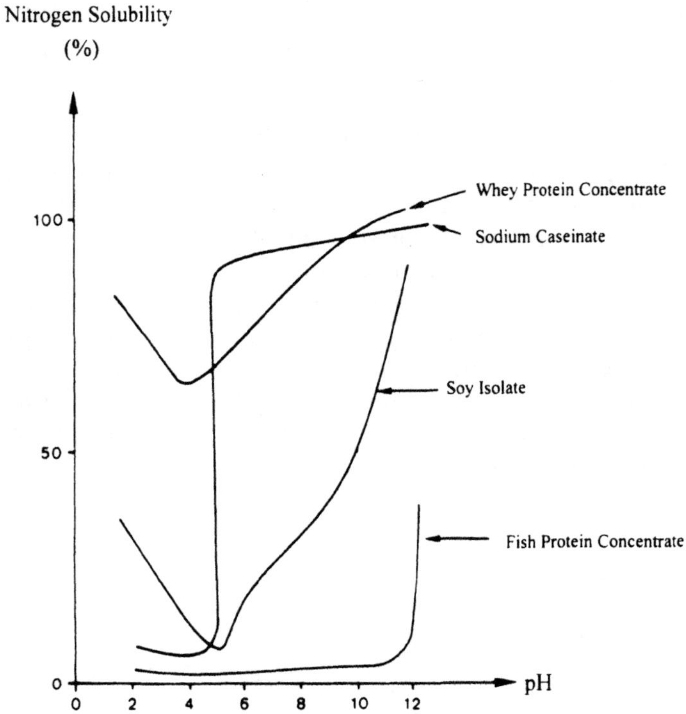

Figure 2.4 Protein solubility as a function of pH in 0.2 M NaCl. Reproduced with permission from Hermansson, A.M. (1973), in *Proteins in Human Nutrition* (eds J.W.G. Porter and B.A. Rolls), Academic Press, London.

2.1.2.2 Temperature. In general, protein solubility is increased with temperature between 0 and 40–50°C. When the temperature of the solution is raised high enough for a given time, the protein is denatured. Proteins are denatured by the effect of temperature on the non-covalent bonds involved in stabilization of secondary and tertiary structure; for example, hydrogen, hydrophobic and electrostatic bonds. When the secondary and tertiary structures of a protein are unfolded, the hydrophobic groups interact and reduce water binding. Hydrophobic interactions lead to aggregation, followed by coagulation and precipitation (Englard and Seifter, 1990). Denaturation decreases protein solubility compared to native protein, and leads to aggregation and difficulty of reversal upon cooling. However, there are many cases of temperature-induced denaturation where biological activity and conformation of the molecules have been restored upon cooling, provided that methods are used to reverse the aggregation, or very dilute solutions of protein are used so that aggregation is avoided.

Li-Chan *et al.* (1984) studied the effect of heating on percent solubility and degree of hydrophobicity. Figure 2.5 shows typical plots of surface hydrophobicity (S_0) and percent protein solubility of meat extracts. Solubility was decreased as the temperature increased from 50 to 85°C, while S_0 increased. At temperatures above 70°C, there was a slight decrease in S_0 due to involvement of part of the hydrophobic residues in the molecular aggregation process and coagulation.

Different proteins require different exposure to heat to denature. The temperature at which a protein denatures, followed by aggregation, also depends on the pH and ionic strength used. For instance, at an extreme pH, denaturation may occur at a lower temperature than at a pH close to the pI. The effect of temperature on the protein solubility at different pH values is shown in Figure 2.6. Protein solubility decreased with temperature between 50 and 90°C, and was more pronounced at pH 2.5 compared to pH 5.5.

The viscosity of a protein solution increases on denaturation and/or gel formation (Hamaguchi, 1992). In the case of gel formation, water absorption is generally increased by entrapment of water in the capillaries of the gel.

Protein denaturation can be detected by sedimentation behavior, viscosity changes, decrease in solubility, optical rotary dispersion (ORD), CD, thermodynamic properties by differential scanning calorimetry,

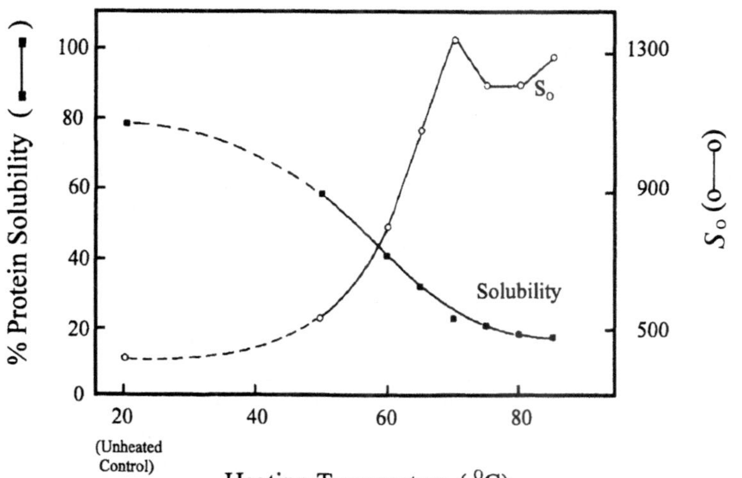

Figure 2.5 Typical plots of surface hydrophobicity (S_0) and percent protein solubility of salt extracts from top round beef as a function of heating temperature. Reproduced with permission from Li-Chan, E., Nakai, S. and Wood, D.F. (1984), Hydrophobicity and solubility of meat proteins and their relationship to emulsifying properties. *J. Food Sci.*, **49**, 345–50.

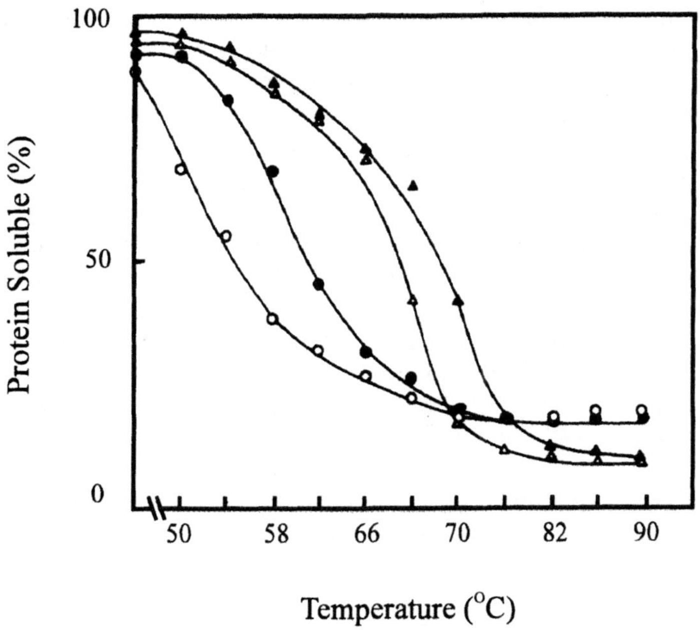

Figure 2.6 Solubility of egg white proteins as a function of temperature (▲, pH 5.5; △, pH 4.5; ●, pH 3.5; ○, pH 2.5). Reproduced with permission from Watanabe, K., Matsuda, T. and Nakamura, R. (1985), Heat-induced aggregation and denaturation of egg white proteins in acid media. *J. Food Sci.*, **50**, 507–10.

fluorescein property and by loss of biological activity. For more detail, there are excellent reviews on protein denaturation (Tanford, 1968, 1970; Kilara and Sharkasi, 1986).

Low temperatures sometimes destabilize proteins when the primary factor stabilizing the tertiary structure is hydrophobic bonds. Hydrophobic bonds, unlike hydrogen and electrostatic bonds, are weaker at lower temperature. Similar phenomena of aggregation, precipitation and coagulation can be induced as heat-treated proteins. If the native structure of a protein is important to functionality, it is very important to choose a temperature where the protein is stable. For most proteins, the ideal temperature range for measuring solubility is 5–24°C, or some cases up to 37°C, depending on the protein. For instance L-threonine deaminase is less stable at 0°C. Certain proteins such as 11S soy protein, gliadins, and some egg and milk proteins denature, aggregate and precipitate when brought to cold or freezing temperature. Low temperature can also result in the dissociation of the subunits of oligomeric proteins (Cheftel *et al.*, 1985). However, most proteins are stable at low temperature.

Protease activity in protein solutions is a problem. In most cases, a lower temperature slows the rate of microbial growth and protease activity.

2.1.2.3 Ionic strength (μ). *In vivo*, proteins are not found in pure water; however, *in vitro*, some proteins (the albumins) are soluble in pure water. In most biological systems, the ionic strength is approximately equivalent to 0.15 M NaCl.

For globulins, and some albumins, the solubility of the proteins increases as the concentration of neutral salts increases, but at some concentration the solubility will reach a maximum and with further increase in salt concentration the solubility decreases. At this point there is competition between protein and salt ions in binding water. The phase where increase in solubility occurs is called salting-in and removal phase is called salting-out. Neutral salts, at a concentration in the order of 0.1–1 M may increase protein solubility depending on the protein, salt, pH and temperature. The salt ions interact with oppositely charged groups on the protein to form a double layer of ionic groups, which decreases electrostatic interaction between protein molecules, causes more protein solvation and thereby increases protein solubility. In this range of salt concentration the logarithm of the solubility is a function of $(\mu)^{1/2}$. The effect of salting-in is dependent on the nature of ions and follows the Hofmeister series. For anions, the Hofmeister series is $SO_4^{2-} > HPO_4^{2-} > CH_3COO^- > Cl^- > Br^- > NO_3^- > ClO_3^- > I^- > ClO_4^- > SCN^-$, and for cations the series is $NH_4^+ > K^+ > Na^+ > Li^{2+} > Mg^{2+} > Ca^{2+} > Ba^{2+}$.

The Hofmeister series of ions at the right side may increase solvation of a protein due to unfolding (perturbation) and dissociation of subunits. Such ions, for example SCN^-, I^-, ClO_4^- and Ba^{2+}, Ca^{2+}, most likely affect the structure of water around the side chain of protein molecules (Creighton, 1984). At concentrations higher than 1 M, the ions to the left of the series decrease protein solubility (salting-out) by increasing hydrophobic interaction and aggregation, and competing with the protein for water, thereby causing the protein to precipitate in the native structure. Such ions stabilize the native conformation of proteins. At the higher concentration of salt ions, most of the water molecules are strongly bound to the salt ions, while there is some reorganization of the water molecules around the protein molecule. This could result in stronger protein–protein interaction (mainly via surface hydrophobic interaction) than protein–water interaction; this results in association, followed by precipitation. The concentration of any given salt or organic solvent necessary to effect precipitation of a given protein is related to the intrinsic properties of the protein, as well as the pH and native or denatured state of the protein.

In some cases, low concentrations of divalent cations can induce the salting-out effect. For instance, calcium changes the protein solubility of caseins by causing the association of the caseins and the formation of large colloidal aggregates which serve to transport calcium and phosphate (Farrell *et al.*, 1988). Another example is tofu-curd formation by the flocculation of protein in soymilk with an addition of 3–10 mM calcium

(Ono *et al.*, 1993). Calcium is especially effective in salting-out processes with phosphorylated proteins, such as α_{s1}- and β-casein and other phosphoproteins; low concentrations of Ca^{2+} (10–20 mM) cause changes in solubility. In experiments with α_{s1}- and β-caseins at concentrations of 10 mg/ml at 37°C, pH 7, the proteins precipitate when $CaCl_2$ exceeds 5 mM (Farrell *et al.*, 1988).

Usually the effect of high salt concentrations on association of proteins is reversible. However, there is an exception in the case of the caseins where high salt concentrations disrupt the micelles.

The ionic strength of the solution can be calculated by:

$$\mu = 1/2\ \Sigma\ C_i Z_i^2 \tag{2.5}$$

where μ is ionic strength, C is concentration of ions and Z is the valence of the ions.

During salting-out, the solubility of the affected protein is as follows (Kilara and Sharkasi, 1986):

$$\log S = -K\mu + \log S_0 \tag{2.6}$$

where S is the solubility of protein in grams per liter of solution, μ is the ionic strength in moles per liter of solution, S_0 is the solubility at zero ionic strength, and K is the salting-out constant, which depends on the protein, pH, temperature and, particularly, on the specific nature of the salt. K, the slope of log solubility versus μ (equation (2.6)) extrapolated to zero ionic strength gives log S_0.

The salting-out action of different salts increases with their hydration energy and with their steric hindrance according to the Hofmeister series. Some commonly used salts for protein precipitation are ammonium sulfate, sodium chloride, potassium phosphate, magnesium chloride, and citrate salts (Ollis and White, 1990).

Urea and chaotropic ions such as guanidine hydrochloride decrease protein–protein interaction and increase protein solvation. These ions, at concentrations in the order of 4–8 M, may cause protein denaturation by disruption of hydrogen bonds and interruption of the tertiary and even secondary structure of the protein to varying degrees, depending on the type of protein. Guanidine hydrochloride forms stronger interactions with protein than does urea, thus it may be more denaturing to the protein.

Salt in solution at low μ relative to that of isotonic solution ($\mu = 0.15$) may cause some proteins to precipitate from solution. For example, myosin (above its critical concentration, 0.2 mg/ml) is insoluble at low salt concentrations ($\mu < 0.15$ of KCl), while at $\mu > 0.3$, it becomes soluble (Margossian and Lowey, 1982).

2.1.2.4 Presence of other components. Other components in the solution, such as certain ions or miscible organic solvents, may also affect the solubility behavior of proteins. Some metal ions such as Zn^{2+}, Cd^{2+} and Ba^{2+}, or bulky organic anions such as picrate, tannate, tungstate, molybdate, trichloroacetate, perchlorate, and sulfosalicylate precipitate proteins, depending on pH, μ and temperature (Ollis and White, 1990). The presence of phytate, Ca^{2+} and Zn^{2+} affects protein solubility, as in the case of soy protein concentrate (Gifford-Steffen and Clydesdale, 1993).

Solubility of proteins is decreased by water-miscible organic solvents such as acetone, ethanol and methanol which denature proteins at high temperature. Therefore, they must be used at 0–5°C to maintain the native state of the protein. Most proteins associate and precipitate at high concentrations of water-miscible organic solvents, primarily by changing the effective concentration of water and decreasing the ionic strength of the solution (Ollis and White, 1990). The salt concentration should be as low as possible when water-soluble organic solvents are used to precipitate proteins. The solubility of proteins is increased by glycine and other dipolar molecules which raise the dielectric constant of water (Haurowitz, 1963).

Polyethylene glycol (PEG) also decreases protein solubility. It is used for protein precipitation and producing protein crystals (Ollis and White, 1990). It affects protein solubility by changing the dielectric properties of the solution and by competing for water.

Lipids interact with proteins in biological systems, such as cellular membranes and plasmatic lipoproteins, or in food system emulsions. Hydrophobic interactions between the apolar aliphatic chain of the lipid and hydrophobic regions on the surface of the protein may reduce the protein–water solubility. Highly hydrophobic proteins interact more strongly with lipids.

Surface-active agents, such as sodium dodecylsulfate (SDS), are strong denaturing agents and solubilize the protein by unfolding the native structure and binding to the hydrophobic regions of the denatured proteins. Anionic detergents, because of the pK_a of the ionizable groups and binding to the proteins, give proteins a large negative charge, increase repulsive forces, and increase unfolding.

Reducing agents, such as dithiothreitol (DTT), β-mercaptoethanol, ascorbic acid and cysteine, may reduce disulfide cross-links and thereby modify the tertiary conformation of proteins (Cheftel *et al.*, 1985). Sometimes it is required to have very low concentrations of reducing agent (0–1 mM) during protein purification to prevent oxidation of –SH groups to S–S groups, causing some aggregation. For example, during myosin or β-lactoglobulin purification 1 mM DTT is added to the buffer to prevent oxidation of the free –SH groups. Reducing agents can be removed by dialysis, if necessary. In food products, a more complex system exists and other components of food interact with protein and contribute to food functionalities. Protein–protein, protein–fat and protein–carbohydrate interactions influence the protein solubility, as well as other physicochemical properties (Schwenke and Raab, 1989).

2.1.2.5 Protein concentration. Solubility is also a function of protein concentration. Most proteins have some solubility in aqueous solutions at any defined condition. At higher concentration, when the solubility product constant is exceeded, some of the protein will be insoluble and will sediment by centrifugation. Figure 2.7 shows the solubility of β-lactoglobulin A and B as a function of the total amount of protein present (Treece *et al.*, 1964). Thermodynamically there is an equilibrium between phases, where the protein in the solid phase is in equilibrium with that in the solution phase.

Usually for protein solubility testing, the protein concentration needs to be in the range of 1–20 mg protein per ml. Figure 2.7 shows that the maximum protein solubility of β-lactoglobulin is 5 mg/ml for type B and less than 1 mg/ml for type A at 6.25 mM NaCl and pH 5.2. The composition of solutions of types A and B β-lactoglobulin are constant above 2 mg/ml total protein.

Protein aggregation increases at high concentration. The problem may be controlled by maintaining the protein concentration at 1–3 mg/ml (Harris and Angal, 1990).

2.1.2.6 Physical parameters. Physical parameters which can affect protein solubilities are mechanical treatment, hydrostatic pressure, irradiation and adsorption at water–air interfaces. Care must be taken to prevent harsh agitation and other physical conditions that denature proteins.

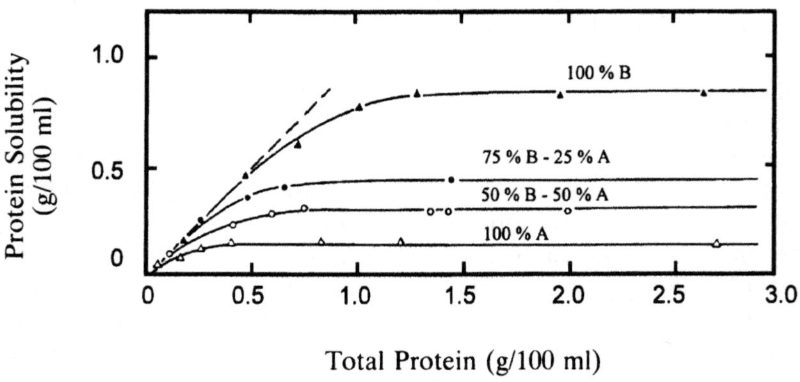

Figure 2.7 Solubilities of β-lactoglobulins A and B and their mixtures in 6.25 mM NaCl as a function of the total amount of protein present at pH 5.2. Reproduced with permission from Treece, J.M., Sheinson, R.S. and McMeekin, T.L. (1964), The solubilities of β-lactoglobulins A, B, and AB. *Arch. Biochem. Biophys.*, **108**, 99–108.

Proteins that adsorb at water–air or water–organic solvent interfaces usually become irreversibly denatured and insoluble (Cheftel *et al.*, 1985; Kinsella and Soucie, 1989).

2.1.3 Definition of terms used for protein solubility

Different terms are used in the literature to express protein solubility, as summarized in Table 2.1. The term dispersibility is very commonly used in the dairy industry to express the properties of milk powder. Dispersibility is a property of a powder describing its uniform distribution in water. It is related to the ease with which aggregates of powder fall apart and disperse when in contact with water. The dispersibility of milk powder is determined by overlaying approximately 40 g of milk powder on 300 ml of water at 24°C to give a surface loading of 4.25 kg/m^2. This mixture is stirred using a kitchen mixer with beaters rotating 175 times for 20 sec, approximately 39 millimeters below the water surface; the dispersion is then drained through a 210 μm screen and the total solids in the filtrate are determined (Kinsella, 1984). Increasing particle size improves dispersibility, whereas dispersibility is decreased as the percentage of fine particles (below 90 μm) increases. An excess of fine particles causes lumpiness. Without the fine particles, water can more easily penetrate and disperse the particles. Factors such as temperature, viscosity and atomizer speed affect the dispersibility (Kinsella, 1984).

As described, dispersibility is a different criterion from solubility. A protein may be dispersed before going into solution. Protein dispersibility index (PDI) and water dispersible protein (WDP) are two terms commonly used to express the degree of dispersed protein (AOCS, 1987). Based on the American Oil Chemists Society (AOCS) definition, PDI is the dispersible protein of a food protein system under the conditions of the test (the original test was developed for soy protein). The protein (20 g dry protein powder) is stirred in distilled water (300 ml) in a Waring blender at 8500 rpm or in a Sorvall Omni-Mixer at 8500 rpm at 25°C until a smooth slurry is obtained. After the slurry has been prepared, a portion is decanted into a 50 ml centrifuge tube and centrifuged for 10 min at 2700 rpm at 24°C. The amounts of protein in 15 ml of supernatant and slurry are then determined by the Kjeldahl method (AOCS Method Aa 5–38, 1987 or AOAC Method 2.049, 1990) and %WDP is calculated and used to calculate PDI (Method Ba 10–65). With a Sorvall Omni-Mixer, 10 g of protein product are stirred in 150 ml of distilled water.

(1) Protein dispersibility index (PDI). This is calculated by:

$$\text{PDI (\%)} = \frac{\text{\% Water dispersible protein}}{\text{\% Total protein}} \times 100$$

(2) Water dispersible protein (WDP). Sometimes the terms WDP and PDI are interchangeably used for dispersible protein. WDP is the percent of water dispersible protein and is calculated as:

$$\%\text{WDP} = \frac{\text{Weight of dispersed protein}}{\text{Weight of sample}} \times 100$$

(3) Nitrogen solubility index (NSI). This method, based on AOCS, Method Ba 11–65 (1987), determines the dispersible nitrogen under the conditions of test. The original method was developed for soy protein. For NSI, in contrast to the fast-stir method for PDI, a slower stirring technique (No. Ba 10–65) is used. The NSI method generally gives lower results than the fast-stir method (PDI).

In this method, protein flour is ground until at least 95% of the sample will go through a 100-mesh screen. In the case of a full-fat sample, the material is ground to a fine flour so that at least 80% of the material will go through an 80-mesh screen and 90% through a 60-mesh screen. A protein sample (5 g) is stirred in distilled water (200 ml) at 30°C, at 120 rpm for 120 min. About 40 ml of the mixture are then decanted into a 50 ml centrifuge tube and centrifuged for 10 min at 1500 rpm at 24°C. After centrifugation, the supernatant is decanted through a funnel containing a plug of glass wool. A 25 ml portion of the clear liquid is used for nitrogen determination by Kjeldahl (AOCS Method Aa 5–38, 1987); percent water soluble nitrogen (%WSN) is calculated and further expressed as nitrogen solubility index (NSI):

$$\%\text{WSN} = \frac{\text{Weight of soluble nitrogen}}{\text{Weight of sample}} \times 100$$

$$\text{NSI} = \frac{\%\ \text{Water soluble nitrogen}}{\%\ \text{Total nitrogen}} \times 100$$

(4) The solubility index (SI). The conventional solubility index reflects the extent of denaturation of the protein in milk powders. This index is measured by dispersing a known amount of protein powder (milk powder) in water under defined conditions and separating the undissolved material by centrifugation. The solubility index is the volume of sediment in milliliters remaining at the bottom of the centrifuge tube (Kinsella, 1984). This solubility index is an inverse of the actual solubility and is used to grade the milk powder. The New Zealand standards for the finest grades of whole dry milk (WDM) and non-fat dry milk (NDM) are 1 and 1.2, respectively, and for the standard grades 1.5 and 2. The lower the solubility index, the better the quality of powder. Apparently the solubility index is

related to denaturation of protein during drying and is directly affected by heat treatments. The method described here indicates the arbitrary nature of some of the methods for assessing the physical properties required in commercial powders and underlines the need for more standardized terminology, and more importantly, for standard methods for determining these characteristics (Kinsella and Soucie, 1989).

(5) Water soluble protein (WSP). This term is an indicator of the solubility properties of proteins in water under a given condition. The term is equivalent to the PDI.

$$\text{WSP} = \frac{\text{Water soluble protein}}{\text{Total sample protein}} \times 100$$

(6) Percent protein solubility (%PS) or percent nitrogen solubility (%NS). These terms are used for solubility measurements of proteins. A given protein is solubilized under a defined condition, centrifuged, and the amount of soluble protein in the supernatant and total amount of protein in the sample are measured. Under the term %PS, the amount of protein can be quantitated by any technique. The percent solubility is then defined as percent of amount of protein in the supernatant to total amount of protein in the sample before centrifugation. Sometimes this is expressed as protein soluble index (PSI) or percent solubility (%S).

For %NS, the amount of protein is measured as soluble nitrogen in the supernatant and in the total sample by a micro-Kjeldahl method (AOCS, 1987). It is generally assumed that a mixture of pure proteins contains 16% nitrogen. The amount of protein is calculated by multiplying %N by the conversion factor 6.25 = (100/16). This factor depends on the type of protein. Factors 6.38, 6.25, 5.70 and 5.46 are used for milk protein, soy protein, wheat protein and peanut protein, respectively. The general factor of 6.25 is used for most foods. There are extremes in nitrogen content ranging from 4.2% in β-lipoproteins to 30% in protamine, but most food contains about 16% nitrogen.

$$\%\text{PS}, (\%\text{NS}; \text{PSI}) = \frac{\text{Amount of soluble protein}}{\text{Total amount of protein}} \times 100$$

(7) Degree of turbidity. The degree of turbidity of a protein solution is the amount of turbidity under a defined condition at 500–540 nm (Kumar Murthy and Narasinga Rao, 1984; Tanimoto and Kinsella, 1988; Ahn *et al.*, 1991). The more turbid the protein solution, the less soluble the protein is in solution. This is an indirect method for expressing the protein solubility, but does not give the same numerical value.

(8) Percent insolubility. Percent insolubility is (100 − percent soluble protein). Percent of protein solubility is measured as in percent solubility ((6) above) and is then subtracted from 100. This term is frequently used to correlate the percent of protein insolubility to functional properties (Hayakawa and Nakai, 1985; Nakai *et al.*, 1986; Li-Chan and Nakai, 1989).

(9) Dispersibility. The term dispersibility is sometimes used to express the degree of dispersed protein. This is equivalent to PDI and sometimes is used to express WDP.

(10) Maximum protein solubility. Other terms, such as maximum solubility, are used; this is a maximum amount of protein which is solubilized under a given condition. In this method, the protein is gradually added to the solvent under the condition of the test, until no further solubilization occurs and additional protein is excluded from the solution. The maximum solubility is mostly used by chemists for expression of solubility.

2.2 Review of methods for measuring solubility

2.2.1 Protein solubilization

Solubility of food proteins is determined in either salt-free water or water containing a specific salt at a constant pH and ionic strength. The pH is usually maintained by a buffer. The choice of a buffer for solubilization of proteins has many important implications besides the creation of a stable pH environment. In general, the lowest possible concentration of buffer required for maintaining the pH is used to minimize the non-specific ionic strength effect.

In most cases where the protein is easily solubilized, a buffer concentration of 20–100 mM is used depending on the pH range. Sen *et al.* (1981) used 0.1 M buffer to solubilize a protein sample of 100 mg/ml, and then adjusted the ionic strength to 0.55 by addition of NaCl. A protein concentration of 10 mg/ml in citrate buffer for pH 3 to 6 and phosphate buffer for pH 6 to 8, and 0.1 M sodium chloride were used by Waniska and Kinsella (1989) for solubility measurements. Townsend and Nakai (1983) also used a 1% protein solution in 0.1 M phosphate buffer, pH 7.0, to test the solubility of several proteins (bovine serum albumin, β-lactoglobulin, trypsin, ovalbumin, conalbumin, lysozyme, κ-casein, and denatured ovalbumin and lysozyme).

However, a higher concentration of specific buffer (0.1–0.5 M) may be needed to solubilize some proteins. In a study of endoplasmic reticulum, concentrations of 0.1–0.5 M phosphate were shown to enhance the solubility of membrane proteins. This effect is not entirely due to the increase in the ionic strength of the buffer, but also to phosphate acting as a surfactant. Solubilization of myosin in a low concentration of pyrophosphate (0.02–0.04 M) is probably due to a specific binding site (Frederiksen and Cunningham, 1982).

Table 2.2 lists the most common buffer salts and the pK_a values of the buffer species. Table 2.3 lists the common buffers mixture and their usable pH ranges. The choice of buffer also depends on the method used for protein determination. Certain buffers are not compatible with specific methods of protein determination.

Some proteins which have stronger hydrophobic interaction need a higher ionic strength, produced by a salt concentration of 0.15–0.5 M in the solubilizing buffer. The most commonly used salts are sodium chloride or potassium chloride. Sodium chloride is more soluble than potassium chloride, but potassium chloride is more effective in solubilizing proteins. Sometimes $CaCl_2$ is used to improve the solubility, but it may have a different solubility effect to NaCl, depending on the type of the protein (i.e. phosphorylated versus unphosphorylated) and the pH of the solution (Figure 2.8; Bera and Mukherjee, 1989). Myosin in 0.001 M phosphate

Table 2.2 Selected buffers and their pK_a values at 25°C. Compiled from Perrin, D.D. and Dempsey, B. (1974) *Buffers for pH and Metal Ion Control*, Chapman & Hall, London.

Trivial name	Buffer name	pK_a
Phosphate (pK_1)	–	2.15
Malate (pK_1)	–	3.4
Formate	–	3.75
Succinate (pK_1)	–	4.21
Citrate (pK_2)	–	4.76
Acetate	–	4.76
Malate (pK_2)	–	5.13
Pyridine	–	5.23
Succinate (pK_2)	–	5.64
Cacodylate	Dimethylarsinic acid	6.27
Dimethylglutarate	3,3-Dimethylglutarate (pK_2)	6.34
Carbonate (pK_1)	–	6.35
Citrate (pK_3)	–	6.40
Bis–Tris	[Bis(2-hydroxyethyl)imino]tris(hydroxymethyl) methane	6.46
Pyrophosphate	–	6.60
Bis–Tris propane	1,3-Bis[tris(hydroxymethyl)methylamino]propane	6.80
Imidazole	–	6.95
Phosphate (pK_2)	–	7.20
Tris	Tris(hydroxymethyl)aminomethane	8.06
Borate	–	9.23
Ammonia	–	9.25
Glycine (pK_2)	–	9.78
Piperidine	–	11.12
Phosphate (pK_3)	–	12.33

Table 2.3 Selected commonly used buffers and their pH ranges at 25°C. Compiled from Perrin, D.D. and Dempsey, B. (1974) *Buffers for pH and Metal Ion Control*, Chapman & Hall, London.

Buffer mixture	pH range
HCl/KCl	1.0–2.2
Glycine/HCl	2.2–3.6
Phosphate (Na_2HPO_4)/citric acid	2.6–7.6
Citric acid/sodium citrate or NaOH	3.0–6.2
Acetic acid/sodium acetate or NaOH	3.7–5.6
Succinic acid/NaOH	3.8–6.0
Phosphate (NaH_2PO_4/NaH_2nPO_4 or NaOH)	5.8–8.0
Bis–Tris/HCl	5.8–7.2
Bis–Tris Propane/HCl	6.3–9.5
Imidazole/HCl	6.2–7.8
Tris/HCl	7.0–9.0
Diethanolamine/HCl	7.8–9.9
Sodium borate/HCl	8.1–9.0
Boric acid/Borax	7.6–9.2
Boric acid/NaOH	8.1–10.1
Glycine/NaOH	8.6–10.6
Borax/NaOH	9.28–10.1
Carbonate/bicarbonate	9.0–10.7
Sodium borate/NaOH	9.3–10.7
Phosphate (Na_2HPO_4)/NaOH	11.0–11.9

buffer, pH 7.0 and above, is soluble when 0.3 M KCl is added. Mattarella and Richardson (1983) used 10% (w/v) protein samples in various buffers of ionic strength 0.55.

To eliminate the effect of inorganic ions on solubility properties, one can solubilize the protein in water and adjust the pH with NaOH or HCl to the desired value using a pH meter (Chobert *et al.*, 1988a,b; Farrell *et al.*, 1988; Tanimoto and Kinsella, 1988; Ahn *et al.*, 1991; Vojdani and Whitaker, 1994). Ahn *et al.* (1991) adjusted the ionic strength to 24 mM with NaCl (1 M) to solubilize phaseolin of red kidney bean in deionized water. Some inorganic ions have a specific binding site on the protein (e.g. the effect of phosphate on Ca-associated protein). Many proteins have high affinity for water and are strongly hydrated and readily soluble in salt-free water. However, if a protein is insoluble in salt-free water, the ionic strength of the solution is adjusted by addition of NaCl, along with pH adjustment (Adler-Nissen and Olsen, 1979; Casella and Whitaker, 1990).

For difficult solubilization problems and solubilization of denatured and aggregated proteins, urea or chaotropic salts such as guanidine hydrochloride may be added to the solubilization buffer to decrease protein–protein interaction, as discussed in section 2.1.2.3.

Some proteins need strong acidic or basic conditions for solubilization. For instance, glutelins (i.e. wheat glutenins) are soluble in acid (pH 2) or

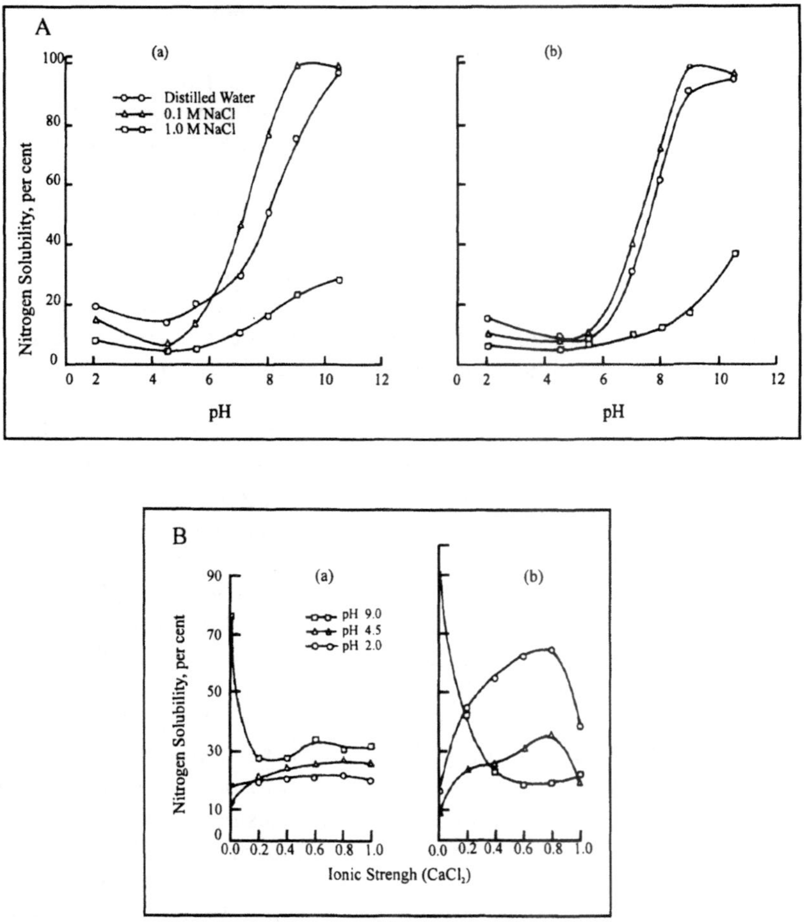

Figure 2.8 Effect of pH and ionic strength on the solubility of full-fat (a) and defatted (b) rice bran protein concentrate: A, in the presence of NaCl, and B, in the presence of $CaCl_2$ (one unit of ionic strength of $CaCl_2 = 0.33$ M). Reproduced with permission from Bera, M.B. and Mukherjee, R.K. (1989), Solubility, emulsifying, and foaming properties of rice bran protein concentrates. *J. Food Sci.*, **54**, 142–5.

alkali (pH 12) (Chefel *et al.*, 1985). If a protein is stable above pH 11 or below pH 3, solubilization by alkali or acid addition can be achieved. The protein suspension is allowed to equilibrate for about 30 min, after which the pH is adjusted to the desired values. However, at extreme pH values, strong electrostatic repulsions of ionized groups occur, which can cause unfolding (denaturation) of the protein. In some cases, the protein may recover its native structure when the pH is readjusted to the initial stable range. During the pH adjustment, amino acid residues may change conformational orientation. When the pH is readjusted, not all the original structures may be regained, leading to a mixture of structures. Some peptide bonds (particularly aspartyl bonds) may be hydrolyzed when a protein is held at an acidic pH for long periods of time; deamidation of asparagine 21 of insulin also occurs (Schein, 1990). Soy isolate in 0.2 M NaCl is only 50% soluble at pH 10, while at pH 11 and 12 it is 70% and 90% soluble, respectively. Fish protein concentrate produced by organic solvent, which is almost insoluble at pH 2 to 11, is soluble at pH 12.

In general, the solubility of a protein is decreased when the temperature is lowered. In most cases, 24°C is used for the protein solubilization test. However, temperatures ranging from 5°C to 37°C or even to 60°C, are used for protein solubility studies (Mattarella and Richardson, 1983; Campbell *et al.*, 1992; Chin *et al.*, 1994; Vojdani and Whitaker, 1994).

Some researchers have used a higher temperature for initial solubilization, then equilibrated the samples at a lower temperature before measuring the solubility. Mattarella and Richardson (1983) used 1 h at 37°C and then 48 h at 4°C. Sen *et al.* (1981) initially solubilized chemically modified casein for 1 h at 60°C, and then incubated for 48 h at 4°C.

For initial protein extraction and purification, chemical, physical or mechanical steps may be required to release and solubilize the proteins. There are excellent reviews by Harris and Angal (1989, 1990) and Deutscher (1990) on practical approaches for protein extraction and purification. Li-Chan *et al.* (1984) extracted meat samples in 100 mM $NaPO_4^-$, pH 7.0, containing 0.6 M NaCl, 1 mM $MgCl_2$ and 0.02% sodium azide (to prevent microbial growth), with stirring in a water bath to internal temperatures ranging from 50 to 85°C. After reaching the desired temperature, the solutions were rapidly cooled in an ice water bath, then mixed with a homogenizer at 3300 rpm for 15 sec. Aliquots of extracts and heated samples were then centrifuged for solubility testing. Protein precipitation techniques, other than salt precipitation (precipitation by salt is reversible and the protein structure is not significantly changed by reversible aggregation) may be used for protein extraction and purification. In such cases, a plot of protein in solution as a function of pH will depend on whether the protein has already precipitated (Schein, 1990).

2.2.2 *Separating the undissolved protein (clarification)*

The best technique to separate the undissolved protein from the solution is centrifugal sedimentation of the insoluble protein. The theory and practice of centrifugation is discussed by Rickwood (1984). The sedimentation velocity (V_g) of a particle in a less dense fluid under the influence of gravity is a function of particle size, density of particle and liquid, and fluid viscosity, as given by Stokes' law:

$$V_g = D^2\,[(\varrho_1 - \varrho_2)/18\,\eta]g \tag{2.7}$$

where D is particle diameter, ϱ_1 is the density of particle, ϱ_2 is the density of liquid, η is the fluid viscosity and g is the gravitational acceleration (980.7 cm/sec^2). In centrifugation, the acceleration term g is replaced by $\omega^2 r$, where ω is the angular velocity of the centrifuge (radian per sec) and r is the distance of the particle from the center of rotation. Centrifugal forces increase the sedimentation rate to V_s (velocity of particle moving through the liquid phase) where:

$$V_s = V_g\,(r\omega^2/g) \tag{2.8}$$

from which the velocity of a particle can be calculated at a distance r from the axis of rotation. In addition, increasing the time of centrifugation will cause the particles to move further from the center of rotation (Whittington, 1989):

$$V_s t = \text{distance settled} \tag{2.9}$$

Another useful parameter is:

$$\text{RCF} = r\omega^2/g \tag{2.10}$$

where RCF is the relative centrifugal force and refers to the force during centrifugation that moves the particle outward from the center of rotation. The RCF value is determined by the following formula:

$$\text{RCF} = (1.117 \times 10^{-5})(r)(n^2) \tag{2.11}$$

where r is the radius in centimeters from the centerline of the rotor to the point in the tube where the RCF value is required and n is equal to the rotor speed in rpm. Whittington (1989) has discussed the use of centrifuge equipment and types of rotors for operation.

Normally, at the concentrations of protein and buffer used for the solubility test, there is a low resistance to sedimentation of protein particles. In this case, the limiting factor is the particle size of undissolved or aggregated proteins.

Many researchers have used centrifugal forces of 105 000 *g* (about 34 000 rpm) to precipitate the undissolved proteins and this is more common for membrane, cytoskeletal and physiologically active proteins (Frederiksen and Cunningham, 1982). Speeds of about 40 000–60 000 rpm are common for sedimentation velocity measurements (Gilbert and Gilbert, 1973; Schwenke *et al.*, 1990). In the literature, centrifugal forces in the range 1600–43 800 *g* were used for 5 to 30 min (Kinsella, 1976; Hayakawa and Nakai, 1985; Farrell *et al.*, 1988; Casella and Whitaker, 1990; Ono *et al.*, 1993). In many cases, 27 000–30 000 *g* forces were used to sediment all of the undissolved protein (Sen *et al.*, 1981; Mattarella and Richardson, 1983; Li-Chan *et al.*, 1984). Kinsella (1984) recommended 30 000 *g* for this purpose. Forces of 20 000 *g* were used by Morr *et al.* (1985) in studies designed to develop a standardized food protein solubility procedure. In many cases, using a small bench Eppendorf-type centrifuge at its maximum speed (12 000–14 000 rpm) is adequate as this generates maximum forces of 13 600–18 500 *g*. In practice, a preliminary test may be required to find the centrifugal forces which sediment all of the undissolved protein from the solution in 15 to 30 min.

After centrifugation, the supernatant should carefully be separated without disturbing the pellet. Morr *et al.* (1985) used Whatman filter paper No. 1 to filter the supernatant after centrifugation. Microfilter-containing nylon filter of pore size 0.22 μm was used for globular proteins of less than 80 000 mol. wt. (Vojdani and Whitaker, 1994). Campbell *et al.* (1992) used 0.45 μm filters in the study of soy protein isolate. Others did not filter the protein sample after centrifugation; the supernatant was removed and used for protein determination.

Microfilterfuge tubes containing nylon or cellulose acetate with no protein binding and different pore sizes are available commercially. For filamentous proteins, microfilters of pore size larger than 0.22 μm are required. For example, myosin has a diameter of 165 nm, which is difficult to filter through a 0.22 μm filter. In some cases, a pore size of 10 μm may be required. However, it is recommended that the researcher finds information on the radius of gyration of the protein under test, and also runs a preliminary test on the binding properties of the filter before use. The filtration step must ensure that all insoluble protein particles are removed from the supernatant.

2.2.3 Methods for measuring protein concentration

In this section different procedures used for measuring protein concentration will be discussed, to be of help in the accurate measurement of the amount of protein. The mechanism of reaction for each method, the range of sensitivity for protein concentration, and interfering substances will be discussed. The final step is very crucial, since in the absence of compatibility between the chosen method and substances present in the sample, errors arise. The choice of the method depends on (Stoscheck, 1990):

1. Amount of protein available to assay
2. Concentration of protein
3. Presence of compounds which may interfere with assay
4. Specificity of assay
5. Ease and reliability, accuracy and precision of performing the assay

For solubility testing, one should have enough protein at a suitable concentration to be able to use the most accurate and precise procedure. Sometimes, due to protein composition, a specific method cannot be used.

Another important factor is simplicity and ease of the procedure, so that many tests can be performed in a day.

2.2.3.1 Measurement of protein nitrogen. Nitrogen is determined by the Kjeldahl method of the AOAC for an amount of protein in the range of about 2 g; however, for 10–40 mg protein, the official micro-Kjeldahl method is used (AOAC method 960.52, 1990). All of the reagents need to be N-free. This procedure is not applicable to material containing N–N or N–O linkages.

Interference of nucleic acids and other N-containing organic compounds can be removed by trichloroacetic acid (TCA) precipitation of proteins.

2.2.3.2 Measurement at 280 nm (UV). The presence of aromatic amino acid residues (phenylalanine, tryptophan, and tyrosine) means that most proteins absorb in the UV region of 260–292 nm with maximum absorption at 280 nm. Knowing the extinction coefficient (E) at 280 nm and the absorbance at 280 nm, the protein concentration can be calculated using Lambert–Beer law:

$$A = E_{280}Cl \tag{2.12}$$

where A is absorbancy at 280 nm, E_{280} is extinction coefficient, C is the protein concentration in %, mg/ml or molarity depending on E units, and l is the cell pathlength in cm. E is usually expressed either as 1%, or 1 mg/ml. For most proteins values of $E_{280}^{1\,\text{mg/ml}}$ are in the range of 0.4–1.5, depending on the aromatic amino acid content. Extreme $E_{280}^{1\,\text{mg/ml}}$ values include some paravalbumins and related Ca^{2+}-binding protein (0.0) and lysozyme (2.65) (Dunn, 1989). A relatively pure protein is needed for accurate measurement, and the extinction coefficient must be known. For measuring the absorbancy at 280 nm, 0.1–2 mg of protein are required. Other compounds which absorb at 280 nm interfere, especially nucleic acids and nucleotides which contain purine and pyrimidine rings. In this case the absorbance of sample is measured at both 280 and 260 nm and a correction formula is applied:

$$\text{Protein (mg/ml)} = 1.55\,A_{280} - 0.76\,A_{260} \tag{2.13}$$

The equation was derived for enolase ($A_{280}/A_{260} = 1.75$) in the presence of yeast nucleic acid ($A_{280}/A_{260} = 0.49$) and thus may not be precise for other proteins and nucleic acids (Segel, 1976).

To determine the extinction coefficient of an unknown protein with known concentration, the following procedure is used. A pure protein is dried and a solution of 1 mg/ml is made in water or a suitable buffer with low absorbancy at 280 nm (Good's buffer has minimum absorbancy; Good *et al.*, 1966; Ferguson *et al.*, 1980). The absorbancy is measured against a blank. Using equation (2.12) gives the value of $E_{280}^{1\,\text{mg/ml}}$. For unknown concentrations of a protein, the procedure described in the next section can be used.

2.2.3.3 Absorbance at far-UV. The peptide bonds of proteins absorb the far-UV with maximum absorption at 191–194 nm. This is much more sensitive than absorbancy at 280 nm since there are many peptide bonds in proteins, and the bonds are essentially constant for all proteins. Consequently, the $E^{1\ \mathrm{mg/ml}}$ values are the same for all proteins ($E_{205}^{1\ \mathrm{mg/ml}} = 31$) (Scopes, 1974). At 191–194 nm, a concentration of 0.01–0.05 mg/ml protein can be measured. Routine spectrophotometers cannot be used for measurement at this wavelength because of UV absorption by oxygen (primarily in water). However, measurements can be made at 205 nm by a spectrophotometer with accurate wavelength setting at 205 nm and in the absence of stray light. Very clean quartz cuvettes and buffer with minimal UV absorption at this wavelength are needed. Including 0.01% polyoxyethylene lauryl ether (Brij 35) in the buffer prevents protein losses on vessel surfaces (Stoscheck, 1990). Protein concentration is estimated by:

$$\text{Concentration (mg/ml)} = 31\ (A_{205}) \qquad (2.14)$$

To correct for absorption of tyrosine and tryptophan at 205 nm, the following equation is used for measurements at 280 nm and 205 nm (Scopes, 1974):

$$E_{205}^{1\ \mathrm{mg/ml}} = 27.0 + 120\ (A_{280}/A_{205}) \qquad (2.15)$$

At normal cellular concentrations of proteins and nucleic acids, nucleic acids contribute relatively little toward the absorbance at 205 nm. At concentrations of 10 μg/ml DNA and RNA have an absorbancy of about 0.3, while for a protein concentration of 15 μg/ml in 5 mM phosphate buffer, pH 7, containing 0.01% Brij 35 and 0.1 M K_2SO_4, the absorbancy is approximately 0.7 and 0.8 for gelatin and bovine immunoglobulin G, respectively (Stoscheck, 1990). However, only a pure protein can give an accurate measurement. To measure the extinction coefficient of a protein of unknown concentration, the absorbancy of the protein at 205 and 280 nm is determined (Scopes, 1974). About a 30-fold dilute protein is needed for a 205 nm reading relative to the 280 nm reading.

Equation (2.15) is used (corrected for dilution factor reading at 205 nm) and the concentration of protein is determined using ($A = E_{205}Cl$). Knowing the concentration of protein in a 1 cm pathlength, the extinction coefficient at 280 nm can be determined using the following equation:

$$E_{280} = \text{concentration}/A_{280} \qquad (2.16)$$

An accuracy of 2% for most proteins with an average amino acid composition and normal phenylalanine content is claimed.

2.2.3.4 Lowry protein assay (Folin phenol method). Protein measurement by the Lowry method (Lowry *et al.*, 1951) is based on the reaction of the Folin–Ciocalteu reagent with protein. While the mechanism of the Lowry reaction is not well understood, it is theorized that in the Folin phenol reagent, the molybdate is reduced to molybdium blue by tyrosine and, to a lesser extent, by cystine, cysteine, histidine and the peptide bonds. The molybdium blue (which may involve tungstate) gives a characteristic blue color with maximum absorbancy at 745–750 nm. Copper ions (Cu^{2+}) in alkaline solution facilitate the reduction process by forming a complex with the peptide bond through the peptide nitrogen and being reduced to Cu^{+}. The Cu^{+}, as well as the R groups of reactive residues, then react with Folin reagent, producing an unstable product which is slowly reduced to become molybdate/tungsten blue (Smith *et al.*, 1985).

Non-protein substances may interfere with color development or produce the blue color. Glycine (0.5%) decreases the color with protein by 50%. Most phenols, except nitrophenols, reduce the reagent; therefore thymol and to a lesser degree sulfosalicylic acid interfere, while picric acid up to 0.1% is permissible. Plant tissues contain high concentrations of phenol so crude protein preparation would be a problem. Agents or buffers which acidify the solution, ammonium sulfate, copper chelators (i.e. EDTA, EGTA) or reducing agents (i.e. β-mercaptoethanol, DTT, phenols) will interfere with the assay. Different proteins produce different degrees of color intensity depending primarily on their tyrosine contents. Table 2.4 shows the effect of selected potential interfering compounds on the Lowry reaction (Smith *et al.*, 1985). Many researchers have modified the Lowry method for their specific needs (Peterson, 1979). Here, the technique as modified to make it less sensitive to interfering buffer components and more sensitive to low levels of protein is described (Stoscheck, 1990). This is similar to the assay reported by Peterson (1979) and Markwell *et al.* (1978).

Reagents: 2 × Lowry reagent concentrate. (a) Copper reagent, mixture of 7.7 w/v Na_2CO_3, 2% $CuSO_4.5H_2O$ and 1% sodium potassium tartrate in water. (b) 1% SDS. (c) 4% NaOH (1 M). (d) Immediately before use, 3 parts copper reagent (a) are mixed with 1 part (b) and 1 part (c). 2 × Lowry concentrate (d) is stable for 2 to 3 weeks. In the case of appearance of a black precipitate, the solution is discarded. If a white precipitate is formed, it is warmed at 37°C. (e) 0.2 N Folin reagent. This solution is stable for several months at room temperature if stored in an amber bottle.

Procedure: 400 μl of 2 × Lowry concentrate (d) are added to a 400 μl sample containing 10–100 μg protein and incubated for 10 min at room temperature. 200 μl of reagent (e) are then added and the mixture is vortexed immediately after each addition. The rapid mixing is very

Table 2.4 Effect of selected potential interfering compounds on the Lowry and BCA assay methods. Reproduced with permission from Smith, P.K., Krohn, R.I., Hermanson, G.T. *et al.* (1985) Measurement of protein using Bicinchoninic Acid. *Anal. Biochem.*, **150**, 76–85.

Sample (50 μg BSA) in	BCA assay (μg BSA found)		Lowry assay (μg BSA found)	
	Water blank corrected	Interference blank corrected	Water blank corrected	Interference blank corrected
Water (reference)	50.00	–	50.00	–
0.1 M HCl	50.70	50.80	44.20	43.80
0.1 M NaOH	49.00	49.40	50.60	50.60
0.2% Sodium azide	51.10	50.90	49.20	49.00
0.02% Sodium azide	51.10	51.00	49.50	49.60
1 M Sodium chloride	51.30	51.10	50.20	50.10
100 mM EDTA (4 Na)	No colour		138.50	5.10
50 mM EDTA (4 Na)	28.00	29.40	96.70	6.80
10 mM EDTA (4 Na)	48.80	49.10	33.60	12.70
50 mM EDTA, pH 11.25	31.50	32.80	72.30	5.00
4 M Guanidine–HCl	48.30	46.90	Precipitated	
3 M Urea	51.30	50.10	53.20	45.00
1% Triton X-100	50.20	49.80	Precipitated	
1% SDS	49.20	48.90	Precipitated	
1% Brij 35	51.00	50.90	Precipitated	
1% Lubrol	50.70	50.70	Precipitated	
1% Chaps	49.90	49.50	Precipitated	
1% Chapso	51.80	51.00	Precipitated	
1% Octyl glucoside	50.90	50.80	Precipitated	

40% sucrose	55.40	48.70	4.90	28.90
10% Sucrose	52.50	50.50	42.90	41.10
1% Sucrose	51.30	51.20	48.40	48.10
100 mM Glucose	245.00	57.10	68.10	61.70
50 mM Glucose	144.00	47.70	62.70	58.40
10 mM Glucose	70.00	49.10	52.60	51.20
0.2 M Sorbitol, pH 11.25	40.70	36.20	68.60	26.60
1 M Glycine	No colour		7.30	7.70
1 M Glycine, pH 11.25	50.70	48.90	32.50	27.90
0.5 M Tris	36.20	32.90	10.20	8.80
0.25 M Tris	46.60	44.00	27.90	28.10
0.1 M Tris	50.80	49.60	38.90	38.90
0.25 M Tris, pH 11.25	52.00	50.30	40.80	40.80
20% Ammonium sulphate	5.60	1.20	Precipitated	
10% Ammonium sulphate	16.00	12.00	Precipitated	
3% Ammonium sulphate	44.90	42.00	21.20	21.40
10% Ammonium sulphate, pH 11.25	48.10	45.20	32.60	32.80
2 M Sodium acetate, pH 5.5	35.50	34.50	5.40	3.30
0.2 M Sodium acetate, pH 5.5	50.80	50.40	47.50	47.60
1 M Sodium phosphate	37.10	36.20	7.30	5.30
0.1 M Sodium phosphate	50.80	50.40	46.60	46.60
0.1 M Caesium bicarbonate	49.50	49.70	Precipitated	

important, since the Folin reagent decomposes rapidly. The mixture is incubated for 30 min at room temperature. The absorbancies are read at 750 nm against the blank of sample buffer processed in the same way as the protein sample. If the absorbancies are too high, they may be read at 500 nm. The developed color is very stable. A standard curve is generated in the same range of protein (0–100 μg).

2.2.3.5 Bicinchoninic acid procedure (Smith et al., *1985).* Bicinchoninic acid (BCA), in the form of sodium salt, is water soluble, stable and a highly specific and sensitive reagent for Cu^+ in alkaline solution. It is similar to the Lowry assay in that in alkaline solution peptide bonds and amino acid side chains of mainly cysteine, cystine, tryptophan, and tyrosine react with Cu^{2+} and reduce it to Cu^+ (Biuret reaction). The Cu^+ then forms a complex with two molecules of BCA to produce a purple color which is water soluble and exhibits a strong absorbance at 562 nm. Figure 2.9 shows the formation of the purple complex with BCA and Cu^+ from the Biuret reaction (Smith *et al.*, 1985; Wiechelman *et al.*, 1988).

When the BCA method is compared to the Lowry method based on the results reported, the BCA reagent is less sensitive to such commonly encountered interferences as nonionic detergents and buffer salts. The stability of the reagent and the resulting chromophore also allows for a simplified, one-step analysis and enhanced flexibility in protocol selection. The BCA method retains the high sensitivity and low protein-to-protein variation associated with the Lowry technique (Smith *et al.*, 1985).

BCA is stable under alkaline conditions, while the Folin reagent is

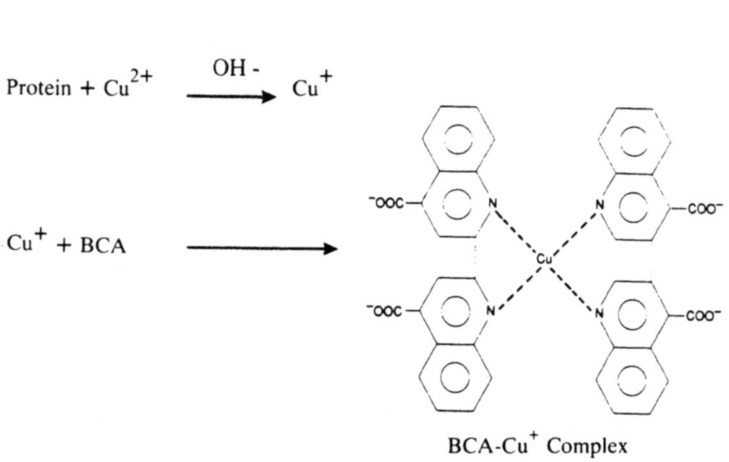

Figure 2.9 Formation of purple complex with BCA and cuprous ion generated from the Biuret reaction. Reproduced with permission from Smith, P.K., Krohn, R.I., Hermanson, G.T. *et al.* (1985), Measurement of protein using Bicinchoninic Acid. *Anal. Biochem.*, **150**, 76–85.

unstable. Thus, the BCA can be added to the alkaline copper solution in a one-step procedure. Table 2.4 shows the effect of interfering compounds on the Lowry and BCA assays (Smith *et al.*, 1985). The BCA assay is more tolerant to the presence of other compounds, which interfere with the Lowry assay. Copper-chelating reagents such as EDTA, EGTA, etc. interfere with both the Lowry and BCA assays by competing with BCA for binding copper. To eliminate the effects of interfering compounds, the protein solution may be precipitated by TCA or by acetone (Bensadoun and Weinstein, 1976; Cabib and Polacheck, 1984; Pohl, 1990). The standard BCA protocol is sensitive for proteins in the range of 0.02–1.0 mg/ml (1–40 μg protein per 1 ml assay). A micro-BCA procedure has also been developed for protein concentrations in the range of 0.5–10 μg/ml of protein (0.2–5 μg per 1 ml assay).

1. Standard BCA method

Reagents: (a) Reagent A, aqueous solution of 1% BCA-Na_2, 2% $Na_2CO_3.H_2O$, 0.16% Na_2 tartrate, 0.4% NaOH and 0.95% $NaHCO_3$. The pH is adjusted to 11.25 with 10 M NaOH. (b) Reagent B, 4% $CuSO_4.5H_2O$ in water. Solutions A and B are stable indefinitely at room temperature. These solutions are commercially available from Pierce Chemical Co. (c) Standard working reagent (SWR) is made by mixing 100 vols of Reagent A and 2 vols of Reagent B. This SWR is green in color and stable for one week.

Procedure: 1 vol. of sample (standard of unknown) is mixed with 20 vols of SWR (for convenience 100 μl sample and 2 ml SWR). The mixtures are incubated for 30 min at 60°C. The samples are cooled to room temperature and the absorbance at 562 nm is measured. The readings are stable for at least 1 h.

2. Micro-BCA method

Reagents: (a) Micro-reagent A (MA), 8% $Na_2CO_3.H_2O$, 1.6% Na_2 tartrate, 1.6% NaOH and sufficient $NaHCO_3$ to adjust the pH to 11.25 (it can be adjusted with 10 M NaOH). (b) Micro-reagent B (MB), 4% BCA-Na_2 in deionized water. (c) Micro-reagent C (MC) 4% (aq.) $CuSO_4.5H_2O$. (d) Micro-working reagent (MWR), 1 vol. of MC and 25 vols MB. 26 vols MA are then added to the mixture of MC and MB. MA and MB are stable indefinitely at room temperature, but MWR should be prepared as needed.

Procedure: 1 vol. of sample or standard is mixed with 1 vol. of MWR. The mixture is incubated for 60 min at 60°C, cooled to room temperature, and the absorbance is measured against a blank at 562 nm.

3. Microtiter assay. For low protein volumes ($\leq$100 μl) and concentration

(1–10 μg/ml), a microtiter plate can be used (Redinbaugh and Turley, 1986; Sorensen and Brodbeck, 1986; Pierce, 1992). 100 μl of each standard, blank or unknown sample are pipetted into the appropriate microtiter plate wells, 100 μl of working reagent are added to each well, and then mixed for 30 sec on a microtiter plate shaker. The microtiter plate is covered and samples are incubated at 37°C for 2 h. The absorbance is recorded at 562 nm with a microtiter plate reader. Overnight incubation improves the sensitivity. The lower detection limit for a 2 h incubation is approximately 1 μg/ml protein.

2.2.3.6 Biuret reaction. A dilute solution of cupric sulfate in strongly alkaline tartrate is added to the protein solution of 1–6 mg/ml. A purplish–violet color is formed, with maximum absorbance at 540–560 nm. The color is due to the interaction of Cu^{2+} ions with adjacent peptide nitrogens of peptide bonds, and formation of a tetracoordinated cupric ion complex (Creighton, 1984). Thus, the reaction is not dependent on specific amino acid composition as the Cu^{2+} reacts with the peptide chain itself rather than with side groups (Gornall *et al.*, 1949). Some buffers, particularly Tris buffers, interfere with the assay. Ammonia complicates the measurement of protein precipitated by ammonium sulfate and must be removed.

Reagent: Mixture of 0.3% copper sulfate ($CuSO_4.5H_2O$), 1.2% sodium potassium tartrate and 3% NaOH. Addition of 1% potassium iodide to the solution inhibits reduction of copper and stabilizes the reagent indefinitely.

Procedure: 3 vols of Biuret reagent are added to 1 vol. of sample or standard containing up to 3 mg of protein, mixed and left to stand for 30 min. The absorbance is measured at 540 nm. To increase the sensitivity, a 5 × concentrated Biuret reagent is made. 4 vols of protein solution and 1 vol. of the reagent are mixed and incubated for 30 min at room temperature. With this procedure, one can measure one-fifth of the concentration of protein as with the standard Biuret method.

2.2.3.7 Dye-binding method (Bradford, 1976). Coomassie Brilliant Blue G250 (Serva, Westbury, NY) forms strong, non-covalent complexes with proteins, at pH 1. Binding to protein shifts the absorption maximum of the dye in acidic solution from 465 nm (red form) to 595 nm (blue form) due to stabilization of the anionic form of the dye. Interaction with the protein may involve electrostatic interactions of the anionic form of the dye with NH_3^+ groups, van der Waals forces (Creighton, 1984) and hydrophobic interaction. The dye principally binds to basic amino acid residues, such as arginine and lysine, and aromatic amino acids such as tyrosine, tryptophan and phenylalanine residues (Compton and Jones, 1985). The dye binding to proteins is based on specific properties of the protein rather than simply on the weight of protein (Kley and Hale, 1977). The dye is prepared as a stock solution in phosphoric acid (Bradford, 1976; Spector, 1978) or perchloric acid (Sedmak and Grossberg, 1977); it is more stable in perchloric acid. The method is a one-step procedure; the dye reagent is added to a protein solution of 0.2–1.4 mg/ml for standard assay and 5–100 μg/ml for microassay. It is very tolerant to high concentrations of salt (i.e. 1 M KCl, 5 M NaCl, 1 M $MgCl_2$, 1 M ammonium sulfate), but is sensitive to EDTA, glycerol, sucrose, β-mercaptoethanol, phenol, SDS and Tritons. Table 2.5 shows the effect of various reagents on determination of protein by the Bradford procedure (Bradford, 1976).

The Bradford method is precise, but has large protein-to-protein variability (Pierce and Suelter, 1977). A modified procedure has been developed which has less protein-to-protein variation in color yield (Stoscheck, 1990). Some proteins do not give a linear relationship between absorbance and protein concentration. Measuring protein concentration with a standard based on the same protein may reduce the potential error (Kley and Hale, 1977). For quantitative measurements, it is recommended that the result be compared with another procedure before use. The following protocol is the modified method of Stoscheck (1990).

Reagent: (a) 0.01% dye (Serva-Blue G), 4.7% ethanol and 8.5% phosphoric acid. Dye is first homogenized in 50 ml of 95% ethanol and

Table 2.5 Effect of various laboratory reagents on the Bradford (Coomassie Brilliant Blue G250–protein complex) assay.[a] Reproduced with permission from Bradford, M.M. (1976) A rapid and sensitive method for the quantitation of microgram quantities of protein utilizing the principle of protein–dye binding. *Anal. Biochem.*, **72**, 248–54.

Substance	Change in absorbance at 595 nm	Equivalent BSA (μg)
1 M KCl	0.000	0.00
5 M NaCl	0.000	0.00
1 M $MgCl_2$	0.000	0.00
2 M Tris	0.026	2.34
0.1 M EDTA	0.004	0.36
1 M $(NH_4)_2SO_4$	0.000	0.00
99% Glycerol	0.012	1.08
1 M 2-Mercaptoethanol	0.004	0.36
1 M Sucrose	0.013	1.17
95% Ethanol	0.000	0.00
Acetone	0.069	6.21
5% Phenol	0.046	4.14
0.1% Triton X-100	0.013	1.17
1% Triton X-100	0.590	53.10
0.1% SDS	0.011	0.99
1% SDS	0.495	44.55
0.1% Hemosol	0.004	0.36
1% Hemosol	0.108	9.72

[a]Values were obtained when 0.1 ml of each substance were assayed in the standard assay.

then mixed with 100 ml of 85% phosphoric acid. After the dye is completely dissolved, it is diluted with water to 1 l.

Procedure: 100 μl of 1 M NaOH are added to a 40 μl sample containing 0.4–40 μg protein. 2 ml of the dye reagent are then added, mixed and incubated for 5 min. The absorbance is measured at 590 nm in glass or polystyrene cuvettes. The sensitivity of the assay may be increased by reading the difference between the absorbance of dye bound to protein (at 595 nm) and unbound dye (at 465 nm) (Bearden, 1978).

Depletion of free dye occurs at high concentrations of protein. To linearize the assay, the formula A_{595}–A_{465} is used, in which the depletion of the free dye is corrected (Bearden, 1978; Stoscheck, 1990).

2.2.3.8 *General considerations*

1. Protein standards. Most methods for protein measurement involve comparison of results obtained with an unknown protein, with a known protein standard curve. Consequently, the results obtained do not give an absolute measure of protein. The set of protein standards of known concentration should be in the same diluent as the unknown samples, and should cover the range of concentrations suitable for the assay. Different proteins, because of variability in their specific amino acid composition, produce different levels of reactivity. In preparing a standard protein, significant care should be taken if the protein is being weighed, since salts and moisture add weight to the protein. It is recommended that highly purified protein which has been extensively dialyzed and lyophilized should be used. The moisture content of the protein must be determined by drying under vacuum until no further changes are observed. Another method of choice is the use of highly purified standard proteins which are available commercially. Yet another option is to determine the protein concentration of the standard protein using trusted extinction coefficient values from the literature. In general, it is best to use the same standard protein as unknown protein. In some cases, this is not possible. In such a case, BSA should be used as it is universally accepted as a standard protein and makes the evaluation of results readily comparable.

2. Removal of interfering compounds. For solubility testing, it is recommended that a relatively pure protein is used, unless there is a need to determine the degree of solubility in a mixture (for example, for industrial use).

In the first step of measuring the protein concentration, strong acids, bases or buffers should be neutralized or adjusted to the pH of the assay system (Lowry, pH 10; BCA, pH 11; Bradford, pH 1) by addition of HCl or NaOH. Table 2.6 shows the neutralization compounds which can be used to eliminate the interference (Stoscheck, 1990).

Table 2.6 Neutralization compounds for interfering substances

Interfering substances	Neutralization compounds
Strong acid Strong acidic buffer	NaOH
Strong base Strong basic buffer	HCl
EDTA	Equimolar $CaCl_2$[a]
DTT	Iodoacetic acid[b]
Mercaptoethanol	Evaporation[c]
Cysteine	Iodoacetic acid[b]

[a]The $CaCl_2$ precipitates in alkaline solutions and must be removed by centrifugation.
[b]2 M iodoacetic acid dissolved in 2.5 M NaOH (Ross and Schatz, 1973).
[c]Makkar *et al.* (1980).

Contaminants can be removed by precipitating the protein with TCA, perchloric acid (PCA) or acetone and then centrifuging and solubilizing the pellet in a known volume of suitable solution for assay. In general, most proteins will be precipitated at 5–10% v/v TCA. Smaller proteins, of mol. wt. less than 20 000 dalton, may require up to 20% TCA. Excess TCA can be removed from the protein pellets by washing with buffer. TCA does not precipitate low concentrations of proteins, in the range of 1–5 μg. 100% recovery is achieved by using deoxycholate and TCA. A 2% solution of sodium deoxycholate (>pH 8) is added to the protein sample to a final concentration of 125–200 μg/ml. After 30 min at 4°C, a 20% solution of cold TCA is added to give a final concentration of 6%. After 1 h standing on ice, the mixture is centrifuged at 2500 *g* for 45 min at 4°C (or 3300 *g* for 30 min). This procedure works satisfactorily for precipitation of protein samples for protein determination (Bensadoun and Weinstein, 1976; Cabib and Polacheck, 1984; Harris and Angal, 1990; Ozols, 1990).

Using the above method, lipids, ions, reducing agents, some detergents, phenols, etc. can be removed. Detergents sometimes interfere with protein precipitation. Some detergents (e.g. deoxycholate, Triton X-100) precipitate at acidic pH values and thus cannot be removed by addition of TCA or PCA. Acid precipitation may result in pellets which are difficult to dissolve unless a coprecipitant ion is used. Sonication and incubation in 1 M NaOH will promote dissolution of the pellet (Stoscheck, 1990). Sometimes, heating or prolonged incubation in alkaline solution (0.1–1 M NaOH) facilitates complete solubilization. More details are given by Bensadoun and Weinstein (1976), Polacheck and Cabib (1981) and Wessel and Flügge (1984).

Dialysis and ultrafiltration may be required to remove the interfering substances. Dialysis of small volumes in large pore size dialysis bags can be done in 1–2 h. Ultrafiltration may take as little as 30 min. For example, to filter a 3 ml volume of a 1 mg/ml sample in 500 mM KCl, using an Amicon filter of 30 000 daltons molecular weight cut off, takes about 30 min to reduce the volume to 40–100 μl at 5000 rpm. The time of filtration depends on the viscosity of the sample and the pore size of the filter.

Interfering substances also can be removed by binding the protein sample to DEAE cellulose or affinity supports, washing the support and then eluting the protein in one step by using a buffer of high ionic strength to prevent protein dilution. It may be necessary to dialyze or lyophilize for long-term storage. This technique can also be used to concentrate the protein sample, if the protein concentration is not high enough for solubility testing.

Lipids may be removed by centrifugation for about 30 min at 4°C and 10 000 *g* or higher. Lipids accumulate on top of the solution and can be removed. The sample can then be filtered.

3. Moisture determination. All of the calculations for solubility testing are based on dried protein samples. Consequently, the amount of moisture in the sample needs to be determined. The AOCS Method Aa 3–38 (1987) or the AOAC Method 927.05 (1990) can be used. In general, to determine the moisture and volatile matter of the sample, an accurately weighed amount of protein, in duplicate, is dried in pre-dried and weighed dishes in an oven at 130°C for 3 h or to constant weight. The dishes are removed from the oven, covered immediately and cooled in a desiccator to room temperature and weighed. The percent moisture is calculated as percent loss in weight. The weight of protein sample is then corrected for the percent moisture.

Another approach is to dry the sample and use the dried sample for protein determination. In such cases, the sample is dried in vacuum oven at 70–100°C under reduced pressure (≤100 mm (4″) of Hg) to constant weight. During drying, a slow current (*c.* two bubbles/sec) of air, dried by passing through H_2SO_4, is admitted into the oven. Samples need to be covered immediately, before removal from the oven, and cooled in a desiccator to prevent moisture uptake. Percent loss in weight is calculated as moisture.

An alternative procedure is to desiccate the sample under vacuum over P_2O_3 until constant weight is achieved. A microbalance is used to weigh the amounts of proteins in mg.

2.3 Protocols for methods

In this section the practical protocols based on the most appropriate method for a particular protein type and application will be outlined and discussed. These protocols are based on the methods discussed in section 2.2, and experiences gained during working on solubility determination with many proteins.

2.3.1 Protein dispersion and solubilization

Proteins are often most soluble and stable under conditions mimicking their natural environment. Serum proteins are soluble in a pH and salt range (pH 6.8 and 0.15 M) where as mature insulin, which is stored in acidic granules in the cell, precipitates (Markussen *et al.*, 1988).

A prime consideration in developing a methodology for protein dispersion and solubilization is the stability of the protein. In their native folded structures, proteins are more stable to aggregation and proteolysis than are unfolded denatured structures. There is a close relationship between protein solubility and protein stability. The free energy of stabilization of proteins in aqueous solution is very low (*c.* 12 kcal/mol at 30°C compared with 8 kcal/mol at 50°C and 6 kcal/mol at 60°C; Privalov, 1979); consequently proteins totter on the verge of denaturation (Schein, 1990). Protein stability can be increased by selection of proper solvent, addition of stabilizer, high concentration of proteins, or by alteration of the protein structure. However, at high concentrations the danger of protein association exists. This problem can be avoided in the presence of 0.5 to 1.0 M NaCl or KCl (Jazwinski, 1990).

Appropriate choice of the solution in which the proteins will be solubilized is important. This choice is dictated by compatibility with the method of solubilization, the subsequent protein determination step and the utilization of the protein solution. Proteins also need to be dispersed or solubilized under the conditions that stabilize them. Besides pH and temperature, the ionic strength of the solution is very important and needs to be carefully monitored.

2.3.1.1 Protein concentration. Solubility depends on the initial concentration of protein as discussed in section 2.1.2.5. Concentrations of 1–10 mg/ml have been used for solubility testing (Kinsella, 1976; Townsend and Nakai, 1983; Morr *et al.*, 1985; Waniska and Kinsella, 1989; Vojdani and Whitaker, 1994). In general, protein solubilities decrease as the protein concentration increases and protein aggregation may be introduced at concentrations more than 3 mg/ml (Harris and Angal, 1990).

A concentration of 1–5 mg/ml is recommended for solubility procedures, unless a specific concentration need be tested.

2.3.1.2 Ionic strength of the solvent. To disperse a protein, four categories of protein are considered, based on the required ionic strength for their solubilization. (a) Proteins in the first category are readily soluble in salt-free medium or at very low ionic strength. Examples of these types of proteins are albumins (i.e. serum albumin, ovalbumin, α-lactalbumin) and some globular proteins such as enzymes (i.e. lysozyme, trypsin, chymotrypsin, etc). (b) Proteins in the second category are those that need a moderate ionic strength (0.1–0.15 M) for their solubilization. Such proteins are globulins (i.e. β-lactoglobulin, arachin and conarachin of peanuts, glycinin of soybeans, legumin of peas), leucosine and glutelins of wheat, oryzenin of rice, hordenine of rye, avenin of oats and some glycoproteins, etc. Histones (found in animals and characterized by high contents of basic amino acids) are also soluble in water, dilute acids and alkali. Most of the proteins in biological systems are in this second category. The ionic strength of blood and eukaryotic cytoplasm is of the order of 0.15 M. (c) The third category of proteins are those that need a relatively high ionic strength in the order of 0.3–0.6 M, for their solubilization and stability. This category includes membrane, cytoskeletal, myofibril, cytoplasmic yeast and bacterial proteins (Dinnbier *et al.*, 1988; Fargin *et al.*, 1988; Lopez-Bote *et al.*, 1989) and prolamins, which are soluble in 50–90% alcohol but insoluble in water, absolute alcohol and other neutral solvents. The prolamins occur in plants such as wheat, rye (gliadin), corn (zein), and barley (hordein) (Altschul, 1964). Zein is soluble in 0.1 M NaCl at pH 12, and is almost 40% as soluble at pH 11 (Casella and Whitaker, 1990). It is also soluble in 60% aqueous acetone. (d) The fourth category of proteins may even need higher salt concentrations (0.5–1 M) for their stabilization and solubilization. This category includes membrane proteins and proteins forming inclusion bodies in yeast and bacteria.

Most food proteins can be classified within the first three categories. Based on this classification, three different solubilization conditions are devised. These require aqueous solvents with conditions of:

(1) ionic strength 0.0–0.025 M;
(2) ionic strength 0.15 M;
(3) ionic strength 0.5 M.

2.3.1.3 pH of the solvent. Maintaining a constant pH of a protein solution is the most critical point in solubility testing. Solubility behavior of most of proteins is a function of pH (Vojdani and Whitaker, 1994). To maintain such a pH, one has a choice of appropriate buffers (Tables 2.2 and 2.3). However, the choice of buffer is also dictated by compatibility with the subsequent step of protein determination and utilization methods. The specific nature of the ions may affect protein solubility by specific interaction with the protein. For example, myosin in phosphate buffer of ionic strength lower than 150 mM, pH 6.8–8.0, is insoluble, while in pyrophosphate buffer of ionic strength 40 mM, pH 8.0, it is soluble (critical concentration ≥200 μg/ml). This may be due to specific interaction of the pyrophosphate ion with myosin (Frederiksen and Cunningham, 1982). Another limitation is that one buffer system cannot provide buffering over the whole range of pH 2–11 (Tables 2.2 and 2.3). This may cause an error in studying the solubility properties as a function of pH, because of the need to choose different buffer systems in different pH ranges. Sometimes a mixture of buffers is used for a wider range of pH. However, careful choice of buffer is recommended, whether one is studying the solubility behavior in one specific pH or over a range of pH values. Another important feature is the choice of buffer concentration. This mainly depends on the buffering capacity of the system, the type of buffer and ionic strength required, and the concentration of protein.

A second option for maintaining a constant pH is to adjust the pH with 0.1–1 M HCl or NaOH, monitoring with a pH meter, while stirring the protein. The pH needs to be monitored for at least 30 min, to make sure a constant pH is maintained. Proteins are excellent buffers. Three protocols for maintaining pH are outlined below; the last protocol is most strongly recommended since the concentration of extraneous ions is low, and, as already stated, proteins themselves are excellent buffers and provide buffering capacity to the solutions.

Protocol A: Buffer concentration of 10 mM for low protein concentrations, in the range of 1–10 mg/ml.

Protocol B: Buffer concentration of 50 mM for high protein concentrations, in the range 10–100 mg/ml.

Protocol C: Adjust the pH by addition of NaOH or HCl.

For any of the three protocols, the final ionic strength (maximum of 0.025, 0.15 and 0.5 M; section 2.3.1.2) is adjusted by addition of NaCl.

2.3.1.4 Mixing procedure and duration of mixing

Procedure A. After choosing the ionic strength (condition 1, 2 or 3 in section 2.3.1.2) of the buffer system (protocol A, B or C in section 2.3.1.3), an accurate amount of dried protein or an aliquot of protein stock suspension with known concentration of protein (final concentration 1–10 mg/ml) is dispersed in the solvent. This is by gradual addition of solvent to protein powder, with stirring to disperse the protein evenly and to facilitate solubilization. The protein mixture is stirred at a low speed to prevent aeration and foaming, since foam formation leads to protein denaturation. The pH is adjusted (protocol C in section 2.3.1.3) by addition of 0.1–1 M HCl or NaOH and monitoring the pH with a pH meter for a period of 30 min to make sure it has reached a constant pH. The mixture is then incubated for an additional 30 min at 24°C to ensure equilibrium and, at the end of incubation, the pH again is measured. When protein is solubilized in a buffer system, the pH is measured prior to and after incubation periods, to maintain a constant pH. An appropriate stirring device is required to prevent heating of the solution during the stirring. One device is an open-mesh sink matting (Morr *et al.*, 1985) or any device that prevents direct contact of the protein container with the stirring plate.

Procedure B. The same procedure as A is followed except that the duration of mixing is longer, 60 min. This procedure is used for categories (b) and (c) proteins (section 2.3.1.2) which are more difficult to disperse and solubilize (i.e. proteins which are more hydrophobic and have less ionic groups) than category (a) proteins.

2.3.1.5 Additional comments. Sodium azide (NaN_3, 0.02%) can be included in the solvent to inhibit possible microbial growth during dispersion and solubilization. Protease inhibitors are included to protein solutions from sources which are highly susceptible to proteolysis, such as bacterial and yeast protein extract. It may be necessary to perform the solubilization and following steps at a lower temperature (0–5°C).

It is essential that after equilibrium at 24°C, the protein solution is not stored in a refrigerator. Protein solutions have different solubilities at different temperatures, therefore protein precipitation may occur. If the solubility test needs to be performed at a lower temperature, all of the steps must be performed at that temperature.

A slightly higher temperature during the initial steps of dispersing may improve the solubilization process for categories (b) and (c) protein. However, it is necessary to keep the temperature constant during pH adjustment, equilibration and all subsequent stages of the solubility test.

A long incubation time may result in protein oxidation and/or polymerization. Polymerized proteins are usually not soluble and precipitate. This is a technical error, rather than intrinsic behavior of the protein itself. It is especially true for proteins with the free –SH groups. In the presence of air, the –SH groups are oxidized, forming S–S bonds; intermolecular –S–S– bonds result in a polymerized complex. To prevent oxidation, 0.5 mM DTT or β-mercaptoethanol are added.

2.3.2 Clarification and phase separation

Clarification of the solution is the next procedure after the protein solution has reached a state of equilibrium. Centrifugation, followed by microfiltration, are used to clarify the protein suspension. They are used in conjunction to ensure that all particles are removed. The main factors controlling the centrifugal forces for the separation of protein particles (undissolved and precipitated forms) is the size and density of the aggregates or particles involved. The size and density of the particles also determines the ease of removal.

In Table 2.7, the size and density of some biological cells are given in comparison to protein precipitates, for determining suitable centrifugation forces required to remove the protein precipitates. As shown, protein precipitates have large size differences, dependent on the type of protein in a specific environment and the form of aggregation. However, most protein precipitates are larger than 0.5 μm.

Typical bacteria, in low-viscosity media such as saline, are usually sedimented at 2000–3000 *g* for 10–15 min. In higher viscosity solutions, higher centrifugal forces and/or longer times are required. For cell debris, or bacterial or fungal spores, higher speeds (up to 12 000 *g*) and longer times (30–45 min) are required, because of smaller size and lower density, respectively. For protein precipitates, 15 000 *g* for 10 min or 5000 *g* for 30 min is usually sufficient. Theoretical calculations can be used to give an indication of the required time and speed, although in practice there is no substitute for preliminary test work (Whittington, 1989).

Two types of centrifuge rotors are available: the angle or conical rotor and the swinging bucket rotors, which can be used in high-speed centrifuges. As discussed in section 2.2.2, the centrifugal force is

Table 2.7 Features of biological cells. Reproduced with permission from Whittington, P.N. (1989) Clarification and extraction, in *Protein Purification Methods. A Practical Approach*, IRL Press, Oxford.

Cell type	Shape	Size (μm)	Density (kg m^3)
Bacteria	Rods Cocci	0.5–3	1050–1080 1050–1090
Yeast	Sphere	5–10	1050–1090
Filamentous fungi; Actinomycetes	Long filaments	1 × 100's	1050–1090
Plant	Irregular	20–100	1050–1090
Animal	Non-defined	10–40	1017–1130[a]
Cell flocs	Assorted	10–100's	1010–1080
Cell debris	Assorted	0.4	1010–1200
Protein precipitates	Assorted	0.1–100's	1010–1200

[a]From Pertoft, H. and Laurent, T.C. (1982)

proportional to the radius of rotation, r (cm). In a swinging bucket rotor, the tubes swing out to a horizontal position. In this type of rotor, a particle travels a longer distance to settle at the bottom. To compensate, prolonged centrifugation and higher speeds are required. In the angle rotor, where the tubes are held at a fixed angle of about 45° in holes in the rotor, the particles travel a distance determined by the diameter of the tube times the secant of the angle of inclination from the vertical. The sedimenting particles deposit at an angle. The particles that strike the tube wall aggregate and slide to the bottom (Pomeranz and Meloan, 1978).

In an Eppendorf-type centrifuge, a particle travels a short distance of less than 10 mm. At its maximum speed (12 000–14 000 rpm) centrifugation for 20–30 min is very effective in sedimentation of most protein precipitates. This speed generates a maximum force of 13 500–18 500 g (DuPont, 1993).

Protocol A. After equilibration, the protein solution is immediately centrifuged to separate the precipitated and undissolved protein from the protein suspension. The solution is transferred to a 1.5–2 ml Eppendorf tube and centrifuged at 16 000 g for 30 min in a high-speed, small centrifuge. The force of 16 000g is equivalent to 13 000–14 000 rpm (maximum speed). Immediately after centrifuging, the supernatant is transferred to a new tube. The following precautions should be taken: do not disturb the pellet during the transfer of supernatant; do not remove all of the supernatant, since by doing so, there is more probability of transferring the sediment materials. The supernatant is then filtered through a microfilter with pore size of 0.22–0.45 μm. For globular proteins, with mol. wt. approximately 300 000 daltons and less, a pore size of 0.22 μm is used (Stokes' radius of 11S globulins from safflower seed and some oilseeds of mol. wt. approximately 300 000 dalton is about 6 nm; Zirwer *et al.*, 1989). Microfilter tubes containing 0.22 μm membrane filters are commercially available (such as Microfilterfuge 0.22 μm Nylon-66 membrane filters). This type of filter has basically no protein-binding affinity. Microfiltration removes any protein particles transferred in the above procedures.

For fibrous, tubular or rod-shaped proteins (i.e. actin or myosin), a much larger pore size is needed. Tropomyosin, a contractile protein, is a coiled coil α-helix with mol. wt. 66 000–70 000 daltons (α–α and α–β subunits) and a length of 400 Å (Harrington, 1979). Myosin is a 150 nm long coiled coil α-helix terminating in two globular regions (Margossian and Lowey, 1982). For these types of proteins a 0.45–15 μm (10 to 20 times the gyration radius) microfilter is appropriate. Some food proteins form microstructures (spheres, granules, particles and micelles), with diameters of 0.1–40 μm (Heertje, 1993). Micelles of casein have diameters of 30–300 nm (milk caseins exist with calcium phosphate in a spherical complex of micelles with particle weights from 10^7–10^9; Brunner, 1977); spheres and granules of egg yolk proteins are 20–40 μm and 1.0–1.7 μm, respectively (Powrie and Nakai, 1985). For testing the solubility of such microstructures, a Whatman filter No. 1 or equivalent is used. The filtrate is saved for determining the soluble protein concentration.

Protocol B. In cases where undissolved and/or aggregated proteins are not precipitated properly at 16 000 *g*, higher forces of 30 000 *g* are required. Larger high-speed centrifuges are used and a fixed angle rotor is recommended. After centrifugation, samples are filtered as in protocol A.

Comment. The best approach is to centrifuge a sample at different *g* forces of 16 000, 20 000 and 30 000 *g* for 30 min and determine the efficiency of separation of the undissolved protein. However, in most cases a 16 000 *g* force is adequate (Whittington, 1989). In such cases, a small centrifuge with a speed of 12 000–14 000 rpm for 30 min, especially for small volumes (1.5–2 ml), is sufficient. Most laboratories are equipped with this type of centrifuge, which is very easy to operate.

Large high-speed centrifuges, for instance the Sorvall-DuPont centrifuge with SS-34 or SA-600 rotors, accommodate tubes with a maximum volume of 50 ml. Different rubber adapters are available for centrifuging volumes ranging from 1–50 ml in such rotors. However, rotors to centrifuge microsamples (≤1 ml) in large ultra- or high-speed centrifuges (i.e. type 25, 42.2 Ti and JA 18.1 rotors, Beckman), and tabletop ultracentrifuges for small volumes (≥0.2 ≤5 ml) are commercially available.

For 15 000 *g*, special toughened glass (Corex) is used. Polypropylene tubes withstand higher *g* forces than 15 000 *g* and are chemically very resistant. Precautions should be taken to ensure that shape of the tube is the same as the hole in the adaptor and rotor.

2.3.3 Determination of protein concentration

Protocol A. Determination of protein concentration by a micro-Kjeldahl method gives accurate and reliable results, providing that a correction is made for non-protein nitrogen in the sample. This method is very suitable for most commercial food proteins and feed proteins, since low amounts of lipid and even high amounts of carbohydrate and other contaminants do not interfere with the procedure. Technically, however, it is laborious and time-consuming, especially when the laboratory is not equipped with an automated system for micro-Kjeldahl determination.

The amount of proteins in the supernatant and in the sample before centrifugation is determined using appropriate aliquot volumes. The percent solubility of the protein product is calculated and expressed as percent nitrogen solubility.

Table 2.8 Concentration limits of chemicals in protein assays.[a] Reproduced with permission from Stoscheck, C.M. (1990) Quantitation of proteins, in *Methods in Enzymology*, Vol. 182 (ed. M.P. Deutscher), Academic Press, New York.

Substance[b]	Concentration limits				
				UV[f]	
	Enhanced copper[c]	BCA[d]	Dye[e]	280 nm	205 nm
Acids and bases					
HCl		0.1 M	0.1 M	>1 M	0.5 M
NaOH		0.1 M	0.1 M	>1 M	25 mM
PCA	>1.25%	<1%		10%	1 M
TCA	>1.25%	<1%		10%	<1%
Buffers					
Acetate		0.2 M	0.6 M	0.1 M	10 mM
Ammonium sulfate	>28 mM	20%	1 M	>50%	9%
Borate		10 mM			>100 mM
Citrate	2.5 mM	<1 mM	50 mM	5%	<10 mM
Glycine	2.5 mM	1 M	0.1 M	1 M	5 mM
HEPES	2.5 μM	100 μM	100 mM		<20 mM
Phosphate	250 mM	250 mM	2 M	1 M	50 mM
Tris	250 mM	0.1 M	2 M	0.5 M	40 mM
Detergents					
Brij 35		1%		1%	1%
CHAPS		1%		10%	<0.1%
Deoxycholate	625 μg/ml		0.25%	0.30%	0.1%
Digitonin				10%	
Lubrol PX		1%		10%	
Octylglucoside		1%		10%	
SDS	1.25%	1%	0.10%	0.10%	0.10%
Triton X-100	0.25%	1%	0.10%	0.02%	<0.01%
Triton X-100(R)				>10%	2%
Tween 20	0.10%	1%		0.30%	0.10%
Reductants					
Dithiothreitol	50 μM	<1 mM	1 M	3 mM	0.1 mM
2-Mercaptoethanol	1.8 μM	<1%	1 M	10 mM	<10 mM
Miscellaneous					
DNA/RNA	0.2 mg	0.1 mg	0.25 mg	1 μg	
DMSO	>6.2%	5%		20%	<10%
EDTA	125 μM	10 mM	0.1 M	30 mM	0.2 mM
Glycerol	25%	10%	100%	40%	5%
KCl	30 mM	<10 mM	1 M	100 mM	50 mM
NaCl	1.75 M	1 M	5 M	>1 M	0.6 M
Sucrose	50 mM	40%	1 M	2 M	0.5 M
Urea	>200 mM	3 M	6 M	>1 M	<0.1 M

[a]This table is a general guide. Test buffer mixtures as described in the text. Figures preceded by (<) or (>) symbols indicate that the tolerable limit for the chemical is unknown but is, respectively, less than or greater than the amount shown. Blank spaces indicate that the data were unavailable.

[b]PCA, perchloric acid; TCA, trichloroacetic acid; HEPES, *N*-2-hydroxyethylpiperazine-*N*′-2-ethanesulfonic acid; CHAPS, 3-[(3-cholamidopropyl)dimethylammonio]propanesulfonic acid; SDS, sodium dodecyl sulfate; (R), reduced; DMSO, dimethyl sulfoxide; EDTA, ethylenediamine-tetraacetic acid.

[c]Figures indicate the final concentration of the chemical in the assay.

[d]Figures indicate the concentration of the chemical in a 50 μl sample.

[e]Figures indicate the concentration of the chemical in a 25 μl sample.

[f]Figures indicate the final concentration of the chemical which does not produce an absorbance of 0.5 over water.

Protocol B. Among other methods for protein quantitation, the BCA method (Smith *et al.*, 1985) is a very sensitive and accurate method for most proteins (section 2.2.3.5). As little as 2 μg protein can be measured. The BCA microtiter plate protocol requires even smaller amounts of protein.

The BCA method is similar to the Lowry method, but is more stable and less sensitive to interference. The BCA method can tolerate up to 40% sucrose and higher amounts of glucose (up to 30–40 mM), while the Lowry method cannot be used in this range of sucrose and glucose concentrations. The Lowry reaction compound precipitates in 1% octyl glucoside while the BCA method is almost unaffected (Table 2.4). However, the Lowry method is accurate for protein quantitation when the protein is relatively pure and no interfering substances are present. Both the BCA and Lowry methods show some protein-to-protein variability but this is significantly less than in the Coomassie dye-binding method.

In a study of protein solubility by Morr *et al.* (1985), the Biuret method gave poorer reproducibility than the micro-Kjeldahl method and there was lack of agreement between the two methods. It was concluded that the lipid was the primary interfering substance with the Biuret method.

In the present protocol, therefore, the BCA method is used for protein determination. If, because of protein structure or the presence of interfering substances, the BCA method is not applicable, other methods can be used.

Table 2.8 shows the concentration limits of chemical in protein assays by the Lowry (enhanced copper), BCA, dye (Bradford), and UV measurement methods. It includes substances such as acids and bases, buffers, detergents, reductants and others (Stoscheck, 1990). Although not a broad range of substances, this gives a guideline for finding the best procedure for specific conditions in the presence of the other substances investigated.

As long as other methods of protein determination give comparable results with the micro-Kjeldahl method, they can be used for the solubility test. A primary test comparing a method with the micro-Kjeldahl method needs to be done before any decision on the choice of method is made.

Acknowledgement

The author gratefully acknowledges Professor John R. Whitaker for his critical review of this manuscript, his suggestions and constant interest.

References

Adler-Nissen, J. and Olsen, H.S. (1979) The influence of peptide chain length on taste and functional properties of enzymatically modified soy protein, in *Functionality of Protein Structure* (ed A.P. Peaco), ACS Symposium Series 92, American Chemical Society, Washington D.C., pp. 125–46.

Ahn, J.K., Sen, L.C. and Whitaker, J.R. (1991) Stability of tertiary structure of phaseolin of red kidney bean (*Phaseolus vulgaris*) as limiting factor in proteolysis. *J. Food Biochem.*, **15**, 263–78.

Altschul, A.M. (1964) Seed proteins, in *Proteins and their Reactions* (eds H.W. Schultz and A.F. Anglemier), AVI Publishing Company, Inc., Westport, Conn., pp. 295–313.

AOAC (1990) *Official Methods of Analysis*, 15th edn, Association of Official Analytical Chemists Inc., Arlington, VA.

AOCS (1987) *Official Methods and Recommended Practices of the American Oil Chemists' Society*, 3rd edn, American Oil Chemists' Society, Champaign, IL.

Arakawa, T. and Timasheff, S.N. (1985) Theory of protein solubility, in *Methods in Enzymology*, Vol. 114 (eds H.W. Wyckoff, C.H.W. Hirs and S.N. Timasheff), Academic Press, Inc. New York, pp. 49–77.

Bearden, J.C., Jr. (1978) Quantitation of submicrogram quantities of protein by an improved protein–dye binding assay. *Biochem. Biophys. Acta*, **533**, 525–9.

Bensadoun, A. and Weinstein, D. (1976) Assay of proteins in the presence of interfering materials. *Anal. Biochem.*, **70**, 241–50.

Bera, M.B. and Mukherjee, R.K. (1989) Solubility, emulsifying, and foaming properties of rice bran protein concentrates. *J. Food Sci.*, **54**, 142–5.

Bradford, M.M. (1976) A rapid and sensitive method for the quantitation of microgram quantities of protein utilizing the principle of protein–dye binding. *Anal. Biochem.*, **72**, 248–54.

Brunner, J.R. (1977) Milk proteins, in *Food Proteins* (eds J.R. Whitaker and S.R. Tannenbaum), AVI Publishing Company, Inc., Westport, Conn., pp. 175–208.

Cabib, E. and Polacheck, I. (1984) Protein assay for dilute solutions, in *Methods in Enzymology*, Vol. 104 (ed. W.B. Jakoby), Academic Press, Inc., New York, pp. 415–16.

Campbell, N.F., Shih, F.F. and Marshall, W.E. (1992) Enzymatic phosphorylation of soy protein isolate for improved functional properties. *J. Agric. Food Chem.*, **40**, 403–6.

Casella, M.L.A. and Whitaker, J.R. (1990) Enzymatically and chemically modified zein for improvement of functional properties. *J. Food Biochem.*, **14**, 453–75.

Cheftel, J.C., Cuq, J.L. and Lorient, D. (1985) Amino acids, peptides, and proteins, in *Food Chemistry*, 2nd edn (ed. O.R. Fennema), Marcel Dekker, Inc., pp. 245–369.

Chin, J.T., Wheeler, S.L. and Klibanov, A.M. (1994) Communication to the editor on protein solubility in organic solvents. *Biotechnol. Bioeng.*, **44**, 140–5.

Chobert, J.M., Sitohy, M. and Whitaker, J.R. (1987) Specific limited hydrolysis and phosphorylation of food proteins for improvement of functional and nutritional properties. *J. Amer. Oil Chem. Soc.*, **64**, 1704–11.

Chobert, J.M., Bertrand-Harb, C. and Nicolas, M.G. (1988a) Solubility and emulsifying properties of caseins and whey proteins modified enzymatically by trypsin. *J. Agric. Food Chem.*, **36**, 883–92.

Chobert, J.M., Sitohy, M. and Whitaker, J.R. (1988b) Solubility and emulsifying properties of caseins modified enzymatically by *Staphylococcus aureus* V8 protease. *J. Agric. Food Chem.*, **36**, 220–4.

Compton, S.J. and Jones, C.G. (1985) Mechanism of dye response and interference in the Bradford protein assay. *Anal. Biochem.*, **151**, 369–74.

Creighton, T.E. (1984) *Proteins. Structures and Molecular Principles*, W.H. Freeman and Company, New York.

Damodaran, S. (1989) Interrelationship of molecular and functional properties of food proteins, in *Food Proteins* (eds J.E. Kinsella and W.G. Soucie), The American Oil Chemists' Society, Champaign, IL, pp. 21–51.

Deutscher, M.P. (1990) Guide to Protein Purification, in *Methods in Enzymology*, Vol. 182 (ed. M.P. Deutscher), Academic Press Inc., New York.

Dinnbier, U., Limpinsel, E., Schmid, R. *et al.* (1988) Transient accumulation of potassium glutamate and its replacement by trehalose during adaptation of growing cells of *Escherichia coli* K-12 to elevated sodium chloride concentrations. *Arch. Microbiol.*, **150**, 348–57.

Dunn, M.J. (1989) Determination of total protein concentration, in *Protein Purification Methods. A Practical Approach* (eds E.L.V. Harris and S. Angal), IRL Press, Oxford, pp. 10–18.

DuPont (1993) *Instruction Manual. Sorvall FA-Micro. Microcentrifuge*, DuPont Company, Wilmington, DE.

Englard, S.and Seifter, S. (1990) Precipitation techniques, in *Methods in Enzymology*, Vol. 182 (ed M.P. Deutscher), Academic Press Inc., New York, pp. 285–300.

Fargin, A., Faye, J.C., Le Maire, M. *et al.* (1988) Solubilization of a tamoxifen-binding protein. *Biochem. J.*, **256**, 229–360.

Farrell, H.M. Jr., Kumonsinski, T.F., Pulaski, P. *et al.* (1988) Calcium-induced associations of the caseins: a thermodynamic linkage approach to precipitation and resolubilization. *Arch. Biochem. Biophys.*, **265**, 146–58.

Ferguson, W.J., Braunschweiger, K.I., Braunschweiger, W.R. *et al.* (1980) Hydrogen ion buffers for biological research. *Anal. Biochem.*, **104**, 300–10.

Frederiksen, D.W. and Cunningham, L.W. (eds) (1982) Structural and Contractile Proteins. Part B, *Methods in Enzymology*, Vol. 85, Academic Press, New York.

Gifford-Steffen, S.R. and Clydesdale, F.M. (1993) Effect of varying concentrations of phytate, calcium, and zinc on the solubility of protein, calcium, zinc, and phytate in soy protein concentrate. *J. Food Protec.*, **56**, 42–6.

Gilbert, L.M. and Gilbert, G.A. (1973) Sedimentation velocity measurement of protein association, in *Methods in Enzymology*, Vol. 27 (eds C.H.W. Hirs and S.N. Timasheff), Academic Press Inc., New York, pp. 273–96.

Good, N.E, Winget, G.D., Winter, W. *et al.* (1966) Hydrogen ion buffers for biological research. *Biochemistry*, **5**, 467–77.

Gornall, A.J. Bardawill, C.J. and David, M.M. (1949) Determination of serum proteins by means of the Biuret reaction. *J. Biol. Chem.*, **177**, 751–66.

Green, A.A. (1932) Studies in the physical chemistry of the proteins. *J. Biol. Chem.*, **95**, 47–66.

Halling, P.J. (1981) Protein-stabilized foams and emulsions. *CRC Crit. Rev. Food Sci. Nutr.*, **13**, 155–203.

Hamaguchi, K. (1992) *The Protein Molecule, Conformation, Stability and Folding*, Springer-Verlag, New York.

Harrington, W.F. (1979) Contractile proteins of muscle, in *The Protein*, Vol. IV, 3rd edn (eds H. Neurath and R.L. Hill), Academic Press, New York, pp. 245–409.

Harris, E.L.V. and Angal, S. (1989) *Protein Purification Methods. A Practical Approach*, IRL Press, Oxford.

Harris, E.L.V. and Angal, S. (1990) *Protein Purification Applications. A Practical Approach*, IRL Press, Oxford.

Haurowitz, F. (1963) *The Chemistry and Function of Proteins*, 2nd edn, Academic Press, New York and London.

Hayakawa, S. and Nakai, S. (1985) Relationships of hydrophobicity and net charge to the solubility of milk and soy proteins. *J. Food Sci.*, **50**, 486–91.

Heertje, I. (1993) Structure and function of food products: a review. *Food Struct.*, **12**, 343–64.

Hermansson, A.M. (1973) Determination of functional properties of protein foods, in *Proteins in Human Nutrition* (eds J.W.G. Porter and B.A. Rolls), Academic Press, London, pp. 407–37.

Jazwinski, S.M. (1990) Preparation of extract from yeast, in *Methods in Enzymology*, Vol. 182 (ed. M.P. Deutscher), Academic Press Inc., New York, pp. 154–74.

Kilara, A. and Sharkasi, T.Y. (1986) Effect of temperature on food proteins and its implications on functional properties. *CRC Crit. Rev. Food Sci. Nutr.*, **23**, 323–95.

Kinsella, J.E. (1976) Functional properties of proteins in foods: A survey. *CRC Crit. Rev. Food Sci. Nutr.*, **7**, 219–80.

Kinsella, J.E. (1984) Milk proteins: physicochemical and functional properties, *CRC Crit. Rev. Food Sci. Nutr.*, **21**, 197–262.

Kinsella, J.E. and Soucie, W.G. (1989) *Food Proteins*. The American Oil Chemists' Society, Champaign, IL.

Kirkwood, J.G. (1943) The theoretical interpretation of the properties of solutions of dipolar ions, in *Proteins, Amino Acids and Peptides* (eds E.J. Cohn and J.T. Edsall), Reinhold Publishing Corp., NY, pp. 276–303.

Kley, H.V. and Hale, S.M. (1977) Assay for protein dye binding. *Anal. Biochem.*, **81**, 485–7.

Kumar Murthy, N.V.K. and Narasinga Rao, M.S. (1984) Acid denaturation of mustard 12S protein. *Int. J. Peptide Protein Res.*, **23**, 94–103.

Lee, B. and Richards, F.M. (1971) The interpretation of protein structures: estimation of static accessibility. *J. Mol. Biol. Chem.*, **55**, 379–400.

Li-Chan, E. and Nakai, S. (1989) Effect of molecular changes (SH groups and hydrophobicity) of food proteins on their functionality, in *Food Proteins* (eds J.E. Kinsella and W.G. Soucie) The American Oil Chemists' Society, Champaign, IL, pp. 232–51.

Li-Chan, E., Nakai, S. and Wood, D.F. (1984) Hydrophobicity and solubility of meat proteins and their relationship to emulsifying properties. *J. Food Sci.*, **49**, 345–50.

Lopez-Bote, C., Warriss, P.D. and Brown, S.N. (1989) The use of muscle protein solubility measurements to assess pig lean meat quality. *Meat Sci.*, **26**, 167–75.

Lowry, O.H., Rosebrough, N.J., Farr, A.L. *et al.* (1951) Protein measurement with the folin phenol reagent. *J. Biol. Chem.*, **193**, 265–75.

Makkar, H.P.S., Sharma, O.P. and Negi, S.S. (1980) Assay of proteins by Lowry's method in the presence of high concentrations of β-mercaptoethanol. *Anal. Biochem.*, **104**, 124–6.

Margossian, S.S. and Lowey, S. (1982) Preparation of myosin and its subfragments from rabbit skeletal muscle, in *Methods in Enzymology*, Vol. 85 (eds D.W. Frederiksen and L.W. Cunningham), Academic Press, New York, pp. 55–71.

Markussen, J., Diers, I., Hougaard, P. *et al.* (1988) Soluble, prolonged-acting insulin derivatives. III. Degree of protraction, crystalizability, and chemical stability of insulins substituted in positions A21, B13, B27 and B30. *Protein Eng.*, **2**, 157–66.

Markwell, M.A.K., Haas, S.M., Bieber, L.L. *et al.* (1978) A modification of the Lowry procedure to simplify protein determination in membrane and lipoprotein samples. *Anal. Biochem.*, **87**, 206–10.

Mattarella, N.L. and Richardson, T. (1983) Physicochemical and functional properties of positively charged derivatives of bovine β-lactoglobulin. *J. Agric. Food Chem.*, **31**, 972–8.

Morr, C.V. and Ha, E.Y.W. (1993) Whey protein concentrates and isolates: processing and functional properties. *CRC Crit. Rev. Food Sci. Nutr.*, **33**, 431–76.

Morr, C.V., German, B., Kinsella, J.E. *et al.* (1985) A collaborative study to develop a standardized food protein solubility procedure. *J. Food Sci.*, **50**, 1715–18.

Nakai, S., Li-Chan, E. and Hayakawa, S. (1986) Contribution of protein hydrophobicity to its functionality. *Nahrung.*, **30**, 327–36.

Ollis, D. and White, S. (1990) Protein crystallization, in *Methods in Enzymology*, Vol. 182 (ed. M.P. Deutscher), Academic Press Inc., New York, pp. 646–59.

Ono, T., Katho, S. and Mothizuki, K. (1993) Influences of calcium and pH on protein solubility in soybean milk. *Biosci. Biotech. Biochem.*, **57**, 24–8.

Ozols, J. (1990) Amino acid analysis, in *Methods in Enzymology*, Vol. 182 (ed. M.P. Deutscher), Academic Press Inc., New York, pp. 587–613.

Perrin, D.D. and Dempsey, B. (1974) *Buffers for pH and Metal Ion Control*, Chapman & Hall, London.

Pertoft, H. and Laurent, T.C. (1982) Sedimentation of cells in colloidal silica (Percoll) in *Cell Separation Methods and Selected Applications*, Vol. 1 (eds T.G. Pretlow II and T.P. Pretlow), Academic Press, New York, pp. 115–52.

Peterson, G.L. (1979) Review of the folin phenol protein quantitation method of Lowry, Rosebrough, Farr and Randall. *Anal. Biochem.*, **100**, 201–20.

Pierce, A. (1992) *Micro BCA Protein Assay Reagent Kit. Instructions*, Pierce Chemical Company, Rockford, IL.

Pierce, J. and Suelter, C.H. (1977) An evaluation of the Coomassie Brilliant Blue G-250 dye-binding method for quantitative protein determination. *Anal. Biochem.*, **81**, 478–80.

Pohl, T. (1990) Concentration of proteins and removal of solutes, in *Methods in Enzymology* Vol. 182 (ed. M.P. Deutscher), Academic Press Inc., New York, pp. 68–83.

Polacheck, I. and Cabib, E. (1981) A simple procedure for protein determination by the

Lowry method in dilute solutions and in the presence of interfering substances. *Anal. Biochem.*, **117**, 311–14.

Pomeranz, Y. and Meloan, C.E. (1978) *Food Analysis: Theory and Practice*, AVI Publishing Company, Inc., Westport, Conn.

Powrie, W.D. and Nakai, S. (1985) Characteristics of edible fluids of animal origin: eggs, in *Food Chemistry*, 2nd edn (ed. O.R. Fennema), Marcel Dekker, Inc., New York and Basel, pp. 829–55.

Privalov, P.L. (1979) Stability of proteins, small globular proteins. *Adv. Protein Chem.*, **33**, 167–241.

Redinbaugh, M.G. and Turley, R.B. (1986) Adaptation of the bicinchoninic acid protein assay for use with microtiter plates and sucrose gradient fractions. *Anal. Biochem.*, **153**, 267–71.

Rickwood, D. (ed.) (1984) *Centrifugation: A Practical Approach*, 2nd end, IRL Press, London and Washington D.C.

Ross, E. and Schatz, G. (1973) Assay of protein in the presence of high concentrations of sulfhydryl compounds. *Anal. Biochem.*, **54**, 301–6.

Schein, C.H. (1990) Review/solubility as a function of protein structure and solvent components. *Bio/Technology*, **8**, 308–17.

Schwenke, K.D. and Raab, B. (eds) (1989) *Interaction in Protein Systems*, Proceedings of the 3rd Symposium on Food Proteins, Akademie-Verlag, Berlin.

Schwenke, K.D., Zirwer, D., Gast, K. *et al.* (1990) Changes of the oligomeric structure of legumin from pea (*Pisum sativum* L.) after succinylation. *Eur. J. Biochem.*, **194**, 621–7.

Scopes, R.K. (1974) Measurement of protein by spectrophotometry at 205 nm. *Anal. Biochem.*, **59**, 277–82.

Sedmak, J.J. and Grossberg, S.E. (1977) A rapid, sensitive, and versatile assay for protein using coomassie brilliant blue G250. *Anal. Biochem.*, **79**, 544–52.

Segel, I.H. (1976) *Biochemical Calculations*, 2nd edn, John Wiley and Sons, Inc., New York.

Sen, L.C., Lee, H.S., Feeney, R.E. *et al.* (1981) In vitro digestibility and functional properties of chemically modified casein. *J. Agric. Food Chem.*, **29**, 348–54.

Smith, P.K., Krohn, R.I., Hermanson, G.T. *et al.* (1985) Measurement of protein using Bicinchoninic Acid. *Anal. Biochem.*, **150**, 76–85.

Sorenson, K. and Brodbeck, U. (1986) A sensitive protein assay method using micro-titer plates. *Experientia*, **42**, 161–2.

Spector, T. (1978) Refinement of the coomassie blue method of protein quantitation. *Anal. Biochem.*, **86**, 142–6.

Stoscheck, C.M. (1990) Quantitation of Proteins, in *Methods in Enzymology*, Vol. 182 (ed. M.P. Deutscher), Academic Press Inc., New York, pp. 50–68.

Tanford, C. (1968) Protein denaturation, Part C, *Adv. Protein Chem.*, **24**, 1–95.

Tanford, C. (1970) Protein denaturation, Part A and B, *Adv. Protein Chem.*, **23**, 121–282.

Tanimoto, S.Y. and Kinsella, J.E. (1988) Enzymatic modification of proteins: effects of transglutaminase cross-linking on some physical properties of β-lactoglobulin. *J. Agric. Food Chem.*, **36**, 281–5.

Townsend, A.A. and Nakai, S. (1983) Relationships between hydrophobicity and foaming characteristics of food proteins. *J. Food Sci.*, **48**, 588–94.

Treece, J.M., Sheinson, R.S. and McMeekin, T.L. (1964) The solubilities of β-lactoglobulins A, B, and AB. *Arch. Biochem. Biophys.*, **108**, 99–108.

Van Holde, K.E. (1985) *Physical Biochemistry*, 2nd edn, Prentice-Hall, Inc., Englewood Cliffs, NJ.

Vojdani, F. (1992) *Effects of Specific Enzymatic And Chemical Modification on Functional Properties of Proteins: α-Lactalbumin and β-Lactoglobulin*. Ph.D. Dissertation, University of California, Davis, CA.

Vojdani, F. and Whitaker, J.R. (1994) Chemical and enzymatic modification of proteins for improved functionality, in *Protein Functionality in Food Systems* (eds N.S. Hettiarachchy and G.R. Ziegler), Marcel Dekker, Inc., New York, pp. 261–309.

Waniska, R.D. and Kinsella, J.E. (1989) Glycosylation of β-lactoglobulin and surface active properties, in *Food Protein* (eds J.E. Kinsella and W.G. Soucie), The American Oil Chemists' Society, Champaign IL, pp. 100–31.

Watanabe, K., Matsuda, T. and Nakamura, R. (1985) Heat-induced aggregation and denaturation of egg white proteins in acid media. *J. Food Sci.*, **50**, 507–10.

Wessel, D. and Flügge, U.I. (1984) A method for the quantitative recovery of protein in dilute solution in the presence of detergents and lipids. *Anal. Biochem.*, **138**, 141–3.
Whittington, P.N. (1989) Clarification, in *Protein Purification Methods. A Practical Approach*, (eds E.L.V. Harris and S. Angal), IRL Press, Oxford, pp. 69–87.
Wiechelman, K.J., Braun, R.D. and Fitzpatrick, J.D. (1988) Investigation of the bicinchoninic acid protein assay: identification of the groups responsible for color formation. *Anal. Biochem.*, **175**, 231–7.
Zirwer, D., Prakash, V., Raab, B. *et al.* (1989) Characterization of the major protein fraction carmin from safflower seeds (*Carthamus tinctorius* L.) by dynamic light scattering and circular dichroism spectroscopy. *Nahrung.*, **33**, 929–33.

3 Viscosity

T.W. SCHENZ and C.V. MORR

3.1 Introduction

Viscosity is one of the most important functional properties of food proteins. It is important for providing physical stability to emulsions and other suspended particles in foods and contributes to the mouthfeel of foods directly and by controlling sugar crystal size in candy and confectionery products. Processing-induced changes in proteins such as polymerization, aggregation and hydrolysis affect the viscosity of food products. Examples of processing-induced protein changes that affect viscosity of foods include:

1. heat-induced milk protein aggregation due to protein denaturation and subsequent protein–protein interaction via calcium ion and disulfide bond formation during sterilization of milk and infant formula products;
2. heat-induced polymerization of milk and/or egg proteins due to denaturation and protein–protein interaction via ionic and disulfide bond formation during the manufacture of candy and other confectionery products;
3. age-thickening of UHT sterile milk concentrates during storage due to protein–protein polymerization via calcium ion and disulfide bond formation;
4. rennet hydrolysis of κ-casein molecules on the surface of milk casein micelles that results in their subsequent polymerization via calcium ion bonding to form cheese curd;
5. a greatly reduced viscosity of extensively hydrolyzed protein solutions (hydrosylates); and
6. the effect of pH, temperature, and ionic composition on the viscosity of commercial caseinate and isolated soy protein solutions.

The concentrations and inherent physicochemical properties, i.e. molecular weight, polydispersity, hydrophobicity, and conformation of each protein species, affect the viscosity of the protein solution. The concentrations and physicochemical properties of other ionic and nonionic solutes also exert an important influence on the viscosity of protein solutions by contributing directly to viscosity and also by their tendency to interact and modify the physicochemical properties of the proteins. All of these factors tend to confound the underlying inverse relationship of protein solubility and viscosity (Shen, 1981).

This chapter considers the definitions of viscosity and the factors that can affect the viscosity of protein solutions. A detailed treatment is given on the various types of viscometers and the kinds of information that each instrument provides.

3.2 What is viscosity?

In 1687, Isaac Newton proposed in his *Principia* that "The resistance which arises from the lack of slipperiness of the parts of the liquid, other things being equal, is proportional to the velocity with which the parts of the liquid are separated from one another". This lack of slipperiness is what we call viscosity.

The concept of viscosity can easily be visualized by considering that a material between two plates (Figure 3.1) is being sheared, i.e. the plates are being moved in opposite directions with a force, F, that results in a velocity of the top plate, V. Intuitively, we know that if the material is 'thin', then a given force will move the top plate farther and faster than if the material is 'thick'. This 'internal resistance' is synonymous with viscosity and is a measure of 'resistance to flow' (Barnes *et al.*, 1989). Mathematically and experimentally, one can measure the force being applied over the area of the plate, F/A. This is denoted as the *stress*, σ. The stress applied to the sample is proportional to the speed of movement, or 'velocity gradient' (or *shear rate*), which is the ratio of the velocity to sample thickness, d. Thus, if one doubles the force, the velocity gradient is also doubled. The constant of proportionality, η, is called the coefficient of viscosity. Thus we have:

$$\sigma = \eta \frac{V}{d} \tag{3.1}$$

This is the mathematical expression for Newton's postulate. Water and glycerin are common liquids that obey it, having viscosities of about 1 mPa·s and 1 Pa·s, respectively. Liquids that follow equation (3.1) are called Newtonian liquids.

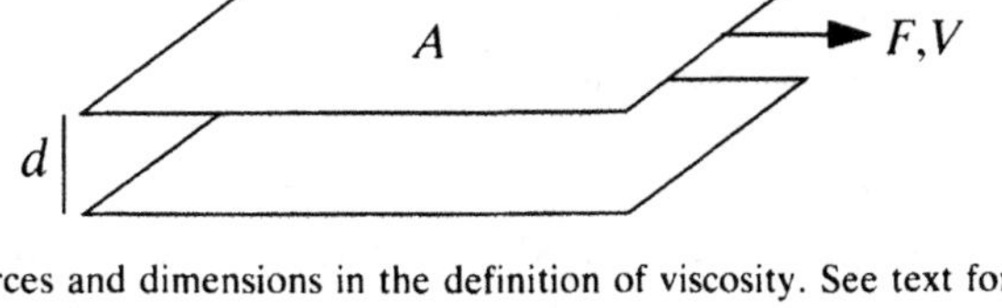

Figure 3.1 Forces and dimensions in the definition of viscosity. See text for explanation.

The velocity gradient V/d, or shear rate, is often denoted by $\dot{\gamma}$, so that shear viscosity is the ratio of shear stress to shear rate:

$$\eta = \sigma/\dot{\gamma} \tag{3.2}$$

Thus to measure viscosity accurately, we need to know the shear stress and shear rate, both of which involve a knowledge of the geometry of our measuring system, i.e. we need to know the area over which the liquid is being sheared and the thickness of the sample. This consideration will have implications as we examine instruments for the measurement of viscosity.

The SI unit for viscosity is the Pascal-second which is abbreviated to Pa·s. Another unit that is still in use today, especially in the food industry, is the cgs unit, the Poise. It is smaller than the Pa·s by a factor of 10. The most common unit is the centipoise, abbreviated cP. The cP is equal to 1 mPa·s (milliPascal-second). Table 3.1 shows the wide range of viscosities encountered in familiar materials at room temperature (Barnes *et al.*, 1989).

3.2.1 Factors which affect viscosity

Obviously variables such as concentration (Southward and Goldman, 1975), dispersion conditions (Southward and Goldman, 1978), pH (Modler and Emmons, 1977) and ionic strength (Girdhar and Hansen, 1974) can affect the viscosity of protein solutions. Figure 3.2 shows the effect of concentration on the Newtonian viscosity of various casein solutions (Korolczuk, 1982). The viscosity of proteins is highly sensitive to the history of the protein and the techniques used to measure it (Shen, 1981). Other factors which affect the viscosity of proteins include processing and

Table 3.1 The viscosity of some familiar materials at room temperature

Liquid	Approximate viscosity (Pa·s)
Glass	10^{40}
Molten glass (at 500°C)	10^{12}
Asphalt	10^{8}
Molten polymers	10^{3}
Corn syrup	10^{2}
Honey	10^{1}
Glycerol	10^{0}
Olive oil	10^{-1}
Milk	10^{-2}
Water	10^{-3}
Air	10^{-5}

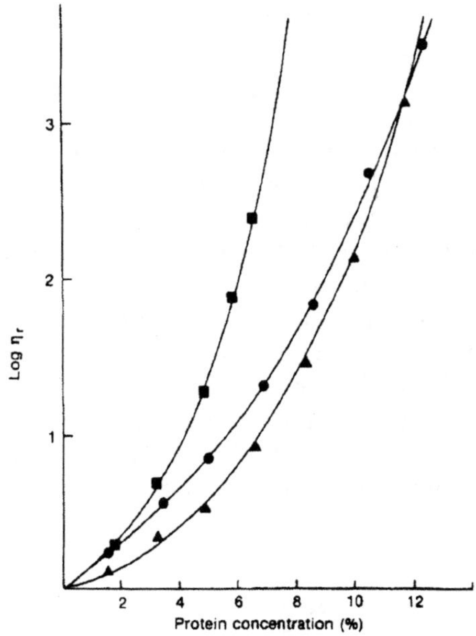

Figure 3.2 The effect of concentration on the relative viscosity, η_r, of casein solutions: (●) sodium caseinate in 2.5% borax; (▲) acidified caseinate in 2.5% borax; (■) acidified casein in deionized water. From Korolczuk (1982).

compositional factors, e.g. temperature and pH of precipitation, temperature of drying, calcium content and the extent of its replacement by sodium, sequestration of calcium by chelating agent, and the presence of reducing agents (Mulvihill and Fox, 1983). Here we will concentrate on the rheological experimental parameters such as shear rate (Johns and Ennis, 1982), temperature (McDonough *et al.*, 1974), time of shearing (Shen, 1981) and pressure that can affect viscosity. We need to understand over what ranges these variables will affect the measurement and to what extent.

3.2.1.1 Relationship of shear rate and processing. The shear rate for any process can be estimated by dividing the velocity of the flowing material by the thickness of the material (or a characteristic dimension of the geometry in which the liquid is flowing, e.g. the radius of a pipe). Table 3.2 shows typical shear rates encountered in a variety of industrial and technological processes.

3.2.1.2 Effect of temperature. For most fluids, viscosity decreases with temperature and can be estimated using an Arrhenius relationship:

$$\eta = Ae^{-B/T} \tag{3.3}$$

Table 3.2 Typical shear rates in processes

Process	Typical range of shear rates (s^{-1})	Example
Lubrication	10^3–10^7	Gears, engines
Rubbing	10^4–10^5	Applying creams and lotions to the skin
Spraying and brushing	10^3–10^4	Spray drying, painting
Pipe flow	10^0–10^3	Pumping, blood flow
Mixing and stirring	10^1–10^2	Batch tanks, mixers
Chewing and swallowing	10^1–10^2	
Extruders	10^0–10^2	Processed foods, polymers
Draining under gravity	10^{-1}–10^1	Painting, foams
Leveling	10^{-2}–10^{-1}	Paints, printing inks
Sedimentation	10^{-6}–10^{-4}	Settling of particles during storage

where T is the absolute temperature (°K) and A and B are constants for a given liquid.

For Newtonian liquids, the greater the viscosity, the greater is the temperature dependence. For water, though, the temperature sensitivity is about 3% per °C at room temperature. Thus, in order to obtain a viscosity reading to within 1%, one must control the temperature to within 0.3°C. Temperature control, then, is one of the most important factors in viscometry.

3.2.2 Non-linear viscosity

Newtonian liquids are those that follow Newton's postulate: viscosity is invariant over shear rate or stress. The viscosity of many liquids, however, is not constant over ranges of shear rates or stresses. These non-Newtonian liquids can have a variety of flow profiles and flow curves, which can be described mathematically. However, almost all liquids will show the flow behavior described in Figure 3.3. The curve indicates that in the limit of low stresses (or rates) the viscosity approaches a constant value, or lower Newtonian region or zero-shear viscosity. At high stresses (or rates), the viscosity will generally approach another constant (lower value) viscosity plateau, called the upper Newtonian region (*Newtonian* referring to the invariance of viscosity with stress or rate, and *upper* and *lower* referring to shear rate or stress). While such behavior can be modeled mathematically using the Cross model, very seldom is one able experimentally to cover the wide ranges of stress and rates necessary to discern the entire model. One can cover more limited ranges of rates and stress and model the resulting flow regimes with appropriate models. It is important, though, always to be aware that extrapolating flow behavior to lower and higher stress or rate regions is difficult at best and dangerous at worst.

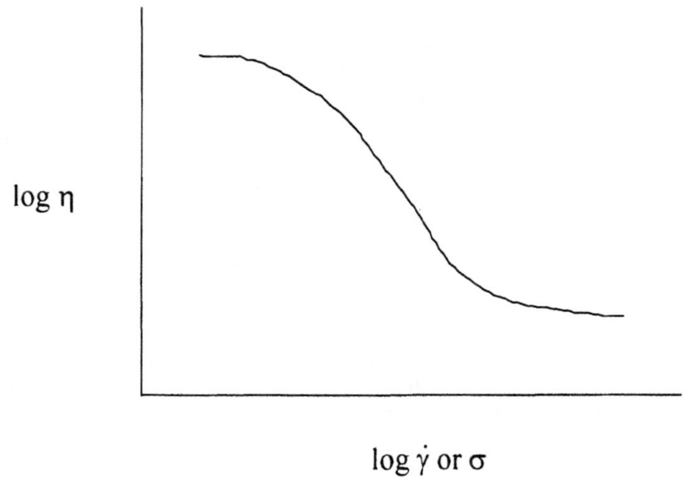

Figure 3.3 Generalized flow curve for most liquids plotted as log viscosity, η, versus log shear rate, $\dot{\gamma}$ (or stress, σ).

While not an exhaustive list, the following is a description of the types of non-Newtonian flow encountered most frequently.

3.2.2.1 Shear-thinning flow. Many dispersions, emulsions, polymer solutions and colloids show departures from Newtonian behavior, i.e. their viscosity decreases with increasing rate or stress. These 'shear-thinning' liquids have also been called 'pseudoplastic', but such terminology is generally avoided. Figure 3.4a shows shear flow curves for Newtonian and non-Newtonian liquids. Curve **1** shows the flow curve of a Newtonian liquid while curve **2** demonstrates a shear-thinning liquid.

Shear-thinning behavior can be attributed to the shear-induced breakdown of any structure in the liquid. Such structures may arise from interparticle interactions or from a randomization of non-spherical particles at low shear leading to an alignment of particles at high shear (lower viscosity).

For a simple shear-thinning liquid, the curvature of the flow curve can be expressed by several mathematical models. The most often used is the *power law model*:

$$\eta = K\dot{\gamma}^{n-1} \tag{3.4}$$

or

$$\sigma = K\dot{\gamma}^{n} \tag{3.5}$$

where K is called the 'consistency' and n is called the power law index. For shear-thinning liquids, n is less than 1. (Equations (3.4) and (3.5) reduce to

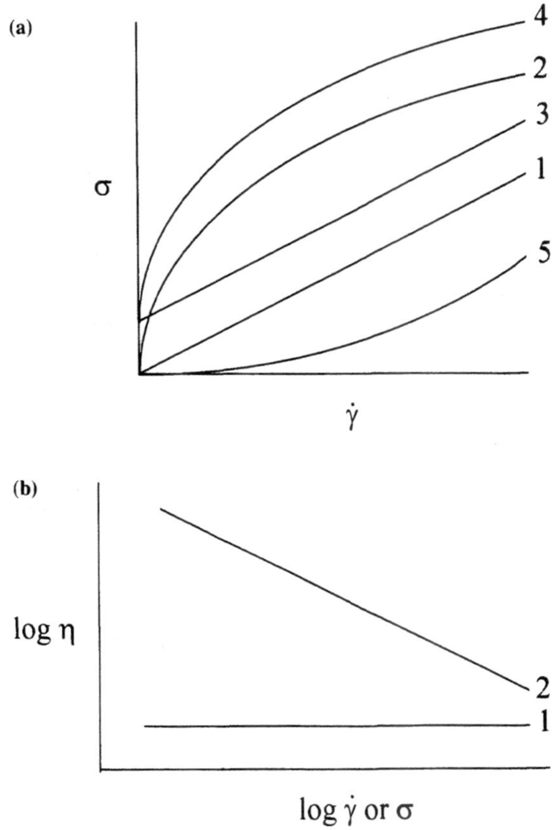

Figure 3.4 (a) Flow curves for various fluid types: **1**, Newtonian; **2**, shear-thinning; **3**, Newtonian with yield stress; **4**, shear-thinning with yield stress; **5**, dilatant. (b) Log viscosity–log shear rates plots of some of the curves in Figure 3.4a.

the Newtonian case for $n = 1$.) Thus a plot of log viscosity versus shear rate yields a straight line of intercept K and slope $n - 1$ (Figure 3.4b). As an example of shear-thinning protein solutions, Figure 3.5 shows the shear-thinning behavior of Na, K and Ca caseinates (Bastier *et al.*, 1993).

3.2.2.2 The yield stress. Certain materials will not flow until a certain stress has been applied to them, at which point they begin to flow. That stress is called the 'yield stress' or 'yield point'. Curve **3** of Figure 3.4a shows a Newtonian liquid with a yield stress, while curve **4** shows a yield stress in a shear-thinning liquid. The former can be modeled with the *Bingham model*:

$$\sigma = \sigma_y + \eta\dot{\gamma} \tag{3.6}$$

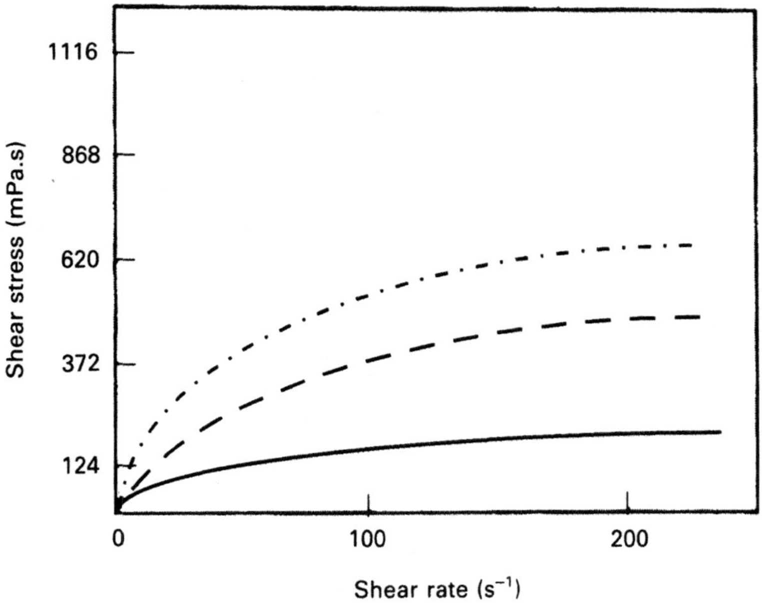

Figure 3.5 Shear stress versus shear rate curves of 150 g/kg aqueous caseinate dispersions at 25°C: (———) Ca caseinate; (- - - -) Na caseinate; (-·-·) Na/Ca caseinate. Adapted from Bastier *et al.* (1993).

where σ_y is the yield stress and η is the viscosity. This is simply equation (3.2) with a yield stress added to it.

Similarly, for a shear-thinning liquid with a yield stress, the *Herschel–Bulkley model* is the power law model (equation (3.5)) with a yield stress added:

$$\sigma = \sigma_y + K\dot{\gamma}^n \quad (3.7)$$

Care must be taken when fitting these yield stress equations to data. While a yield stress may be implied from extrapolation, it can quite likely be due to instrumental artifact. The only valid verification of a yield stress is to make sure that the liquid does not flow at a given stress. For self-leveling aqueous liquids, any measurement of a yield stress above about 0.01 Pascal (the stress on a unit volume of water due to gravity) may be spurious.

3.2.2.3 Shear-thickening flow. Some concentrated dispersions will show flow behavior as illustrated in curve **5** in Figure 3.4a. This shear-thickening, or *dilatant*, flow is generally due to particles colliding with one another as the shear increases; the rapid rate of movement does not allow enough time for one particle to get out of the way of the next. This type of

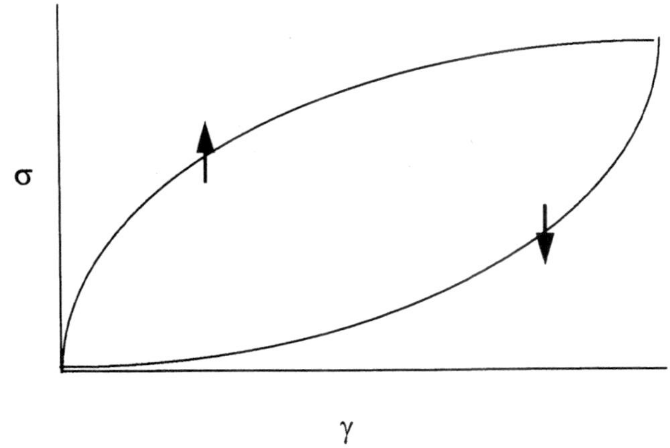

Figure 3.6 Generalized flow curve showing time-dependent behavior, in which the up curve (↑) and down curve (↓) do not coincide.

flow can be modeled using equations (3.4)–(3.7) with the exception that n, the power law index, will now be greater than 1.

3.2.2.4 Time-dependent flow. Food materials often exhibit flow behavior that changes with time or with applied shear stress. For example, Figure 3.6 shows typical data from performing a flow curve with increasing shear, followed by decreasing shear. The fact that there is a hysteresis loop indicates that structure is being destroyed in a time-dependent manner. Such behavior is called *thixotropy*. It is also seen as a decrease in viscosity under constant stress or shear rate. The opposite effect, increasing viscosity under constant shear, is called *anti-thixotropy* or *rheopexy*. Experimentally, this means that previous shear history, including sample preparation, may affect the results. Similarly, flow of a thixotropic material down a long pipe is affected by the fact that the viscosity may change with distance down the pipe.

3.2.3 Viscosity of suspensions

Since proteins are either colloidal or particulate suspensions, it is important to understand the general shape of the viscosity/shear curve for suspensions (Figure 3.7). Note that the axes are designated as logarithmic so that one could cover many orders of magnitude in shear rate over the entire curve. The first (lower) Newtonian region at low shear rate is followed by a power law shear-thinning region and then by a leveling out to the upper Newtonian region. After more shear is applied, the suspension will generally tend to become shear-thickening (dilatant).

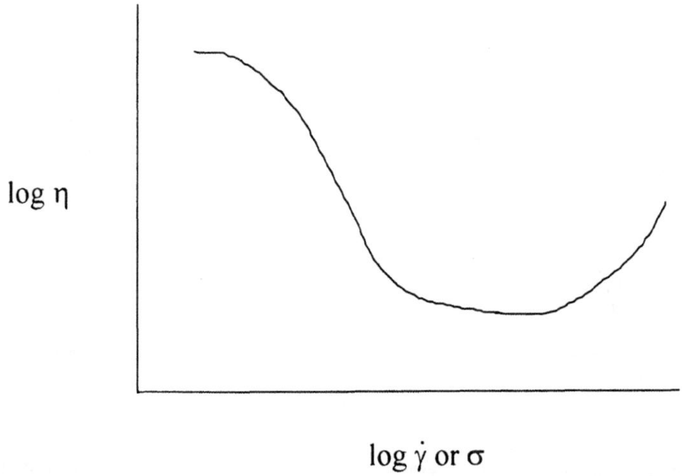

Figure 3.7 Generalized flow curve for colloidal dispersions over a wide range of shear rates plotted as log viscosity, η, versus log shear rate, $\dot{\gamma}$ (or stress, σ).

In actual practice, the lower Newtonian region may be so high as to be experimentally unreachable. The resulting low shear rate behavior is often seen as an apparent yield stress. The important point to consider is that depending on the concentration of the suspension and the experimental shear rate ranges employed, one may find that the same material produces quite different rheological results, ranging from Newtonian to shear-thinning to shear-thickening, from fluid flow to flow with a yield stress. A proper appreciation of one's instrumental capabilities and a knowledge of the nature of the material are most important in interpretation of viscometric data.

3.3 Viscometers for measuring shear viscosity

3.3.1 Rotational viscometers

Rotational viscometers are the workhorses of laboratory viscometry. Their ease of use makes them ubiquitous in food science laboratories. The principle of operation is that a simple rotational motion produces the necessary shear. To measure viscosity, two approaches can be used. In the first, one can impart a known shear rate (rotational speed) to the viscometer and measure the resulting torque (stress) that the material transmits to the measuring system. This class of viscometers is called *rate-controlled* and includes viscometers produced by Brookfield, Haake, and Mettler, to name a few. In the second approach, one imparts to the sample a known torque or stress and then measures the resulting speed or rate with which the sample moves. These viscometers are called *stress-controlled* and are manufactured by TA Instruments, Haake and Rheometrics. The merits of each will be discussed in the corresponding section.

In both types of rotational viscometer, common geometries are used to accurately measure or calculate shear rate (which depends on sample thickness) and shear stress.

3.3.1.1 Rate-controlled viscometers. For many years, rate-controlled rheometers were the only instrument type available to researchers. Their advantages include rugged design, high upper shear rate limit and relatively low cost. The disadvantages of rate-controlled rheometers are that they are generally less sensitive than the stress-controlled rheometers and that the principle of operation limits the ability to collect low stress data. That is, if one wishes to measure structure (e.g. a yield stress) at low stresses (or the corresponding low rate), the operation of the rate-controlled rheometer dictates that one first applies a specified low shear rate to the sample. If the sample possesses some structure having a low yield stress, the rate-controlled rheometer will force the sample to move at a low shear rate regardless of the nature of the structure present. One will probably break the structure in the first instant of applied shear rate. Thus, the best that one can hope for is to extrapolate the rate–stress data to zero shear rate to estimate the corresponding yield stress. Often, such extrapolations are not correct.

One of the most common rate-controlled viscometers in use in the food industry is the Brookfield viscometer. This viscometer comes with a variety of 'spindles' to cover various viscosity ranges, and later models have digital readouts. While their presence is ubiquitous in the lab, the following limitations should be kept in mind:

1. Except for the Ultra Low Viscosity adapter and Low Volume Adapter, the spindles of the Brookfield viscometer do not provide a precise measuring geometry for viscometric measurement. This presents a problem especially in thixotropic materials, where the material being sheared is constantly being replaced by new material.
2. The maximum shear rate for the viscometer is limited (60 rpm = 13.7 sec^{-1}). Often, manufacturing and sensory processes occur at much higher shear rates.

Haake, Mettler and Bohlin manufacture rate-controlled rheometers that provide precise measuring geometries and offer high upper shear rate limits. Of course, such features increase the cost of the viscometer.

3.3.1.2 Stress-controlled viscometers. With the advent of computer control and data acquisition, stress-controlled rheometers can be produced simply and at prices competitive with rate-controlled instruments. The stress-controlled instrument overcomes the yield stress extrapolation problem by the very nature of its construction. If one applies to a material a low stress that is lower than the material yield stress, then the material will not flow and the shear rate will be zero. Not until the applied stress is greater than the yield stress will flow begin. Thus one can accurately measure yield stress with a stress-controlled instrument. One disadvantage of stress-controlled viscometers is that their upper shear rate limit tends to be less than their rate-controlled counterparts.

TA Instruments and Bohlin are among those manufacturing stress-controlled instruments. They are also capable of measuring other rheological properties in modes other than shear, i.e. oscillation and creep modes.

3.3.2 Other techniques

Although shear viscometry is the most commonly used technique for viscosity data, there are several other techniques that should be mentioned for completeness.

3.3.2.1 Capillary viscometry. In capillary viscometry, the sample geometry is a capillary tube through which the sample flows, usually under the stress of gravity. By timing the flow of a given volume of the material through the tube, one can calculate the viscosity and shear rate. By changing the diameter of the capillary tube, one can measure the viscosity at various shear rates. A variety of capillary viscometer tubes is available, e.g. Ubbelohde, Cannon-Fenske, Ostwald Pipet (Cannon, Fisher). Multi-bulb capillary viscometers even allow the researcher to measure the same sample over a range of shear rates in one experiment (Cannon).

Capillary viscometry is time consuming but very accurate and is used extensively to measure the *intrinsic viscosity* of a material. This intrinsic viscosity is the viscosity of a series of increasingly dilute solutions extrapolated to zero concentration.

3.3.2.2 Surface viscometry. Surface viscosity is a two-dimensional analog of the three-dimensional bulk viscosity (Dickinson and Stainsby, 1982). It measures the deformation and flow of films adsorbed or spread at a fluid interface. Surface rheology is important in emulsions and foams because of its sensitivity to macromolecular structure and interactions at the interface.

The experimental details are analogous to the bulk viscosity situation (Dickinson *et al.*, 1989). A biconical disk or similar geometry, attached to a sensing module by a torsion wire (to measure torque), is suspended at an interface with the edge of the disk at the interface. The solution to be measured is contained in a vessel that is larger than the disk, the annulus

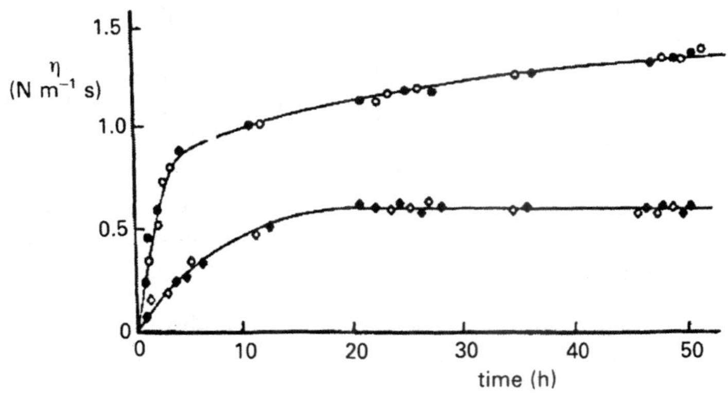

Figure 3.8 Time-dependent surface viscosities of pure whey proteins at the *n*-tetradecane–water interface: (◇, ◆) α-lactalbumin; (○, ●) *β*-lactoglobulin. From Dickinson *et al.*, (1990).

between the disk and the wall of the vessel becoming the measurement gap for the surface active material. The vessel is then rotated at steady slow speeds and the torque recorded.

Both air/water and oil/water interfaces can be measured (Walstra and DeRoos, 1993). Various experiments can be performed, including following the aging of a film or examining the adsorption of proteins injected below the surface. Figure 3.8 shows the time-dependent surface viscosities of whey proteins at an *n*-tetradecane–water interface (Dickinson *et al.*, 1990).

3.4 Protocols for measuring viscosity

3.4.1 Shear viscosity

Several important considerations need to be made before and during the measurement of shear viscosities:

1. Does the protein solution or suspension show Newtonian or non-Newtonian behavior? If the former, then almost any single shear rate or stress will be adequate. If the latter, then flow curves should be reported over a range of rates or stresses.
2. What is the purpose of the viscosity measurement? If it is to show the effect of increased hydration of a protein, then a single shear rate or stress will suffice (Korolczuk, 1982). If, on the other hand, one is making a comparison to some processing conditions, then the viscosity should be reported at the appropriate shear.
3. Is the solution or dispersion thixotropic, i.e. does it show time-dependent behavior? Constantly decreasing viscosity values indicate

thixotropic behavior. One should either wait for equilibrium or report the data at a specified time after the start of the experiment.

4. Would a model (*supra vide*) for a non-Newtonian solution convey more information than reporting a single viscosity at a shear rate? Often, compiling the parameters of a model over the variables of an experiment can give useful information.
5. Is low stress structural information desired? Stress-controlled viscometers give better structural data than do rate-controlled instruments, for which one must rely on extrapolation to low stresses or rates.

Only after critically analyzing the results desired and performing a few range-finding experiments is one able to design the best viscometric experiment.

However the experiment is performed, one must be unambiguous in reporting the data. Details such as the overall rheological nature of the sample are valuable (Newtonian, non-Newtonian, thixotropic), as well as the precise experimental conditions: temperature, shear rate or shear rate range, etc.

3.5 Summary

While the measurement of the viscosity of protein dispersions and solutions is a fairly straightforward task, many factors need to be considered. We have attempted to give a foundation for the measurement of viscosity of proteins that may assist the researcher in exploring protein functionality. An understanding of the principles of viscometry and their application to complex colloidal dispersions such as proteins is important in the proper interpretation of the contribution of protein viscosity to protein functionality.

References

Barnes, H.A., Hutton, J.F. and Walters, K. (1989) *Introduction to Rheology*, Elsevier Science Publishers, New York. (Table 2.1, 2.2)
Bastier, P., Dumay, E. and Cheftel, J.C. (1993) *Lebensm.-Wiss. u.-Technol.*, **26**, 529–37.
Dickinson, E. and Stainsby, G. (1982) *Colloids in Food*, Applied Science Publishers, London.
Dickinson, E., Elverson, D.J. and Murray, B.S. (1989) *Food Hydrocolloids*, **3**, 101–14.
Dickinson, E., Rolfe, S.E. and Dalgleish, D.G. (1990) *Int. J. Biol. Macromol.*, **12**, 189–94.
Girdhar, B.K. and Hansen, P.M.T. (1974) *J. Food Sci.*, **39**, 1237–43.
Johns, J.E. and Ennis, B.M. (1982) *N.Z. Jl Dairy Sci. Technol.*, **15**, 79–86.
Korolczuk, J. (1982) *N.Z. Jl Dairy Sci. Technol.*, **17**, 135–40.
McDonough, F.E., Hargrove, R.E., Mattingly, W.A., Posati, L.P. and Alford, J.A. (1974) *J. Dairy Sci.*, **57**, 1438–43.
Modler, H.W. and Emmons, D.B. (1977) *J. Dairy Sci.*, **60**, 177–84.
Mulvihill, D.M. and Fox, P.F. (1983) IDF Session, Commission F, F-Doc 93.

Shen, J.L. (1981) In *Protein Functionality in Foods* (ed. J.P. Cherry), American Chemical Society, Washington, pp. 89–109.
Southward, C.R. and Goldman, A. (1975) *N.Z. Jl Dairy Sci. Technol.*, **10**, 101–12.
Southward, C.R. and Goldman, A. (1978) *N.Z. Jl Dairy Sci. Technol.*, **13**, 97–105.
Walstra, P. and DeRoos, A.L. (1993) *Foods Reviews International*, **9(4)**, 503–25.

List of manufacturers

Bohlin Instruments Inc., 2540 Rt. 130, Cranbury, NJ 08512 USA.
Brookfield Engineering Laboratories, Inc., 240 Cushing St., Stoughton, MA 02072 USA.
Cannon Instrument Company, P.O. Box 16, State College, PA 16804 USA.
Fisher Scientific, 711 Forbes Ave., Pittsburgh, PA 15219 USA.
Haake Instruments, 53 W. Century Dr., Paramus, NJ 07652 USA.
Mettler-Toledo, Inc., 69 Princeton-Hightstown Rd., P.O. Box 71, Hightstown, NJ 08520 USA.
TA Instrument, Inc., P.O. Box 311, 109 Lukens Dr., New Castle, DE 19720 USA.

4 Gelation

Y. MATSUMURA and T. MORI

4.1 Introduction

Gelation is one of the most important properties of food proteins as many foods are made of protein gels, e.g. yogurt, cheese, pudding, sausage, jelly, etc. Some traditional oriental foods are also based on protein gels, for instance *tofu* (soybean curd), *kamaboko* (fish meat) and *fu* (wheat gluten).

In the mass production process in modern food industries strict control of the processing conditions is necessary to maintain the constant high quality of such traditional foods. To establish the processing conditions properly, fundamental and practical studies must be made on gelling properties of the food proteins which constitute these traditional foods. Furthermore, there is now an increased demand for new protein gel foods in order to cope with the diverse nutritional needs and tastes of people.

In this chapter, we describe the methods of testing the heat-set gelling properties of food proteins which are expected to be of use as gelling agents.

4.2 Scientific basis of protein gelation

4.2.1 Definition of gelation and types of gels

Gel formation can be defined as a protein aggregation phenomenon in which attractive and repulsive forces are so balanced that a well-ordered tertiary network or matrix, capable of holding much water, is formed (Hermansson, 1979). If attractive forces predominate, a coagulum is formed and water is expelled from the gel matrix. If repulsive forces predominate, no network will be formed.

Gelation consists of two steps: (i) conformational change or partial denaturation of protein molecules; and (ii) the following gradual association or aggregation of the individual denatured proteins. The second step should be slow relative to the first one, so that a well-organized gel network is formed. Ferry (1948) speculated that the protein molecule was unfolded completely and that the resultant extended linear polymers

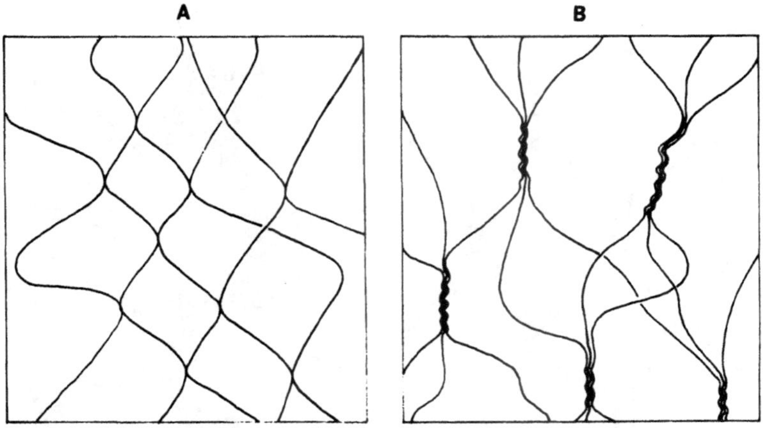

Figure 4.1 Gel network formation by connection of long linear molecules. (A) Network formation by connection of the long molecules at points. (B) Connection of the long molecules at junction zones.

associated to form networks such as those shown in Figure 4.1. It has been found that polysaccharide gels have structures as described by this figure (Morris, 1986). The structure of a gelatin gel is very similar to that shown in Figure 4.1B (Ledward, 1986) while an illustration of a heat-induced myosin gel, which is supported by electron microscopic observation, very closely resembles that shown in Figure 4.1A (Hermansson and Langton, 1988). In the case of fibrous proteins such as gelatin or myosin, the networks are composed of polypeptide chains of random or helical structures, i.e. the threads of networks are composed of peptide chains.

This type of network structure, however, does not describe the gels of globular proteins. Barbu and Joly (1953) reported that horse serum albumin and other protein gels were formed by end-to-end aggregation of proteins that retain globular structure. Tombs (1974) described the gelation of globular proteins by two models: (i) a highly-oriented 'string of beads'; and (ii) random aggregation. These two types of gel networks are illustrated in Figure 4.2.

Gels of the 'string of beads' type (Figure 4.2A) with transparent or translucent appearance have been reported for serum albumin, insulin, lysozyme (Clark *et al.*, 1981), ovalbumin (Doi *et al.*, 1987) and soybean glycinin (Nakamura *et al.*, 1984). Examples of gels made by random aggregation of molecules (Figure 4.2B) with more turbid appearance have been observed for gels of myosin heated at high ionic strength (Hermansson *et al.*, 1986), β-lactoglobulin and whey proteins (Hermansson, 1988). However, most gels of globular proteins take on both network forms depending on the conditions of gelation, such as protein concentration, pH, ionic species and strength, heating temperature, heating time, etc.

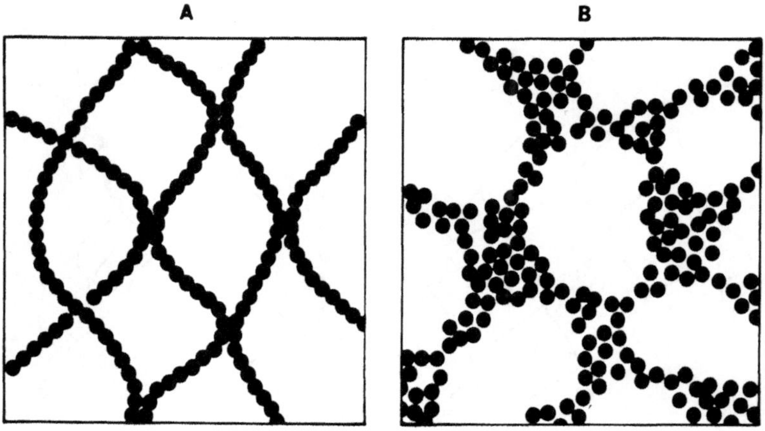

Figure 4.2 Two types of gel network formed by the aggregation of globular proteins. (A) Aggregation of 'string of beads' polymers. (B) Random aggregation of molecules.

4.2.2 Conformational change of proteins

As described in the previous section, a conformational change of protein molecules is a prerequisite for the aggregation of protein molecules, particularly globular proteins. The protein molecule was found to retain a globular conformation in the strands of the network. In this section we will consider the denatured state of globular proteins, since a lot of available and promising food proteins for use as gelling materials have a globular conformation.

In the classical two-state theory of a globular protein's unfolding transition, there are only two protein states: native (N) and highly disordered (D) (Privalov, 1979). It has now been established that a kinetic intermediate clearly exists along the unfolding path from the D state to the N state (Kuwajima *et al.*, 1985). The occurrence of a similar intermediate state has been demonstrated under equilibrium conditions at low (or high) pH, in the presence of moderate concentrations of a denaturant, and at high temperatures (Ptitsyn, 1992). This intermediate state, called 'the molten globule state' can be defined as a compact globular molecule with native-like secondary structure but unfolded tertiary structure (Ptitsyn, 1992). As will be described in section 4.3, accumulated data indicate that globular proteins transform from the native state to the molten globule state upon heating, and that this partially denatured form is mainly involved in the heat-set gelation.

Clark and Lee-Tuffnell (1986) reported that, upon heating, the α-helix content of bovine serum albumin decreased with concomitant increases in the β-sheet structure to an extent approaching that observed in fibrous proteins. This indicates the dramatic alteration of the secondary structure in heat-set gelation, which is contradictory to the concept of the molten globule-like structure. As will also be shown in section 4.3, the increase of the β-sheet content seems to be due to the formation of an intermolecular β-sheet structure induced by association between denatured molecules. That is, β-sheet formation probably has no direct relation to the denaturation step of globular proteins.

There is a special case where the highly unfolded states of the globular protein molecule play important roles in gelation. This is the gelation seen in the presence of high concentration of denaturants such as urea and guanidine hydrochloride (Katsuta and Kinsella, 1990) where gel formation can be ascribed to thiol–disulphide interchange reactions (Xiong and Kinsella, 1990). The high concentration of denaturants facilitates extensively the unfolding of the secondary as well as the tertiary structure of polypeptides, and thus promotes thiol–disulphide interchange reactions.

4.2.3 Chemical forces involved in gelation phenomena

The association of the denatured protein molecules and the formation of the gel network is based on a good balance of attractive and repulsive chemical forces. The important chemical forces involved in the protein gelation phenomena will now be described.

4.2.3.1 Hydrophobic interactions. Hydrophobic interactions would be enhanced between non-polar segments of adjacent polypeptides when the exposure of occluded non-polar polypeptides is induced by heating. Therefore, the average hydrophobicity (i.e. the ratio of hydrophobic amino acids in the protein) should affect the process of the gel network formation. In several studies of proteins in which gel formation was correlated with the content of hydrophobic amino acid residues, Shimada and Matsushita (1980) showed that proteins containing greater than 31.5% mole percent of non-polar amino acid residues formed coagulation-type gels, whereas those that contained less than 31.5% mole percent of the non-polar residues formed translucent-type gels. This systematic approach clearly shows the importance of hydrophobic interactions in gel formation and the possibility of predicting gel properties from the amino acid compositions of proteins.

However, not only amino acid composition but also amino acid sequences and the resultant local conformation should be considered in relation to the protein association. Hydrophobic interactions have been found to occur between hydrophobic 'areas' or 'patches' exposed on the surface of different protein molecules. The appearance of the surface hydrophobic areas is determined primarily by the amino acid sequence, but the areas can be altered by physical/chemical treatments and with buffer conditions. Therefore, the 'actual' surface hydrophobicity of the unfolded molecule determined by the methods described in Chapter 8 should be more important than the average hydrophobicity of the protein for the gel network formation (Nakai and Li-Chan, 1988).

4.2.3.2 Hydrogen bonds. The involvement of hydrogen bonding in gel formation can be demonstrated by the melting of gels at elevated temperature. However, except for gelatin (Ledward, 1986) and a few other examples such as arachin from peanut (Navin Kumar *et al.*, 1980), soy protein (Bikbov *et al.*, 1981), lysozyme (Clark and Lee-Tuffnell, 1986) and whey protein gels (Rector *et al.*, 1991) under special conditions, this phenomenon is difficult to demonstrate.

Nevertheless, the increase in intermolecular β-sheet structure during gel formation suggests the general contribution of hydrogen bonding to gelation. The effects of treatment by formamide and alcohol on the physical properties of gels also supports the contribution of hydrogen bonding (Zheng *et al.*, 1993a).

4.2.3.3 Electrostatic interactions. Electrostatic interactions are thought usually to behave as repulsive forces during the process of protein aggregation, particularly when the system contains a single protein species or different species having similar isoelectric points (pI) (Hermansson, 1979). Since the net charge is zero at pI, the proteins easily form a coagulum as a result of the rapid and random aggregation at pH values near pI. At pH conditions far away from pI, the repulsive forces predominate because of the greater net charge, and so the aggregation of proteins does not occur. At the intermediate regions between pI and extreme pH, the repulsive electrostatic forces and attractive forces (mainly hydrophobic interactions) are well-balanced to form the gel network. The importance of electrostatic repulsive forces in gelation has also been demonstrated by the effect of salt addition to the medium. Even at pH conditions far away from pI, the addition of salts shields the excess charges on the molecules' surface and enables them to associate to form the fine-stranded network. However, the addition of salts at intermediate pH values breaks the balance of attractive and repulsive forces, leading to the formation of turbid gels consisting of large aggregates, or coagulums.

4.2.3.4 Disulphide bonds. Intermolecular covalent disulphide (SS) bonds do not act as initial network formers, because it has been shown for several proteins that gelation is possible even if the sulphydryl (SH)–SS interchange reaction is inhibited (Koseki *et al.*, 1989a). However, the physical integrity of the gel network is attained by the formation of intermolecular SS bonds, as demonstrated by gel solubility experiments (Clark and Lee-Tuffnell, 1986). It remains unclear whether SS bonds are formed in the junction zone in the gel network, or whether they simply contribute to the increase in the effective chain length of the polypeptides. In the latter case, the long polypeptide chains are easily entangled, and consequently the formation of non-covalent bonds is enhanced within the gel network (Beveridge *et al.*, 1984).

4.2.4 Rheological aspects

Methods to evaluate rheological properties are categorized into two groups: (i) the small-deformation test and (ii) the large-deformation mechanical test.

The small-deformation mechanical (dynamic) test is very useful for monitoring sol–gel transition and for characterizing the viscoelastic behaviour of gels in the 'linear region' in which the amplitudes of the stress and strain are adjusted to sufficiently low values that stress is proportional to strain (Ross-Murphy, 1984). In the small-deformation mechanical test, the sample is subjected to oscillating deformation. The viscoelastic parameters of the material are determined by comparing the strain with the resulting oscillating stress. Ideally, an elastic solid will have no lag between stress and strain, while the phase angle for a Newtonian fluid is 90°. With a viscoelastic material, the behaviour lies somewhere in between, that is, there is a phase shift δ between stress and strain ($0 < \delta < 90°$), the value of which depends on the relative amounts of elastic and viscous behaviour at the frequency of oscillation.

In an oscillatory rheological experiment, the relationship between stress and strain is described by the complex shear modulus G^* defined by:

$$G^* = G' + iG''$$

where G' is the storage modulus (ratio of in-phase stress to strain) i a complex number and G'' is the loss modulus (ratio of 90° out-of-phase stress to strain). Each of the quantities G' and G'' is dependent on the deformation frequency). The main advantage of oscillatory measurement is that it leads to minimal disruption in the structure of the material being investigated, because the measurement is carried out at low strain and stress.

The large-deformation mechanical test is used for obtaining the rheological parameters corresponding to the food processing conditions and sensory properties (Clark and Lee-Tuffnell, 1986). The specimen is deformed, usually rapidly, over the range of linear viscoelasticity, and the gel structure is disrupted to a large extent. Depending on the materials and purposes, the test cells contacting the specimen may also be varied (Voisey, 1979). Theoretical analyses are therefore almost impossible, although several fundamental studies have been made on the fracture and yielding behaviour of food gels (Stading and Hermansson, 1991; van Vliet *et al.*, 1991).

4.3 Methods of testing gel properties of proteins

4.3.1 Determination of heating temperature

In the case of heat-set gelation, gelation normally occurs at or above the denaturation temperatures of proteins, therefore data on the denaturation temperature are necessary as a first stage in studies of heat-set gelation. Differential scanning calorimetry (DSC) is most widely used to analyse the thermal behaviour of food proteins (Ma and Harwalker, 1990). The real denaturation temperature can be determined by extrapolating denatured temperature observed in DSC curves at different heating rates to 0°C/min. However, normally the denaturation temperature is estimated by DSC measurement with a constant heating rate. Figure 4.3 shows the results of DSC analyses of broad bean 11S globulin 'legumin' under various NaCl concentrations (Zheng *et al.*, 1993b). With an increase in NaCl concentration, the DSC curve shifted to a higher temperature, indicating the change of denaturation temperature of legumin with ionic strength. The other interesting finding in Figure 4.3 is that the gel hardness is very sensitive to the heating temperature. Here we define T_o, T_d and T_f as the temperatures of onset, peak and end points of denaturation. Temperatures below T_o

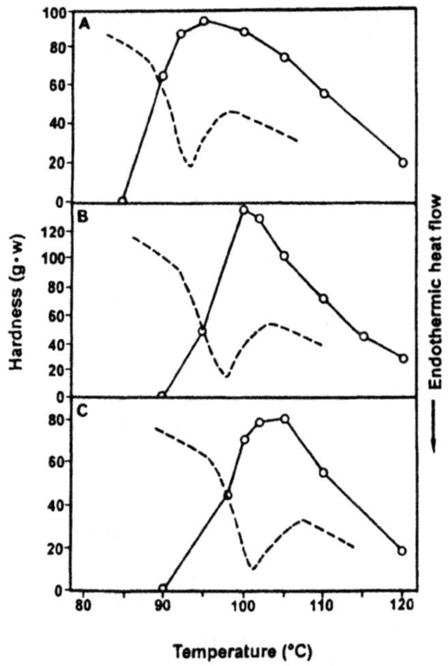

Figure 4.3 Effect of temperature on gel hardness relating to the denaturation temperature of legumin. The protein concentration of the legumin solution was 12.5%. NaCl concentrations: (A) 0.2 M; (B) 0.4 M; (C) 0.8 M. (– – – –), DSC curves with 1°C/min heating rate. (○), gel hardness with 60 min of heating time for gelation. From Zheng *et al.* (1993b).

showed no heat-set gel. In the region from T_o to T_d, a gel was formed and its hardness increased with increasing temperature towards T_d. The highest values of hardness were attained at temperatures from T_d to T_f. Above T_f, the gel hardness progressively decreased with increasing temperature. These results are consistent throughout Figures 4.3A, B and C. The reason for the lower gel hardness in the region from T_o to T_d could be explained by the slow denaturation rate relative to the following aggregation rate (see section 4.2.1). The mechanism of the decrease in gel hardness above T_f remains unclear but is probably due to excess heat damage to the protein molecules. Therefore, it is clear that the gel formation and gel properties are very dependent on small differences of temperature around denaturation points. To obtain a firm gel, the temperature should be set in the region from T_d to T_f.

4.3.2 Detection and analysis of the early stage of gelation

Prior to gel formation, the protein solution becomes more viscous, with the association of protein molecules to form flocs with an open structure. The detection of such an early stage of gelation is important with respect to the prediction of gelling points with the least consumption of the sample proteins. Furthermore, information on the gelling mechanism of the sample proteins may be obtained from the analysis of the early stage of gelation.

For the rheological view, the early stages of gelation can be detected by changes in viscosity of a protein solution or dispersion (Rector *et al.*, 1991). For instance, a whey protein concentrate dispersion (10% w/v) displayed a significant increase in viscosity at about 75°C (Hermansson, 1979). This reflects the predominant thermal behaviour of β-lactoglobulin.

The kinetics of protein aggregate formation have been analysed by chromatography techniques with UV monitoring. The transformation process of ovalbumin monomers to aggregates by heating was followed by gel permeation (GP) high performance liquid chromatography (HPLC) with a TSK G4000 SW column (the exclusion limit of this column is relatively small: 1.0×10^6 in molecular weight for globular proteins) (Koseki *et al.*, 1989b). The rate of polymer formation at 75°C exhibited approximate first-order behaviour, being independent of the protein concentration. Van Mil and Roefs (1993), using a similar method, analysed the time-dependent decrease of monomer α-lactalbumin and β-lactoglobulin by heating whey protein isolate solutions.

A column with small exclusion limits and UV monitoring is not powerful enough to estimate the molecular weights of large protein aggregates. The HPLC–low angle laser light scattering (LALLS) system developed by Takagi (1981) has been applied successfully to the analysis of the molecular weights of such large aggregates, formed by heating ovalbumin (Kato and Takagi, 1987) or by an enzyme-catalysed cross-linking reaction of bean globulins (Chanyongvorakul *et al.*, 1994). In this system, GP is performed through a tandem array of two columns, i.e. the TSK Gel G4000 SW (covering the molecular weight-range for dextran between 5000 and 4.0×10^5) and TSK Gel G6000 PW column (covering the range between 1.0×10^5 and 2.0×10^7). Good separation of aggregrates according to molecular weight is attained by using the two columns. Elution from the columns is monitored with a LALLS photometer and then with a differential refractometer. The molecular weight of the aggregates can be determined directly from the ratio of the output of a LALLS photometer to that of the refractometer.

Light scattering techniques not connected to chromatography systems have been used for the study of the kinetics of protein aggregation. For instance, dynamic light scattering was used to monitor the change in the apparent hydrodynamic radii of aggregates by heating whey protein solutions (Verheul *et al.*, 1995).

Sucrose density gradient centrifugation is normally used for the fractionation of cell organella. Nakamura *et al.* (1984) and Zheng *et al.* (1991) successfully applied this technique to separate aggregates from soybean 11S globulin glycinin and broad bean 11S globulin legumin formed by heating. Electrophoresis techniques can be used to monitor the aggregate to polymer formation. The protein–protein interaction between different species is easily proved using this technique. For instance, Utsumi and Kinsella (1985) showed the interaction of basic subunits of glycinin and β subunits of β-conglycinin (soybean 7S globulin) using the two-dimensional electrophoresis technique.

4.3.3 Detection of the sol–gel transition (determination of gelling points)

The term 'gelling point' normally means the minimum concentration of protein for gelation, gelling temperature or gelling time. The occurrence of gelation can be checked using fundamental or practical rheological techniques.

The sol–gel transition can be monitored by small-deformation mechanical testing (oscillatory rheological method). To discriminate the 'real' gel from 'weak' gel or sol, the frequency dependence of the storage modulus (G') and the loss modulus (G'') should be investigated. For instance, a 'weak' gel such as polysaccharide solution is distinguished from the 'real' gel using this method, because the former shows predominantly liquid-like behaviour ($G'' > G'$) at low frequency (da Silva and Rao, 1992). In practical aspects, however, the sol–gel transition of food protein solution has been usually performed at a constant frequency (often ≤ 1 Hz) with varied temperature and/or time. The gelation temperature (or time) is 'empirically' regarded

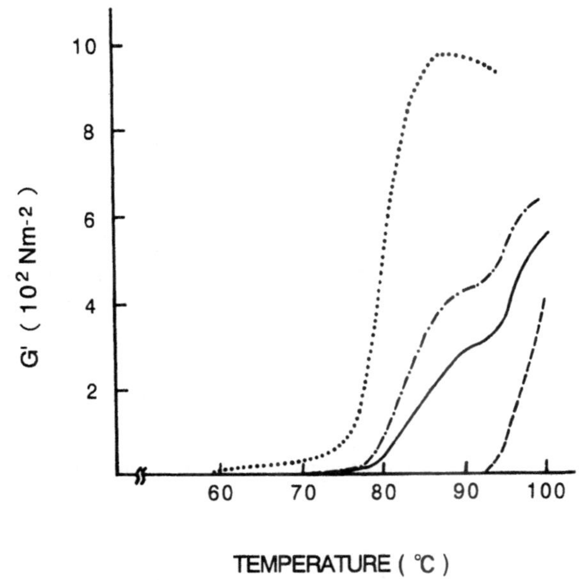

Figure 4.4 Dynamic shear modulus versus temperature curves for gels of soy protein. The protein concentration of the solution was 19%. Measurement was done at a fixed frequency of 2 Hz and ±100 μm displacement. (·····), β-conglycinin; (————), mixture of β-conglycinin (35%) and glycinin (65%); (–·–·–), mixture of β-conglycinin (50%) and glycinin (50%); (– – – –), glycinin. From Kang *et al.*, unpublished data.

as the temperature (or the time) at which G' steeply increases (Hamann *et al.*, 1990).

The temperature dependence of G' of soy proteins measured using an oscillatory rheological apparatus is shown in Figure 4.4 (Kang *et al.*, unpublished data). The shear deformation was done at a fixed frequency of 2 Hz and ±100 μm displacement. The curve of 7S globulin of soybean, β-conglycinin, displayed a steep rise in G' at around 76°C. In the case of soybean 11S globulin, glycinin, G' increased markedly at around 93°C. These temperatures (76°C and 93°C), indicating the occurrence of an viscoelastic gel matrix, coincided with temperatures of denaturation for β-conglycinin and glycinin determined by DSC measurements. The mixture of β-conglycinin and glycinin exhibited two step rises in G' at around 76°C and 93°C. This result may reflect the real gelation behaviour of soybean proteins, since these two globulins are major components of soybean storage proteins.

Simple and more empirical methods without any specialized equipment have been devised to detect the sol–gel transition or to determine the gel points of protein solutions. For instance, Utsumi *et al.* (1982) reported a micro-method to detect the gelation of soybean glycinin. The method is briefly illustrated in Figure 4.5. 15 to 20 μl of the protein solution were

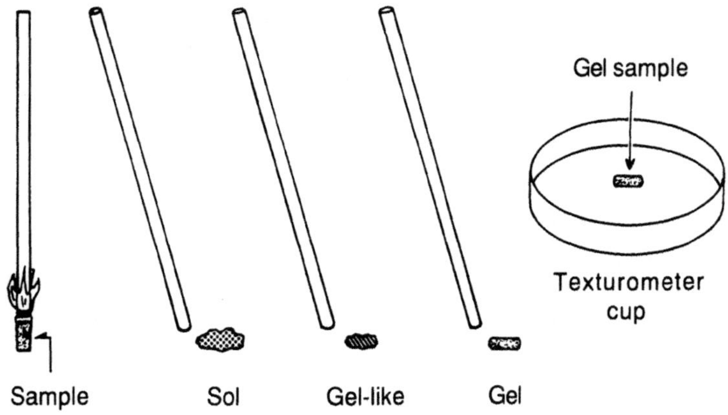

Figure 4.5 Schematic representation of the procedure for the micro-analysis of 11S globulin gel. The details are described in the text. From Utsumi *et al.* (1982).

taken up into micropipettes, which were carefully sealed at one end with polyvinylidenechloride film. After heating, the contents were removed to check the gelation. The formed gels were cut into 5 mm length, after which hardness was measured with a texturometer. For the measurement of gel turbidity, the gels formed in the micropipettes were scanned with a dual wavelength chromatoscanner to detect the light absorbance at 600 nm. Using this method, the degree of gelation, i.e. sol, gel-like but not self-supporting, or self-supporting gel, could be judged macroscopically. Furthermore, the effect of heating and protein concentration on the hardness of gels could be investigated satisfactorily. The advantages of this method are the small amount of sample required for the experiment and the very high heat conductivity because of using micropipettes 1.68 mm in diameter. The disadvantage is the difficulty in handling the micropipettes, particularly sealing with polymer films.

Zheng *et al.* (1991) and Hatta *et al.* (1986) used small caps and moulds made of aluminium and stainless steel, respectively, to form gels from protein solutions upon heating. In these cases, the gel formed was compressed *in situ* using a rheometer, and the generated force was recorded as gel hardness. Compression of the gel inside a vessel is not recommended from a fundamental view point, but is very useful for checking sol–gel transition and comparing gel firmness directly under various conditions.

Recently, Kitabatake *et al.* (1994) devised a simple, rapid, and probably more easily handled method for measuring the turbidity and rheological properties of protein gels and sols. They used this method to investigate the effects of pH and ionic strength on gel properties of whey proteins. However, the method can be also used to detect the gelling point. The unique feature of the method is to use a microwell and/or microreader. These are normally used in the biomedical field for the detection of the formation of antigen–antibody complex, for instance an ELISA reaction. The procedure is as follows.

For measurements of rheological properties, protein solution (250 μl) was poured into each well (6.3 mm in diameter × 12 mm in height) of a 12-well chamber slide, and the chamber was tightly sealed with heat-resistant film. The sealed chamber was placed in a tightly closed box in which distilled water was placed to saturate the interior of the box with water vapour. The box containing the chamber was then heated in a convection oven. After cooling at room temperature the bottom slide glass was removed. The gel was then carefully removed from each well using a Teflon piston (6.3 mm in diameter) and cut to 5 mm in height. The gel was immediately measured for rheological parameters.

For turbidity measurements, test samples (250 μl) were placed in the wells of a 96-well microplate, heated as described above and then cooled at room temperature. Figure 4.6 shows β-lactoglobulin solutions heated in the wells where the turbidity (the absorbance at 590 nm) of each well was measured with a microwell reader.

The disadvantage of this method is the difficulty in controlling the heating time precisely. However, such a simple and rapid method, using low amounts of sample, should be applicable to the determination of good gelling conditions for various food proteins.

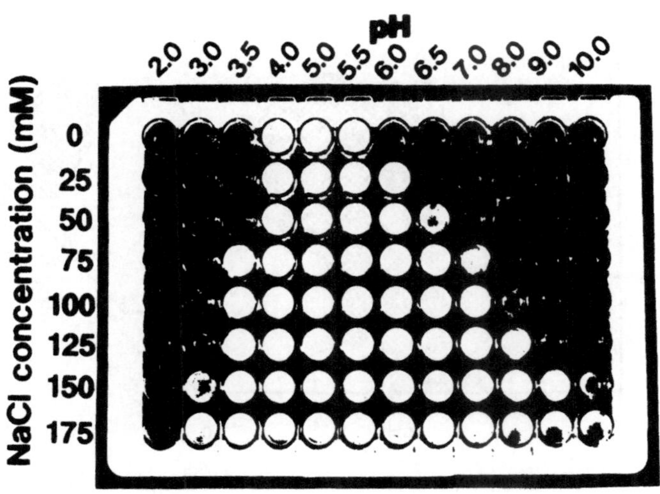

Figure 4.6 β-Lactoglobulin solutions heated in the wells of a 96-well microplate at 80°C for 1 h. Each well contained 250 μl of β-lactoglobulin solution (70 mg/ml). Turbidity (absorbance at 590 nm) was measured with a microwell reader. From Kitabatake *et al.* (1994).

4.3.4 Effects of pH and ionic strength

The importance of balanced electrostatic interactions in the formation of the well-developed network structure of protein gels has been demonstrated by the effects of pH and ionic strength on gel formation and gel properties of various proteins. This type of experiment necessitates simple microscale methods, such as the one of Kitabatake *et al.* (1994) described in section 4.3.3, because of the enormous numbers of the combinations of pH and ionic strength (and/or ionic species). In this section we discuss the case of egg white ovalbumin gel investigated systematically by Doi's group (Doi, 1993).

Hatta *et al.* (1986) investigated the effects of pH and NaCl conditions on the gel turbidity and hardness of ovalbumin gels formed by heating 5% protein solution at 80°C. The results are shown in Figure 4.7. Without added NaCl, ovalbumin solutions did not gel when heated. At low NaCl concentrations (10–20 mM), turbid and soft gels were obtained near the pI of ovalbumin (4.7), while transparent gels were obtained at pH values above or below the pI; gel hardness peaked at around pH 3 and 7. Gel turbidity increased with increasing NaCl concentration; at higher NaCl concentrations (30–80 mM), turbid gels were also obtained for a wide range of pH values at, above and below the pI. The gel hardness at pH 3.5 and 7.5 was maximal at 30 mM and 50 mM NaCl, respectively, and decreased with higher concentrations. Such a relationship between pH, ionic strength, gel turbidity and gel hardness has been observed for serum albumin (Murata *et al.*, 1993), lysozyme (Tani *et al.*, 1993) and whey

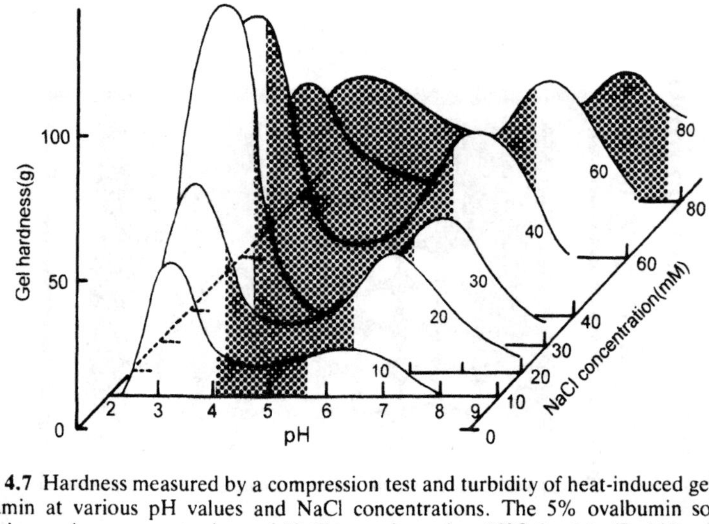

Figure 4.7 Hardness measured by a compression test and turbidity of heat-induced gels from ovalbumin at various pH values and NaCl concentrations. The 5% ovalbumin solutions containing various concentrations of NaCl were heated at 80°C for 1 h. Turbid gels were obtained in the shaded regions. Numerals under the curves indicate the corresponding NaCl concentrations. From Doi (1993).

protein (Mulvihill and Kinsella, 1988; Shimada and Cheftel, 1988; Kinekawa and Kitabatake, 1995), although a transparent and firm gel with a fine strand network structure is not formed for whey proteins at acidic pH regions.

It is generally observed that transparent or translucent gels can be obtained only in a narrow range of pH and salt concentration (Hatta *et al.*, 1986). To overcome this limitation and cope with the requirement for the formation of firm and transparent gels in various buffer conditions, a two-step heating method was devised (Kitabatake *et al.*, 1987). Recently, this method was applied to improve the gelling properties of whey proteins (Kinekawa and Kitabatake, 1995). The first heating step (80°C for 1 h) was done using whey protein isolate solution (74 mg/ml) at neutral pH under salt-free conditions. The resultant transparent solution was named 'process whey protein'. The process whey protein solution was then heated at 80°C for 1 h at various pH values (2.0 to 11.0) and NaCl concentrations (0 to 200 mM). For comparison, the unprocessed whey protein solution (i.e. without the first heating step) was heated under the same conditions. The turbidity and gel properties of the heat products from unprocessed whey protein and process whey protein were measured using the method of Kitabatake *et al.* (1994). Figure 4.8 shows the results of gel turbidity. The whey protein solutions without the first heating step showed high turbidity by heating over a wide range of pH and NaCl conditions. However, for the process whey protein, transparent and/or translucent samples were obtained over a much broader area. The process whey protein gel showed higher values of breaking energy around neutral pH values than did the unprocessed whey protein gels (Figure 4.9), indicating that the process whey protein can produce firmer gels. Therefore, the two-step heating method is very useful for producing a novel food protein material and the desired textural properties.

4.3.5 Spectroscopic analysis of protein structural change through the gelation process

As described in section 4.1, the partially denatured state of protein molecules may play an important role in the gelation process. The denatured state of proteins is normally analysed using spectroscopic techniques such as circular dichroism (CD), fluorescence, UV absorbance difference spectra, and so on.

Zheng *et al.* (1993b) investigated the heat-induced changes in conformation of broad bean legumin molecules using CD and UV difference absorbance spectra. It was shown that the secondary structure, detected by far-UV CD spectroscopy, of legumin molecules during thermal aggregation alters slightly, reflecting the native-like conformation of the denatured form. However, based on the results of UV difference spectra, aromatic

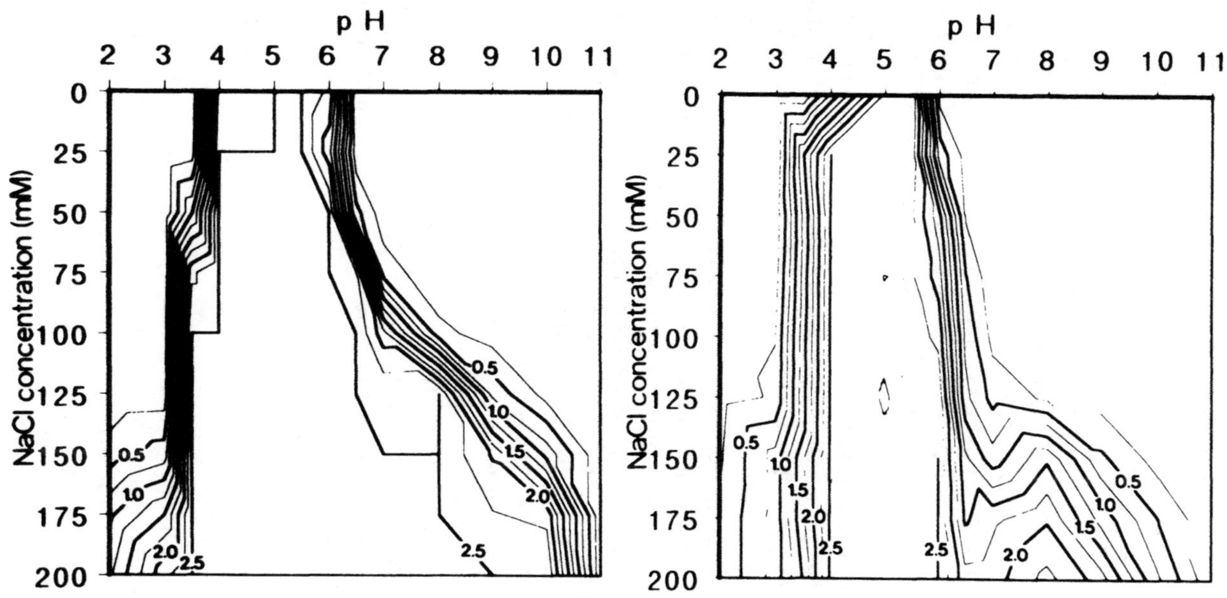

Figure 4.8 Turbidity of the whey protein isolate and the heat-processed whey protein. Each contour curve indicates the absorbance at 590 nm of the whey protein isolate (left) and the heat-processed whey protein (right). From Kinekawa and Kitabatake (1995).

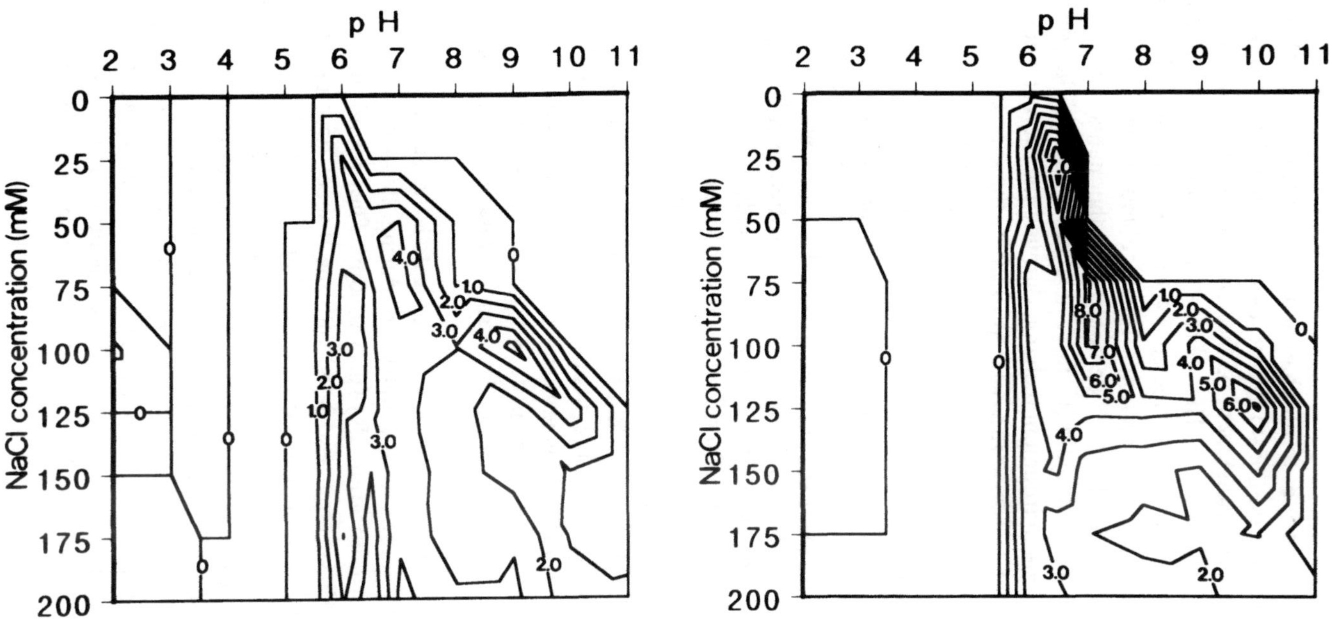

Figure 4.9 Breaking energy of the whey protein isolate and the heat-processed whey protein. Each contour curve indicates the breaking energy values ($\times 10^6$ dyne/cm^2) of the whey protein isolate (left) and the heat-processed whey protein (right). From Kinekawa and Kitabatake (1995).

amino acid (i.e. tyrosine, tryptophan and phenylalanine) side chains were found to be exposed on the surface of legumin molecules. In other words, heating altered the arrangement of the aromatic amino acid residues, i.e. tertiary structural changes. The increase in surface hydrophobicity of heated legumin molecules, detected by fluorescence dye binding, also indicates the partial denaturation of heated molecules (Zheng *et al.*, 1993b). Koseki *et al.* (1989a) demonstrated the native-like secondary structure of heated ovalbumin (by far-UV CD spectra) and the completely irreversible micro-environmental changes around the aromatic amino acid residues (by near-UV spectra). Taken together, these results indicate the close relationship between the molten globule form and heat-denatured states in the gelation process.

The experiments described above usually employ relatively low protein concentrations which are suitable for investigating protein denaturation without the effect of the association of denatured molecules. In real gelation systems, however, protein molecules with relatively high concentrations aggregate and form the network structure following thermal denaturation. The resultant sol or gel is normally turbid or translucent. Because of severe light scattering, the spectroscopic techniques described are not suitable for monitoring the secondary structural changes of protein molecules in a gel network structure.

Clark and Lee-Tuffnell (1986) applied Raman spectroscopy to analyse the secondary structure of polypeptides in heated solutions and gels of α-chymotrypsin, lysozyme, bovine serum albumin and ribonuclease. Evidence was found for β-sheet development in aggregates and gels formed from all the proteins except ribonuclease. The result for lysozyme was confirmed by Li-Chan and Nakai (1991).

Recently, Nagano *et al.* (1994) investigated the secondary structural changes of glycinin and β-conglycinin during heat-set gelation using Fourier-transform infrared (IR) spectroscopy. They found that a band at 1618 cm^{-1} on IR spectra, associated with the β-sheet structure, increased with the formation of heat-induced gels for both proteins. This relationship was not seen in the results of Wang and Damodaran (1991) who determined the secondary structure of heated glycinin and β-conglycinin by CD analysis. The contradiction is likely to be caused by a difference in the use of protein concentration, as the protein concentration used in CD measurements is lower. This confirms the speculation suggested in section 4.2.2 that, at high protein concentrations, the development of a β-sheet in the gel network is due to the association of denatured molecules.

4.3.6 Contribution of chemical bonding to the gelation process

The simplest way to assess the contribution of chemical bonding to gel formation may be to study the effects of additives on gelation behaviour.

The idea is to add reagents such as SH-blocking reagents or salts to the medium prior to heating, and to demonstrate their effect on the gelation process. The problem with this procedure is ambiguity in relation to the time of action of the additives. It is not known whether the additives influence mainly the unfolding step or the subsequent aggregation step, therefore care should be taken when defining the contribution of chemical bonding to the gelation process. The two-step heating method described in section 4.3.4 may be one possible way to discriminate the role of chemical bonding in the unfolding or aggregation process.

The SH and SS interchange reaction plays a crucial role in the formation of a protein gel network, particularly in the case of thermo-irreversible, elastic gels. The effects of addition of an SH-blocking agent such as *N*-methylmaleimide (NEM) on the gelling process has been reported for various proteins. Mori *et al.* (1982) found that the presence of 10 mM NEM inhibited the heat-set gelation of 5% glycinin solution, and that heating only produced linear soluble aggregates without the branched structure necessary for the gel network formation. The effect of NEM addition was also investigated for legumin gelation by observing the rate of protein association using GP HPLC (Zheng *et al.*, 1993b). When NEM was added to the solutions of β-lactoglobulin (Phillips *et al.*, 1994), hardness increased slightly up to 5 mM NEM, and then rapidly decreased when NEM concentrations exceeded 10 mM. This result indicates that the ratio of free SH groups and SS bonds is crucial for the developed network structure and gel hardness.

The effects of the deletion or introduction of cysteine and/or cystine groups on the gelling properties of soybean glycinin (Utsumi *et al.*, 1993) and β-lactoglobulin (Lee *et al.*, 1993) were investigated using protein engineering techniques. The rapid development of such techniques means that in the future, recombinant protein will more frequently be used to assess the relationship between the primary sequence and the gelling properties of food proteins.

The contribution of non-covalent bonding has been investigated by the addition of salts to gelling systems. Experiments using NaCl for assessing electrostatic interactions have been described in section 4.3.4. Other salts consisting of chaotropic anions such as SCN^- have been used to assess hydrophobic rather than electrostatic interactions in gel formation, because chaotropic anions can break the water structure effectively (Hatefi and Hanstein, 1969). Zheng *et al.* (1993c) found that NaSCN addition delayed the aggregate formation of legumin molecules during heat gelation. The effects of different anions on gel hardness were also demonstrated for β-lactoglobulin and bovine serum albumin (Phillips *et al.*, 1994).

In addition to treatment prior to gelation, attempts have been made to swell and dissolve the 'formed' gel in a solvent noted for its ability to attack particular types of linkage. For instance, the critical roles of hydrophobic interactions and hydrogen bonding in maintaining the gel network of legumin were demonstrated by the complete solubilization of gels by 8 M urea (Zheng *et al.*, 1993a). By contrast, mercaptoethanol (10 mM), which is expected to attack the SS linkage, could solubilize only 33% legumin from the gels, but decreased the hardness of the gel dramatically.

Physical methods such as high-pressure treatment could be useful for discriminating between the contribution of hydrogen bonding and hydrophobic interactions in gels. Doi *et al.* (1991) found that agarose or gelatin gels could not be melted by high pressure (600 MPa), whereas heat-set globular protein gels such as transparent ovalbumin gels were melted by the same treatment. It is well known that hydrogen bonding is strengthened under high pressure (Terasawa *et al.*, 1975), but that hydrophobic interactions are weakened (Kauzmann, 1987). Therefore, these results indicate that the predominant forces in the network of transparent ovalbumin gels are hydrophobic interactions.

4.3.7 Rheological properties of gels

The relationship between dynamic moduli and protein concentrations has been analysed using several theories to assess the numbers of physical cross-links and the size of junction zones. Theories and their application to the actual data were reviewed by Clark and Lee-Tuffnell (1986) and Ross-Murphy (1991).

The frequency dependence of the shear modulus for a gel is characteristic of the type of gel (Clark and Ross-Murphy, 1987). G' shows a plateau over a wide range of frequency when the system behaves like a gel (particularly when the network is stabilized by covalent cross-links), but decreases with decreasing frequency when the system behaves like a sol or weak physical gel. Nagano *et al.* (1994) demonstrated that the G' of β-conglycinin gel showed frequency independence when the gel was formed at temperatures above 75°C (the denaturation temperature of β-conglycinin is 65°C). Stading *et al.* (1992) discriminated between homogeneous and inhomogeneous gels of β-lactoglobulin based on data on the frequency dependence of G'.

The concentration dependence of the modulus can be approximated by an exponential function (Clark *et al.*, 1990):

$$G \propto C^n$$

In most cases of biopolymer gels, the modulus is proportional to approximately C^4 at relatively low concentrations and proportional to C^2 over high concentrations (10–20%). Kohyama *et al.* (1992) obtained the exponent 3.4 for the heat-set gel of soybean glycinin (2–7% protein concentration), while an exponent of 3.6 was found for glycinin gels

(4–11% protein concentration) formed by a cross-linking reaction of transglutaminase (Matsumura *et al.*, 1993). The similar exponent values indicate similar association behaviour and similar network structure, although the different chemical forces seem to be predominant in each gel, i.e. non covalent bondings for heat-set gelation and covalent isopeptide bonding for transglutaminase-catalysed gels.

Stading *et al.* (1993) demonstrated the usefulness of fractal dimension for the analysis of the relationship between dynamic modulus and protein concentrations. The fractal dimension D_f was approximately 2.5 for particulate gels of β-lactoglobulin formed near the pI point, whereas D_f was approximately 2.9 for fine-stranded gels formed at pH 7.5, far from the pI point. The lower values of fractal dimension indicate the more condensed packing and more diffusion-limited aggregate mode of particulate gels. It may be interesting to investigate the concentration–modulus relationship using the fractal concept to obtain information on the network dimension of protein gels produced by various methods.

The most popular of the large-deformation tests is the textural profile determined by texturometer (Szczesniak, 1968). From these measurements, several parameters relating to the mastication process can be obtained, such as 'hardness', 'cohesiveness', 'adhesiveness', 'fracturability', 'chewiness' and 'gumminess'.

To assess precisely the gel's rigidity and brittle (or ductile) character, compression or tensile experiments have often been carried out. However, good reproducibility has rarely been obtained, since the methods are very sensitive to the sample shape and the small cracks accidentally occurring within the sample. The methods of sample preparation, cutting and forming, and the method of sample setting are therefore very important. There have been several attempts to minimize the effect of accidental factors on intrinsic gel properties. Normally, square moulds made from metals have been used to prepare gel species (Stading *et al.*, 1993; Chanyongvorakul *et al.*, 1995). The materials of the mould and the sealer were chosen carefully to avoid sticking the samples to the moulds and leakage of the samples during heating. Khoyama *et al.* (1993) obtained reproducible results by applying a ring-shaped gel sample for tensile measurement.

Hermansson and co-workers have succeeded in characterizing physical properties of several types of protein gels, such as myosin, blood plasma proteins, whey proteins, etc., using compression or tensile experiments. In work on the tensile properties of β-lactoglobulin, Stading *et al.* (1993) made artificial notches on samples and analysed the fracture behaviour. They demonstrated clearly that the rigidity and brittleness of β-lactoglobulin gels changed according to pH, and that such physical properties showed a close relationship with the network structure observed by electron microscopy.

Elasticity is also one of the most important rheological properties. It is evaluated by the measurement of tanδ (G''/G') or creep tests but, more practically, by the compression–decompression test. The hysteresis of the compression and decompression curves can be regarded as the index of the deviation from ideal elasticity (Kaletunc *et al.*, 1991). Chanyongvorakul *et al.* (1995) showed that transglutaminase-catalysed gels of bean 11S globulins were more elastic than were heat-set gels.

4.3.8 Microscopic observation

The microstructure of the protein gel network has a close relationship with the gelling behaviour and physical properties of formed gels. The microstructure of protein gels can be characterized by electron microscopy, and microscopic techniques used for the visualization and characterization of gel structures have been reviewed by Hermansson and Langton (1992), and Hermansson (1994).

A negative staining technique and observation by transmission electron microscopy (TEM) were used to characterize the heat-induced soluble aggregates of soybean glycinin (Nakamura *et al.*, 1984) and broad bean legumins (Zheng *et al.*, 1991). In the case of soybean glycinin, the formation by short-time heating of strands 10–12 nm in thickness and 100 nm in length was observed (Figure 4.10a,b). Further heating caused an increase in length to 170–200 nm, and branching of the strands (Figure 4.10c). The thickness 10–12 nm is similar to the diameter of the native glycinin molecule. On heat treatment therefore, glycinin tends to associate regularly to form linear or branched strands while retaining its native subunit structure. However, the strands formed from legumin on heating exhibited an irregular thickness range from 8.3 nm (diameter of native molecule) to 36 nm (Figure 4.10d,e). As a result, the strands were associated more randomly to form a more uneven structure. The irregularity of the legumin gel structure compared with the glycinin gel structure could be responsible for the less firm and less elastic properties of legumin gels, although the two globulins are categorized in the same group of 11S bean globulins. Similar observation and characterization using negative staining and TEM have been carried out for various protein gels such as ovalbumin (Koseki *et al.*, 1989a), bovine serum albumin (Clark and Lee-Tuffnell, 1986), myosin (Hermansson and Langton, 1988), etc.

Structural analyses of the gel network should be carried out using scanning electron microscopy (SEM) and TEM with freeze-etching and thin-sectioning techniques (Hermansson and Langton, 1992). The network structure of particulate gel types can successfully be characterized using SEM observation, whereas TEM is better suited to the visualization of the network structure of a fine-stranded gel.

It has been demonstrated that computer-based image analysis provides a

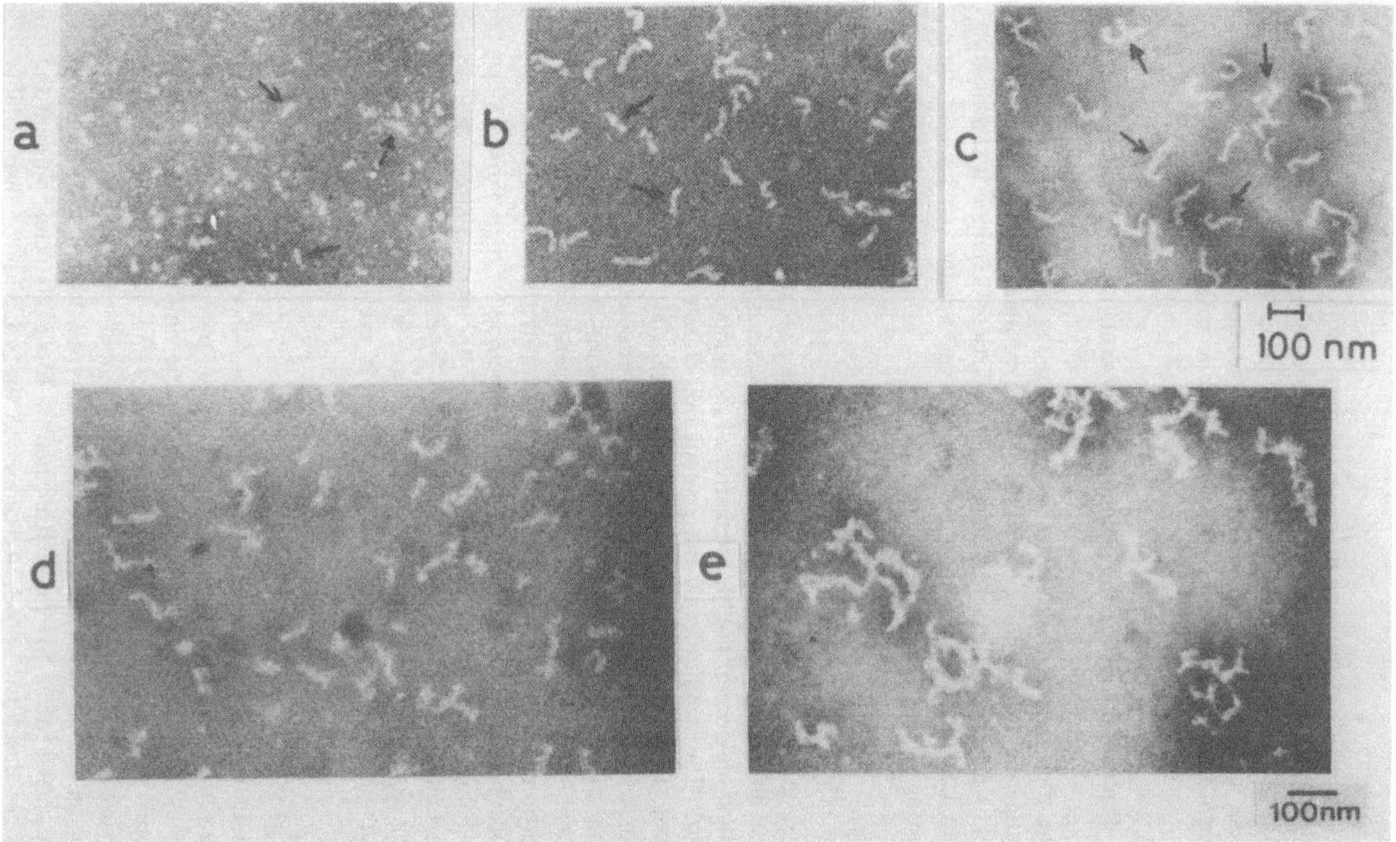

Figure 4.10 Transmission electron micrographs of negatively stained materials. Glycinin solution (5%) and legumin solution (12.5%) were heated at 100°C for various times: (a) glycinin, 15 sec; (b) glycinin, 30 sec; (c) glycinin, 60 sec; (d) legumin, 60 sec; (e) legumin, 3 min. a, b, c from Nakamura *et al.* (1984), d, e from Zheng *et al.* (1991).

rapid means of quantifying many image parameters, including the fractal dimension from micrographs (Bottcher and Foegeding, 1994; Hermansson, 1994). The introduction of image analysis to characterize SEM images of protein gels may aid in understanding the quantitative relationship between functional property parameters and microstructure.

4.4 Protocol for testing the gelling ability of proteins

In the final part of this chapter we put ourselves in the position of one investigating the gelling properties of food proteins, and consider the proper way to get satisfactory results. The protocol described is shown in Figure 4.11. Each step in the flow chart will be explained in detail in the following sections.

4.4.1 Information on samples

Before entering the procedure of functionality testing, it is necessary to collect information on samples as fully as possible. Is the sample familiar as a gelling material or not? Is the source of the sample animal, plant, fish or others? The conditions of the protein preparation (extraction), drying process (if the sample is a powder) and storage are also important. For instance, lipid extraction at high temperatures could cause denaturation and the aggregation of protein molecules. This may lead to a change in the solubility and gelling behaviour (gelling temperature, gelling time, etc.). Data on chemical composition (ash, carbohydrate, lipid, water contents) are also helpful in choosing the appropriate method for the following steps in the flow chart. If the sample is a powder, the particle size is also an important factor in relation to the solubility or dispersibility. When the average particle size is too large, the particle should be ground by a grinding or milling machine to the micrometer level.

4.4.2 Solubilization or dispersion

Prior to the gelation process, sample proteins should be solubilized or dispersed in the aqueous medium. For the purpose of the chemical analyses, proteins can be solubilized by severe buffer conditions such as extreme pH, addition of denaturants, surfactants, or reducing reagents, etc. However, these conditions are not permitted or should be avoided in the actual food systems and it is desirable that food proteins are solubilized or dispersed in the pH range from neutral to weakly acidic without the additives described above. According to the solubility or dispersibility under such buffer conditions, food proteins can be categorized into two groups (A and B in Figure 4.12).

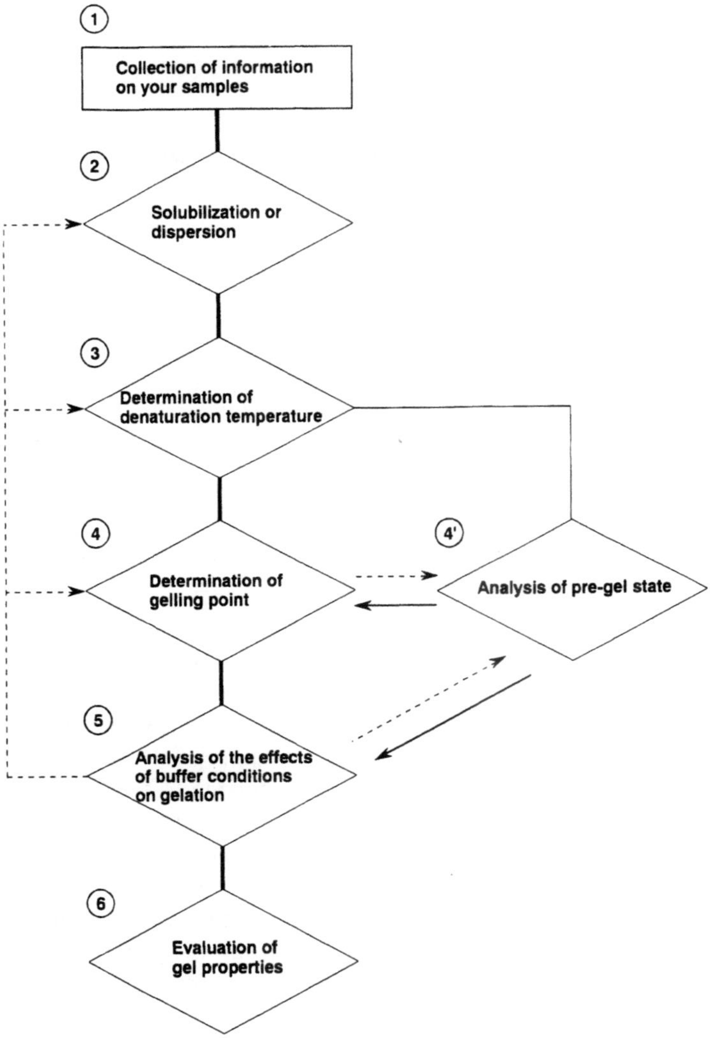

Figure 4.11 Flow chart to find optimum conditions of heat-set gelation of proteins. The solid thick line indicates the route normally followed. The solid thin line indicates how the data of the pregel state can be used to predict the appropriate gelling conditions. The dashed line indicates working back to previous steps.

Group A includes proteins which are soluble or dispersible in the aqueous medium with neutral or weakly acidic pH values; this group is further divided into two groups according to their solubility in water. Gelatin, egg white, milk whey and soy proteins are the most popular gelling materials and belong to the Group A. Blood proteins also belong to the same group and the functional properties of blood plasma proteins

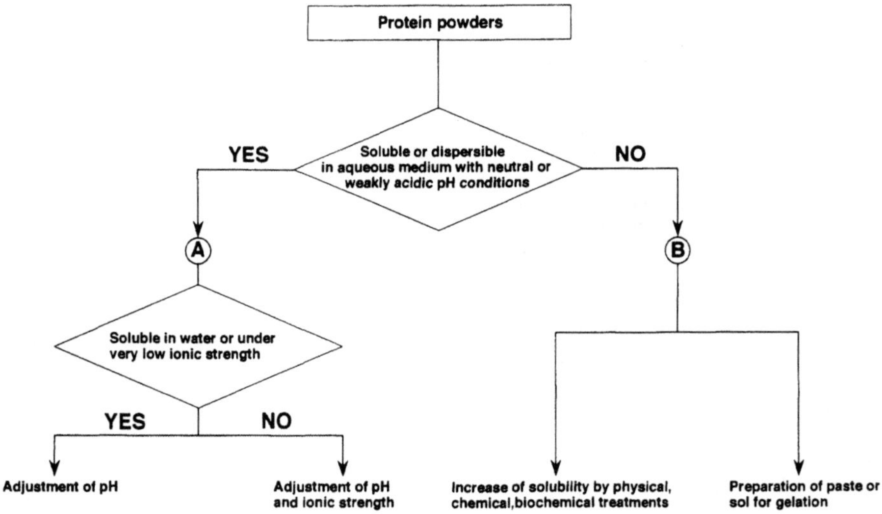

Figure 4.12 Classification of protein powders according to their solubility or dispersibility.

have recently been attracting great interest from fundamental and practical view points (Hermansson, 1986). For Group A food proteins, there is much information on the solubility or dispersibility in the aqueous medium. We can therefore predict the most appropriate buffer conditions for their solubilization or dispersibility by referring to the literature. It is however possible that denaturation, causing the decrease of solubility, occurs during the preparation and/or the storage process of the sample proteins. In addition to the popular food proteins listed above, there may be the opportunity to investigate the gelling properties of novel food proteins. If the novel proteins originate from the legumes, they can probably be categorized in Group A because major fractions of legume seed storage proteins are globulins (Wright, 1987). The major component of muscle proteins from various animals and fishes, myosin, is also categorized in Group A (Ranken, 1984), although whole muscle proteins are very difficult to solubilize in water. If no information is available on the origin and the solubility of the sample it may be categorized in Group A and the solubilization experiment started.

First, the sample should be solubilized or dispersed in water. The protein powder should be added gradually to distilled water with gentle stirring. The alternative way is to add distilled water to the protein powder, stirring with a small glass rod to form a smooth paste. Relatively high protein concentrations are necessary for gelation, i.e. from a few percent to more than 10%. The resultant dispersions are adjusted to the appropriate pH from 3–8, with 1 M HCl or NaOH solution. If the protein powder is difficult to disperse it may be necessary to use several types of disperser or homogenizer (polytron, ultrasonic, etc.), although care should be taken with generation of heat.

The dispersion is stirred for several hours at room temperature (20–25°C). If strong intermolecular hydrophobic interaction is expected to decrease the solubility, as for caseins (Binghan, 1971), the storage temperature should be decreased to around 4°C. By contrast, warming the dispersion sometimes increases the rate of water hydration, as in the case of gelatin. When the sample protein contains a significant amount of ash, the ash should be removed by dialysis.

When testing the real solubility of proteins, the proteins should not be sedimented by moderate centrifugation forces as described in Chapter 2. However, complete solubilization of proteins to give a transparent solution is sometimes impossible in dispersions of the high protein concentrations necessary for the gelation. Furthermore, even if the protein dispersion is a turbid or viscous sol state, gel formation is sometimes possible. Such dispersions normally form turbid, aggregate-type gels by heating, therefore complete solubilization is not necessary for gelation unless the proteins precipitate under natural gravity. It is of course better to try and remove the large colloidal particles of proteins by centrifugation or filtration in order to obtain the most transparent or translucent gels.

Poor solubility or dispersibility of the proteins in water may necessitate solubilization with moderate or high concentrations of salts such as NaCl, KCl, etc. First, the sample proteins are dispersed in distilled water, and the pH adjusted to the desired value as described above. Solutions of high salt concentrations (e.g. 5 M) are added to the dispersion to adjust the ionic strength to the desired values. The subsequent procedure is the same as described above. Most globulins and myosin must be solubilized under these buffer conditions.

Group B of Figure 4.12 contains proteins which are very difficult to solubilize or disperse in aqueous solution under normal conditions, and so have poor gelling properties. Major proteins of this group are cereal seed proteins, which have the greatest potential for sustaining the population. The main components of cereal seed proteins are prolamine and glutelin, which are soluble only in organic solvents (70% ethanol), or at extremely acidic or alkaline pH, or in the presence of denaturants. Addition of a surfactant such as lecithin (and ultrasonication) may improve the solubility of prolamines. However, these buffer conditions may not be applicable to gelation in real food systems.

Chemical modifications such as succinylation are known to be applicable for improving the solubility of wheat gluten (Ma *et al.*, 1986). However, chemical modifications are not desirable tools for application in real food systems from the point of view of safety. Cereal prolamine and glutelin normally contain a high content of glutamine, and hydrogen bonding among glutamine residues is thought to be responsible for the poor solubility. Deamidation under acidic conditions has therefore been attempted to improve the solubility of gluten proteins. Recently, trans-glutaminase has been found to be capable of catalysing the deamidation of gluten proteins (Alexandre *et al.*, 1993). Enzymatic treatments, including protease reaction, should be useful for modifying the solubility as well as other functional properties of cereal prolamines and glutelins.

Fish meats are also promising protein sources which sustain the population. Whole muscle proteins containing connective tissues are not soluble in aqueous media. Traditionally, meat proteins are homogenized to form a sol which changes to a gel upon heating, for example sausage or kamaboko. For raw fish meat, therefore, it is best to choose such traditional methods to make gels. For a fish (or animal) muscle protein powder, distilled water or salt buffer is added dropwise to the powder with gently stirring to form the paste (sol), and then a gel by heating. Alternatively, the proteins may be extracted from the powder with buffers, and the resultant soluble protein solutions (consisting mainly of myosin) used for gelation as in Group A.

The gelation method starting from such a paste or viscous sol is different from the original definition of gelation in section 4.2.1, and investigation of the gelation mechanism of such systems is very difficult. However, when given proteins which are difficult to disperse in aqueous media, but have water-holding capacity, this method should be tried. A gel can be made from cereal prolamine and glutelin; an eastern traditional gel food, fu, is just a hydrated gluten ball.

4.4.3 Determination of the denaturation temperature

In the case of heat-set gelation, determination of the denaturation temperature is necessary under the desired buffer conditions using DSC to determine the heating temperature for gelation. To avoid the effect of the interaction between protein molecules as much as possible, the aggregates should be removed by centrifugation or filtration if there are large particles in the dispersion.

If a single peak is found in the DSC curve for the sample, it is easy to set the heating temperature between T_d and T_f as described in section 4.3.1. However, if several endothermic peaks or poorly defined sharp peaks, i.e. a broad peak over a wide range of temperatures are found, these may be attributed to the mixed components in the sample and/or the denaturation of protein molecules during processing and storage. For mixed components, it may be possible to separate the protein peak from those of other components involved in the sample. This requires data on the chemical composition of the sample powder or the ability to predict the temperature of the phase transition of other components (for instance, the denaturation temperatures of starch and lipids can normally be discriminated from those of proteins). However, practically it is very difficult to pick out the protein peak from such complex DSC curves and it is therefore better to set the heating temperature near the T_d of the final peak (i.e. the peak which appears at the highest temperature) or near the end point of endothermic flow.

The denaturation temperatures of plant seed proteins are normally higher than those of animal proteins. For instance, the T_d of oat globulin is over 110°C (Ma and Harwalkar, 1988); for such proteins it is necessary to prolong heating at the boiling temperature or to increase the heating temperature using an oil bath or autoclave.

4.4.4 Determination of the gelling point

It is natural to investigate the pre-gel state (step 4′ in Figure 4.11) before going on to the gel state. However, because of practical interest and importance, studies on the pre-gel state are often omitted, and determination of the gelling point is the next step.

A laboratory equipped with dynamic rheology apparatus can monitor the change in G' of protein solutions with temperature as shown in Figure 4.4, or with time (at constant temperature). Otherwise the gelling point can be determined using a microscale method, such as that of Utsumi *et al.* (1982), Hatta *et al.* (1986) or Kitabatake *et al.* (1994).

4.3.4.1 Analysis of the pre-gel state. Based on information from the pre-gel state, we can predict the gelling point (direction 4′ → 4 in Figure 4.11). For instance, the temperature of the onset of viscosity increase detected in step 4′ may be helpful in finding the real gelling temperature in step 4 but, as described above, gelling points are often determined before analysing the pre-gel state. It may be more usual to study the pre-gel state after knowing the gel point of the protein in order to characterize the gelation mechanism (direction 4 → 4′ in Figure 4.11). In other words, the mechanism of good gelling ability of the protein detected in step 4 can be interpreted by the results in the pre-gel state (step 4′) such as: the rapid aggregation of protein molecules (data from chromatography, light scattering, viscometry); the developed network structure or dimension of aggregates (light scattering, TEM); the conformational change of protein in aggregates (spectroscopic analyses); and the formation of chemical forces (the effects of chemical additives).

4.4.5 The effects of pH, ionic strength and other conditions

The method of Kitabatake *et al.* (1994) is very useful for investigating the effects of the pH and ionic strength (or ionic species) on the gelation behaviour (section 4.3.4). When the protein sample solutions form soft and turbid gels in a wide range of pH and ionic strength, it may also be useful to attempt the two-step heating method.

To investigate the effects of the enormous numbers of conditions in the experiments, it is very important to process and present the data in the most appropriate way. Contour line graphics of data such as those shown in Figures 4.8 and 4.9, and three-dimensional graphics are very helpful in finding the region of pH and ionic strength for producing the best gels. Combining the statistical analyses or design with a computer program, such as the Mapping Super-Simplex Optimization developed by Nakai *et al.* (1984), facilitates the prediction of the optimum conditions.

Besides pH and ionic strength (species), other conditions may influence the gelling properties in real food systems. For instance, when more than two species of proteins are involved in food systems, the interaction between them could modify the individual gelation behaviour. The effects of other additives such as sugar, polysaccharide (starch), lipid, alcohol, etc. are also of immense practical importance. For the examination of the effects of such additives on gelation behaviour the microscale methods described above are excellent tools.

As with the relationship between steps 4 and 4′ in Figure 4.11, there should be some interaction between steps 5 and 4′. Based on the information on the pre-gel state in step 4′, we can predict the best experimental conditions for gelation in step 5 (direction 4′ → 5). However, like the relationship between steps 4 and 4′, it is usual to try and explain the gelation behaviour under certain buffer conditions found in step 5 using data from the pre-gel state under the same buffer conditions (step 4′). For instance, no gelation of the protein at certain pH values can be explained by little or no aggregation of the protein molecules in the pre-gel state as detected by chromatography, light scattering and TEM observation techniques. By contrast, turbid gel formation may be related to the random and compact aggregation of protein molecules shown by light scattering and TEM. In heterogeneous protein systems, the important role of heterogeneous protein–protein interactions in gelation can be confirmed by the detection of the protein complex by electrophoresis techniques. Working back from step 5 to step 4′ is of fundamental importance.

If the desired gelation behaviour is found in step 5, it is not necessary to work back to step 4′. However, sometimes the effects of other additives are negative, i.e. the additives cause a dramatic decrease in gelling ability of the protein in comparison with the protein single-component system. In such cases, we can investigate the reason for the negative effects of additives by going back to step 4′, or earlier steps if necessary. For instance, if the solubility (step 1) and the denaturation temperature (step 2) of the protein are not changed by the additives in spite of a great change in gelling behaviour (step 5), we can infer that the additives influence the aggregation process after heat denaturation; this is checked by analyses in step 4. If the denaturation temperature is found to be increased by additives, we could consider that the mechanism of the increase is by an increase in the conformational stability of the protein by the additives. In an extreme case, it may be necessary to start again from solubilization step (1).

4.4.6 Evaluation of gel properties

Following the evaluation of good conditions for gelation (step 5), it is appropriate to advance to the next step, i.e. the evaluation of gel properties (step 6). Based on the results obtained in step 5, we can select the appropriate conditions which produce gels with the desired or unique rheological properties. The gel properties can then be evaluated using SEM or TEM observation, spectroscopic analysis, gel solubility experiments, rheological testing or measurements of fat-binding or water-holding capacity (Chapter 7).

Rheological properties are the most important factors determining gel properties and the most important and difficult point is which rheological method to choose. If in doubt, the small-deformation mechanical test should be tried first, because of the good reproducibility and the exact physical significance of the parameters obtained as (section 4.2.4). As discussed in section 4.3.7, it is easy to change the viscoelastic parameters by altering the conditions of gel formation. The other advantage of the small-deformation test is that the gel specimen can be reused for other experiments (for instance, large-deformation tests or gel solubility experiments) after the measurement. This is because the small-deformation test causes minimal disruption of the gel structure.

For more definitive properties concerning real processing and eating conditions the large-deformation mechanical test should be used but care must be taken with several points as described in section 4.3.7 in order to obtain good reproducibility. Large-deformation tests such as texture profile, compression tests and tensile tests give rise to many parameters from which a small number may suit the sample and proposed use. For instance, if the sample gel is expected to show adhesive behaviour, the 'adhesiveness' should be checked using a texturometer. Elasticity is the primary requirement for a surimi gel (*kamaboko*), therefore the elastic behaviour of the surimi gel should be checked by the compression–decompression test. To assess the fracture behaviour of a sample in the mastication process the fracture point and breaking energy are determined by means of a compression curve.

SEM or TEM observations and spectroscopic analyses are not directly linked to practical applications; however, from a fundamental view point, data on the microstructure of the gel should be compiled in order to determine the general relationship between the physical properties and the microstructure of protein gels. SEM observation may be preferable to TEM observation, because of the easier pretreatment of samples and more realistic imaging. Resolution of SEM has been improved such that it can clearly differentiate between the strand thicknesses of the network structures for aggregate and fine-stranded gels (Yuno-Ohta *et al.*, 1992).

Acknowledgement

The authors wish to thank Dr Etsushiro Doi of Kinki University and Dr Naofumi Kitabatake of Kyoto University for providing their original figures and valuable advice.

References

Alexandre, M.-C., Larre, C., Viroben, G., Popineau, Y. and Gueguen, J. (1993) Modification of a wheat gamma gliadin by bovine plasma factor XIII, in *Food Protein. Structure and Functionality*, VCH Publishers, Weinheim, pp. 172–86.

Barbu, E. and Joly, M. (1953) The globular–fibrous protein transformation. *Discuss. Faraday. Soc.*, **13**, 77–93.

Beveridge, T., Jones, L. and Tung, M. (1984) Progel and gel formation and reversibility of gelation of whey, soy beans and albumin protein gels. *J. Agric. Food Chem.*, **32**, 307–13.

Bikbov, T.M., Grimberg, V.Y., Schmandke, H., Chaika, T.S., Vaintraub, I.A. and Tolstogusov, V.B. (1981) A study on gelation of soybean globulin solutions. *Coll. Polym. Sci.*, **259**, 536–47.

Binghan, E. (1971) Influence of temperature and pH on the solubility of αs1, *β* and ϰ-casein. *J. Dairy Sci.*, **54**, 1077–80.

Bottcher, S.R. and Foegeding, E.A. (1994) Whey protein gels: fracture stress and strain and related microstructural properties. *Food Hydrocoll.*, **8**, 113–23.

Chanyongvorakul, Y., Matsumura, Y., Sakamoto, H., Motoki, M., Ikura, K. and Mori, T. (1994) Gelation of bean 11S globulin gels by Ca^{2+}-independent transglutaminase. *Biosci. Biotech. Biochem.*, **58**, 864–9.

Chanyongvorakul, Y., Matsumura, Y., Nonaka, M., Motoki, M. and Mori, T. (1995) Physical properties of soy bean and broad bean 11S globulin gels formed by transglutaminase reaction. *J. Food Sci.*, **60**, 483–8.

Clark, A.H. and Lee-Tuffnell, D.D. (1986) Gelation of globular proteins, in *Functional Properties of Food Macromolecules* (eds J.R. Mitchell and D.A. Ledward), Elsevier Applied Science, London, pp. 203–72.

Clark, A.H. and Ross-Murphy, S.B. (1987) Structural and mechanical properties of biopolymer gels. *Adv. Polymer Sci.*, **83**, 57–192.

Clark, A.H., Judge, F.J., Richards, J.B., Stubbs, J.M. and Sugget, A. (1981) Electron microscopy of network structure in thermally-induced globular protein gels. *Int. J. Peptide Res.*, **17**, 380–92.

Clark, A.H., Ross-Murphy, S.B., Nishinari, K. and Watase, M. (1990) Shear modulus–concentration relationships for biopolymer gels. Comparison of independent and cooperative crosslink descriptions, in *Food Polymers, Gels, and Colloids* (eds W. Burchard and S.B. Ross-Murphy), Elsevier Applied Science, London, pp. 209–29.

da Silva, J.A.L. and Rao, M.A. (1992) Viscoelastic properties of food hydrocolloid dispersions, in *Viscoelastic Properties of Food*, Elsevier Applied Science, London, pp. 285–315.

Doi, E. (1993) Gels and gelling of globular proteins. *Trends in Food Sci. Tech.*, **4**, 1–5.

Doi, E., Koseki, T. and Kitabatake, N. (1987) Effects of limited proteolysis on functional properties of ovalbumin. *J. Am. Oil Chem.*, **64**, 1697–703.
Doi, E., Shimizu, A., Oe, H. and Kitabatake, N. (1991) Melting of heat-induced ovalbumin gel by pressure. *Food Hydrocoll.*, **5**, 409–25.
Ferry, J.D. (1948) Protein gels. *Adv. Protein Chem.*, **4**, 1–78.
Hamann, D.D., Purukayastha, S. and Lanai, T.C. (1990) Applications of thermal scanning rheology to the study of food gels, in *Thermal Analysis of Foods*, Elsevier Applied Science, London, pp. 306–32.
Hatefi, Y. and Hanstein, W.G. (1969) Solubilization of particulate proteins and nonelectrolytes by chaotropic agents. *Proc. Natl. Acad. Sci., USA*, **62**, 1129–36.
Hatta, H., Kitabatake, N. and Doi, E. (1986) Turbidity and hardness of a heat-induced gel of hen egg ovalbumin. *Agric. Biol. Chem.*, **50**, 2083–9.
Hermansson, A.-M. (1979) Aggregation and denaturation involved in gel formation, in *Functionality and Protein Structure* (ed. A. Pour-El), American Chemical Society, Washington DC, pp. 82–103.
Hermansson, A.-M. (1986) Water and fat-holding, in *Functional Properties of Food Macromolecules* (eds J.R. Mitchell and D.A. Ledward), Elsevier Applied Science, London, pp. 273–314.
Hermansson, A.-M. (1988) Gel structure of food biopolymers, in *Food Structure: Its Creation and Evaluation* (eds J.M.V. Blanshard and J.R. Mitchell), Butterworths, Borough Green, Sevenoaks, pp. 25–40.
Hermansson, A.-M. (1994) Microstructure of protein gels related to functionality, in *Protein Structure–Function Relationships in Foods* (eds R.Y. Yada, R.L. Jackman and J.L. Smith), Blackie Academic & Professional, Glasgow, pp. 22–42.
Hermansson, A.-M. and Langton, M. (1988) Filamentous structures of bovine myosin in diluted suspensions and gels. *J. Sci. Food Agric.*, **42**, 335–69.
Hermansson, A.-M. and Langton, M. (1992) Electron microscopy, in *Physical Techniques for the Study of Food Biopolymers* (ed. S.B. Ross-Murphy), Blackie Academic & Professional, Glasgow, pp. 277–342.
Hermansson, A.-M., Langton, M. and Harbits, O. (1986) Formation of two types of gels from myosin. *J. Sci. Food Agric.*, **37**, 69–84.
Kaletunc, G., Normand, M.D., Nussinovitch, A. and Peleg, M. (1991) Determination of elasticity of gels by successive compression–decompression cycles. *Food Hydrocoll.*, **5**, 237–47.
Kato, A. and Takagi, T. (1987) Estimation of the molecular weight distribution of heat-induced ovalbumin aggregates by the low-angle laser light scattering technique combined with high-performance gel chromatography. *J. Agric. Food Chem.*, **35**, 633–7.
Katsuta, K. and Kinsella, J. (1990) Spontaneous gelation of whey protein in urea and guanidine hydrochloride. *Agric. Biol. Chem.*, **54**, 2423–4.
Kauzmann, W. (1987) Thermodynamics of unfolding. *Nature*, **325**, 763–4.
Kinekawa, Y. and Kitabatake, N. (1995) Turbidity and rheological properties of gels and sols prepared by heating process whey proteins. *Biosci. Biotech. Biochem.*, **59**, 834–40.
Kitabatake, N., Hatta, H. and Doi, E. (1987) Heat-induced and transparent gel prepared from hen egg ovalbumin in the presence of a two-step heating method. *Agric. Biol. Chem.*, **51**, 771–8.
Kitabatake, N., Doi, E. and Kinekawa, Y. (1994) Simple and rapid method for measuring turbidity in gels and sols from milk whey protein. *J. Food Sci.*, **59**, 769–72.
Kohyama, K., Yoshida, M. and Nishinari, K. (1992) Rheological study on gelation of soybean 11S protein by glucono-δ-lactone. *J. Agric. Food Chem.*, **40**, 740–4.
Kohyama, K., Iida, H. and Nishinari, K. (1993) A mixed system composed of different molecular weights konjac glucomannan and kappa-carrageenan. *Food Hydrocoll.*, **7**, 213–26.
Koseki, T., Kitabatake, N. and Doi, E. (1989a) Irreversible thermal denaturation and formation of linear aggregates of ovalbumin. *Food Hydrocoll.*, **3**, 123–34.
Koseki, T., Fukuda, T., Kitabatake, N. and Doi, E. (1989b) Characterization of linear polymers induced by thermal denaturation of ovalbumin. *Food Hydrocoll.*, **3**, 135–48.
Kuwajima, K., Hiraoka, Y., Ikeguchi, M. and Sugai, S. (1985) Comparison of the transient folding intermediates in lysozyme and α-lactalbumin. *Biochemistry*, **24**, 874–81.

Ledward, D.A. (1986) Gelation of gelatin, in *Functional Properties of Food Macromolecules* (eds J.R. Mitchell and D.A. Ledward), Elsevier Applied Science, London, pp. 171–201.
Lee, S.-P., Cho, Y. and Batt, C.A. (1993) Enhancing the gelation of β-lactoglobulin. *J. Agric. Food Chem.*, **41**, 1343–8.
Li-Chan, E. and Nakai, S. (1991) Raman spectroscopic study of thermally and/or dithiothreitol induced gelation of lysozyme. *J. Agric. Food Chem.*, **39**, 1238–45.
Ma, C.T., Oomah, B.D. and Holme, J. (1986) Effect of deamidation and succinilation on some physicochemical and baking properties of gluten. *J. Food Sci.*, **51**, 99–103.
Ma, C.Y. and Harwalkar, V.R. (1988) Studies of thermal denaturation of oat globulin by differential scanning calorimetry. *J. Food Sci.*, **53**, 531–4.
Ma, C.-Y. and Harwalkar, V.R. (1990) Instrumentation and techniques of thermal analysis in food research, in *Thermal Analysis of Foods* (eds V.R. Harwalkar and C.-Y. Ma), Elsevier Applied Science, London, pp. 1–15.
Matsumura, Y., Kang, I.-J., Sakamoto, H., Motoki, M. and Mori, T. (1993) Filler effects of oil droplets on the viscoelastic properties of emulsion gels. *Food Hydrocoll.*, **7**, 227–40.
Mori, T., Nakamura, T. and Utsumi, S. (1982) Gelation mechanism of soybean 11S globulin: formation of soluble aggregates as transient intermediates. *J. Food Sci.*, **47**, 26–30.
Morris, V.J. (1986) Gelation of polysaccharides, in *Functional Properties of Food Macromolecules* (eds J.R. Mitchell and D.A. Ledword), Elsevier Applied Science, London, pp. 121–70.
Mulvihill, D.M. and Kinsella, J.E. (1988) Gelation of β-lactoglobulin: effects of sodium chloride and calcium chloride on the rheological and structural properties of gels. *J. Food Sci.*, **53**, 231–6.
Murata, M., Tani, F., Higasa, T., Kitabatake, N. and Doi, E. (1993) Heat-induced transparent gel formation of bovine serum albumin. *Biosci. Biotech. Biochem.*, **57**, 43–6.
Nagano, T., Akasaka, T. and Nishinari, K. (1994) Dynamic viscoelastic properties of glycinin and β-conglycinin gels from soybeans. *Biopolymers*, **34**, 1303–9.
Nakai, S. and Li-Chan, E. (1988) *Hydrophobic Interactions in Food Systems*, CRC Press, Boca Raton, pp. 23–41.
Nakai, S., Koide, K. and Eugster, K. (1984) A new mapping super-simplex optimization for food product and process development. *J. Food Sci.*, **49**, 1143–8.
Nakamura, T., Utsumi, S. and Mori, T. (1984) Network structure formation in thermally induced gelation of soybean glycinin. *J. Agric. Food Chem.*, **32**, 349–52.
Navin Kumar, K.D., Nandi, P.K. and Narasingh Rao, M.S. (1980) Reversible gelation of arachin. *Int. J. Pept. Prot. Res.*, **15**, 67–72.
Phillips, L.G., Whitehead, D.M. and Kinsella, J. (1994) *Structure–Function Properties of Food Proteins*, Academic Press, San Diego, pp. 179–204.
Privalov, P.L. (1979) Stability of proteins. Small globular proteins. *Adv. Protein Chem.*, **33**, 167–241.
Ptitsyn, O.B. (1992) The molten globule state, in *Protein Folding* (ed. T.E. Creighton), W.H. Freeman and Company, New York, pp. 243–300.
Ranken, M.D. (1984) Composition of meat: some structural and analytical implications, in *Developments in Food Proteins-3* (ed. B.J.F. Hudson), Elsevier Applied Science, London, pp. 1–32.
Rector, D., Kella, N. and Kinsella, J. (1991) Reversible gelation of whey proteins: melting, thermodynamics and viscoelastic behavior. *J. Texture Studies*, **20**, 457–71.
Ross-Murphy, S.B. (1984) Rheological method, in *Biophysical Methods in Food Research* (ed. H.W.-S. Chan), Blackwell Scientific Publications, Oxford, pp. 138–99.
Ross-Murphy, S.B. (1991) Concentration dependence of gelation time, in *Food Polymers, Gels and Colloids* (ed. E. Dickinson), The Royal Society of Chemistry, Cambridge, pp. 357–68.
Shimada, K. and Cheftel, J.C. (1988) Texture characteristics, protein solubility and sulfhydryl group/disulfide bond contents of heat-induced gels of whey protein isolate. *J. Agric. Food Chem.*, **36**, 1018–25.
Shimada, K. and Matsushita, S. (1980) Relationship between thermo-coagulation of protein amino acid compositions. *J. Agric. Food Chem.*, **28**, 413–17.
Stading, M. and Hermansson, A.-M. (1991) Large deformation properties of β-lactoglobulin gel structures. *Food Hydrycoll.*, **5**, 339–52.

Stading, M., Langton, M. and Hermansson, A.-M. (1992) Inhomogeneous fine-stranded β-lactoglobulin gels. *Food Hydrocoll.*, **6**, 455–70.
Stading, M., Langton, M. and Hermansson, A.-M. (1993) Microstructure and rheological behaviour of particulate β-lactoglobulin gels. *Food Hydrocoll.*, **7**, 195–212.
Szczesniak (1968) Correlations between objective and sensory texture measurements. *Food Technol.*, **22**, 981–6.
Takagi, T. (1981) Confirmation of molecular weight of *Aspergillus oryzae* α-amylase using the low angle light scattering technique in combination with high pressure silica gel chromatography. *J. Biochem.*, **89**, 363–8.
Tani, F., Murata, N., Higasa, T., Goto, M., Kitabatake, N. and Doi, E. (1993) Heat-induced transparent gel from hen egg lysozyme by a two step heating method. *Biosci. Biotech. Biochem.*, **57**, 209–14.
Terasawa, S., Itsuki, H. and Arakawa, S. (1975) Contribution of hydrogen bonds to the partial molar volumes of nonionic solutes in water. *J. Phys. Chem.*, **79**, 2345–51.
Tombs, M.P. (1974) Gelation of globular proteins. *Faraday Discuss. Chem. Soc.*, **57**, 158–64.
Utsumi, S. and Kinsella, J.E. (1985) Structure–function relationships in food proteins: subunit interactions in heat-induced gelation of 7S, 11S, and soy isolate proteins. *J. Agric. Food Chem.*, **33**, 297–303.
Utsumi, S., Nakamura, T. and Mori, T. (1982) A micro-method for measurement of gel properties of soybean 11S globulin. *Agric. Biol. Chem.*, **46**, 1923–4.
Utsumi, S., Gidamis, A.B., Kanamori, J., Kang, I.-J. and Kito, M. (1993) Effects of deletion of disulfide bonds by protein engineering on the conformation and functional properties of soybean proglycinin. *J. Agric. Food Chem.*, **41**, 687–91.
van Mil, P.J.J.M. and Roefs, S.P.F.M. (1993) Denaturation of whey proteins studied by thermal analysis and chromatography, in *Food Colloids and Polymers: Stability and Mechanical Properties* (eds E. Dickinson and P. Walstra), The Royal Society of Chemistry, Cambridge, pp. 45–54.
van Vliet, T., Luyton, H. and Walstra, P. (1991) Fracture and yielding of gels, in *Food Polymers, Gels and Colloids* (ed. E. Dickinson) The Royal Society of Chemistry, Cambridge, pp. 392–403.
Verheul, M., Roefs, S.P.F.M. and deKruif, C.G. (1995) Heat-induced denaturation and aggregation of β-lactoglobulin: influence of sodium chloride, in *Food Macromolecules and Colloids* (eds E. Dickinson and D. Lorient), The Royal Society of Chemistry, Cambridge, pp. 437–44.
Voisey, P.W. (1979) Recent advances in texture test instrumentation and their application, in *Food Texture and Rheology* (ed. P. Sherman), Academic Press, London, pp. 55–64.
Wang, C.H. and Damodaran, S. (1991) Thermal gelation of globular proteins: Influence of protein conformation on gel strength. *J. Agric. Food Chem.*, **39**, 433–8.
Wright, D.J. (1987) The seed globulins, in *Developments in Food Protein-5*, Elsevier Applied Science, London, pp. 81–157.
Xiong, Y. and Kinsella, J. (1990) Mechanism of urea-induced whey protein gelation. *J. Agric. Food Chem.*, **28**, 413–17.
Yuno-Ohta, N., Maeda, H., Okada, M. and Hasegawa, K. (1992) Formation of transparent gels of sesame 13S globulin: effects of fatty acid salts. *J. Food Sci.*, **57**, 86–90.
Zheng, B.-A., Matsumura, Y. and Mori, T. (1991) Thermal gelation mechanism of legumin from broad beans. *J. Food Sci.*, **56**, 722–5.
Zheng, B.-A., Matsumura, Y. and Mori, T. (1993a) Relationships of molecular forces to rheological and structural properties of legumin gels from broad beans. *Biosci. Biotech. Biochem.*, **57**, 1257–60.
Zheng, B.-A., Matsumura, Y. and Mori, T. (1993b) Relationship between the thermal denaturation and gelling properties of legumin from broad beans. *Biosci. Biotech. Biochem.*, **57**, 1087–90.
Zheng, B.-A., Matsumura, Y. and Mori, T. (1993c) Conformational changes and surface properties of legumin from broad beans in relation to its thermal aggregation. *Biosci. Biotech. Biochem.*, **57**, 1366–8.

5 Foam formation and stability

P.J. WILDE and D.C. CLARK

5.1 Scientific background

5.1.1 Introduction

Foams are to be found in a huge variety of diverse applications including the food industry, large-scale construction (e.g. insulation) and firefighting. Applications involving protein-stabilised foams are more limited, probably due to their shelf life. Their largest field of application is the food industry, where they are widely used in meringues, mousses, a variety of whipped products and beer. The study of foam behaviour has wider implications, for example it may be used for separating and concentrating (Lalchev and Exerowa, 1981; Lalchev *et al.*, 1982). In biotechnology and fermentation, foams can be undesirable products of gas formation (Evans, 1972). Knowledge of the principles underlying foam formation and stabilisation is therefore essential for their efficient control and destruction.

Proteins in foams have been studied as far back as the mid-nineteenth century (Neurath and Bull, 1938). Experiments with protein solutions showed that they formed a stiff viscoelastic layer at the air/water interface, and this was thought to play an important role in foam stabilization. These early observations revealed that when a foam was created by shaking a protein solution, insoluble material was formed as the foam collapsed. This was thought to be the remains of a protein 'skin' surrounding the bubble. Since then, knowledge of the protein behaviour at interfaces has progressed significantly, and has shed light on the manner by which proteins stabilise dispersions such as foams and emulsions.

A thorough knowledge of the mechanism of protein foam formation is essential if a foam of the required characteristics is to be produced, or destruction of an unwanted foam by product is to be achieved. The structure of a typical foam is shown in Figure 5.1. The gas bubbles are caged within a network of Plateau borders. These latter structures form where three thin films or foam lamellae, which constitute the walls of the bubbles, meet. It is the properties of the two air/water interfaces of the thin films which make or break a foam. When a thin film is stretched, perhaps because of local reorganisation of bubbles in the foam or thermal perturbations, the surface tension in the region of the stretch increases

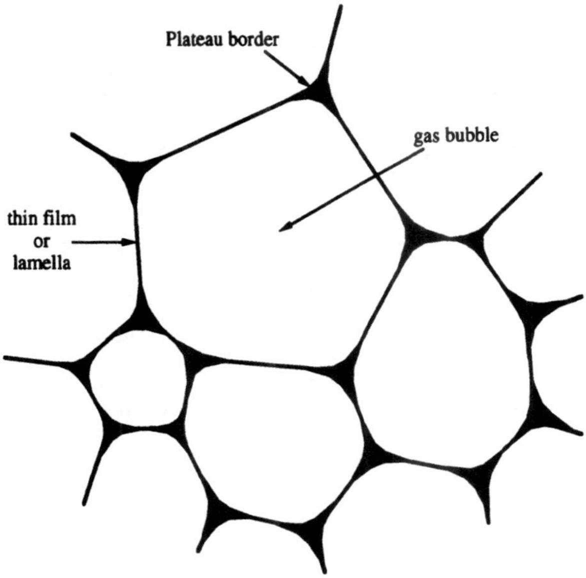

Figure 5.1 Diagrammatic representation of the structure of foam bubbles.

from the equilibrium level due to a local change in surface concentration of the surfactant. Consequently, the film gets thinner. This is an unstable situation. There are two distinct mechanisms for restoring stability in such a system – one is characteristic of surfactant-stabilised films, the other of protein-stabilised films. Surfactants or detergents are fairly small, simple, amphiphilic molecules, which generally do not interact with their neighbours. Equilibrium can be restored if the surfactants can diffuse rapidly and laterally towards the thinnest region of the film, against the gradient in surface tension. If this occurs, the surfactants drag a layer of interlamellar liquid, associated with their headgroup, into the thin part of the stretched film. This serves to thicken the film. The mechanism, shown in Figure 5.2a, is often referred to as the Marangoni effect. By contrast, proteins adsorb at the interface and interact with each other by formation of electrostatic, hydrophobic, hydrogen and covalent bonds (e.g. oxidation of sulphydryl groups) to form a stiff, viscoelastic adsorbed layer (Figure 5.2b). This layer can resist stretching of the film and dampens rapid fluctuations in film thickness within limits. The adsorbed layer acts rather like a rubber sheet. Limited stretching of the film can be dissipated by deformation of the protein molecules in the adsorbed layer. We refer to this as the viscoelastic mechanism of stabilisation, and it is in marked contrast to the Marangoni mechanism of stabilisation. For effective stabilisation by the viscoelastic mechanism it is essential that the protein molecules interact with each other and are not free to diffuse in the plane of the interface as in the

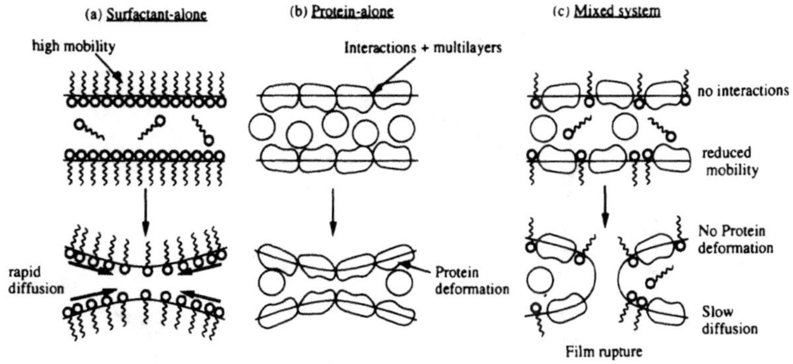

Figure 5.2 Illustration of the different mechanisms of thin film stabilisation. (a) Surfactant-like (Marangoni mechanism); (b) protein-like (viscoelastic); (c) mixture of protein and surfactant (unstable).

Marangoni mechanism. This unique behaviour of adsorbed layers of proteins is often overlooked by physical and interfacial chemists in classical studies of protein foam stabilisation.

Instability can occur if both surfactants and proteins are present in the foaming mixture (Lee and Tynan, 1988; Coke *et al.*, 1990; Pearce *et al.*, 1991) as is often the case with foods. The surfactants can interfere with the formation of protein–protein interactions in the adsorbed layer, by competitive adsorption or by interaction with charged or hydrophobic regions on the protein surface (Figure 5.2c). This can result in a transition in the stabilisation mechanism from viscoelastic to Marangoni as the level of surfactant is increased (Lee and Tynan, 1988; Coke *et al.*, 1990). Foam stability is generally significantly reduced in the region of this transition. As a result, the study of the surface properties of protein films, i.e. surface tension, viscosity, diffusion and drainage properties, indicates which stabilisation mechanism is in operation. This is important, as many real systems contain a complex mixture of surface active species, and although there may be much protein present, it may not necessarily be stabilising the foam. The study of foam formation alone, therefore, is not enough to understand fully foam stability and to progress towards its control. For a full interpretation or understanding of the foaming experiments the mechanisms responsible for stabilisation need to be determined.

There are two distinct phases to protein foaming: (i) the effectiveness of gas encapsulation (foamability); and (ii) the lifetime of the foam (foam stability). The rates at which a protein is transported to the gas/water interface, and adsorbed, are important determinants of foamability (Graham and Phillips, 1976; Kalischewski and Schurgeri, 1979; Mitchell, 1986; Lorient *et al.*, 1989; Closs *et al.*, 1990). Reduction in molecular weight, perhaps by use of peptides rather than proteins, can improve foam properties (Groninger and Miller, 1975; Horiuchi *et al.*, 1978). The adsorbed layer needs to become stable at an early stage for the bubble to remain intact against disruption during the dynamic phase of foam formation. Good foam stability properties result if the adsorbed protein layer remains as an intact structure which can resist collapse. The strength of this viscoelastic interfacial layer of protein, has been linked with good foam stability. Surface shear or dilational viscosity studies of a variety of proteins have revealed that proteins which form highly viscous surface films form the most stable foams (Phillips, 1981; Kim and Kinsella, 1985; Mitchell, 1986; Maeda *et al.*, 1991). It is important that foamability and foam stability are considered separately, as the features of the protein which determine these properties are different. A protein may have excellent foamability, but it may not produce a stable foam, and vice versa. The rest of this section deals with the scientific principles behind these two stages, the physical properties required, factors which affect these properties, mechanisms of destabilisation, and methods which have been employed to improve protein foam formation.

5.1.2 Principles of protein foam formation

Proteins are polymers of amino acids which have hydrophilic, apolar or hydrophobic side chains. The amphipathic character that these side chains confer to proteins is responsible for their adsorption at interfaces. To form a foam efficiently (i.e. to possess high foamability), a protein needs to adsorb rapidly during the transient stage of foam formation (Graham and Phillips, 1976; Kalischewski and Schurgeri, 1979; Mitchell, 1986; Lorient *et al.*, 1989; Closs *et al.*, 1990).

The adsorption of proteins at interfaces has often been described in terms of three processes, namely, transport from bulk solution to the interface, penetration into the surface layer and reorganisation of structure (surface denaturation) of the protein in the adsorbed layer. Figure 5.3 shows these three processes. Proteins are transported to interfaces by diffusion, convection or a combination of the two processes. The case of pure diffusion is the easiest to model, since if there is no barrier to adsorption other than diffusion, simple diffusion theory can be used to predict the rate of adsorption.

In dilute solutions, the kinetics of adsorption can be studied by direct measurement of the change in surface concentration of the protein, using radiocounting of radioactively labelled protein, or by ellipsometry (DeFeijter *et al.*, 1978; Graham and Phillips, 1979a). With simple amphipathic molecules, changes in surface pressure (π) are dependent on surface concentration. The surface pressure, π is equal to $\gamma_o - \gamma$, where γ_o is the surface tension of the solvent and γ is the observed surface

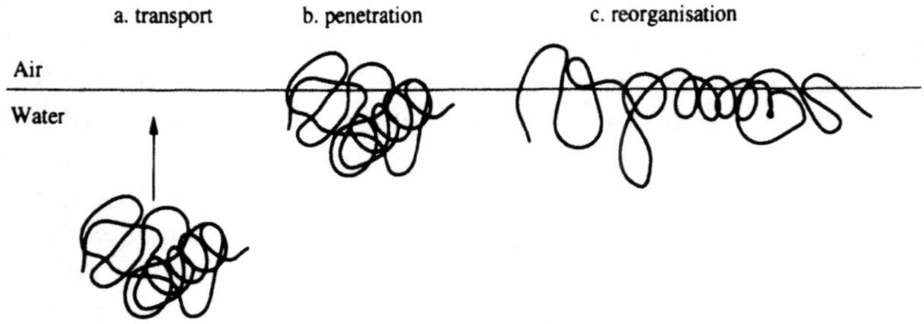

Figure 5.3 Adsorption process of proteins. (a) Diffusion or transport to the interface; (b) penetration or adsorption; (c) structural reorganisation (surface denaturation).

tension. However, care must be taken in the case of proteins since post-adsorption conformational rearrangements (surface denaturation) can also result in changes in π (Graham and Phillips, 1979a). It is possible to correct for this effect by plotting a π versus surface concentration (Γ) curve from quantitative spreading experiments. High foamability has been correlated with a high rate of increase of π (Graham and Phillips, 1976; Kalischewski and Schurgeri, 1979; Kim and Kinsella, 1985). In a diffusion-limited system, a high diffusion coefficient, hence low molecular weight, is important. Consequently, small molecular weight proteins and peptides can often play a major role in foam formation (Grunden *et al.*, 1974).

Ward and Tordai (1946) first derived a general equation for the rate of arrival of solute molecules at liquid interfaces. In deriving this equation, it was assumed that the diffusion coefficient was independent of solute concentration, which is not always true with proteins. In addition, it was assumed that there was no activation barrier for adsorption or desorption, leaving the rate of adsorption diffusion controlled. This approach has been reviewed widely in the literature (MacRitchie, 1978; Damodaran, 1990). The theory has been shown to be valid during the initial stages of adsorption; however, at higher surface coverages, the rate of adsorption decreased below that of the rate of diffusion (MacRitchie and Alexander, 1963). Thus it appears that diffusion-controlled adsorption holds only during the initial stages of protein adsorption. A barrier to adsorption develops as proteins become closely packed at the interface. It has been suggested that the barrier to adsorption is related to surface denaturation of the protein, which is determined by the surface water activity (Saraga, 1981). A protein with high surface hydrophobicity will increase π rapidly, due to its affinity for the interface, and have a low rate of desorption. Hence, foamability has been linked with surface hydrophobicity (Badwan *et al.*, 1977; Kato *et al.*, 1983a; Townsend and Nakai, 1983; Mitchell, 1986).

The adsorption properties of the three conformationally distinct proteins, bovine serum albumin (maleable globular), lysozyme (compact globular) and β-casein (random coil) were studied by Graham and Phillips (1979a,b,c) in an attempt to identify the role of protein conformation in determining the kinetics of adsorption. Changes in π and Γ were shown to occur simultaneously in β-casein. The lack of structure in β-casein means that there cannot be any further denaturation of the protein by adsorption. Changes in π due to surface denaturation will be accelerated if the protein structure is more flexible. This is confirmed in the case of lysozyme where the Γ curve reaches equilibrium before major changes occur in π. Independently, protein flexibility, measured mainly by electron spin resonance, has been shown to correlate with increased foamability (Mitchell, 1986; Closs *et al.*, 1990; Le Meste *et al.*, 1990; Baniel *et al.*, 1992).

The principles underlying the efficient formation of a protein-stabilised foam, rely therefore mainly on the intrinsic properties of the protein itself; that is, the size, surface hydrophobicity and structural flexibility of the protein.

5.1.3 Principles of protein foam stability

Once a protein foam is formed, its lifetime is, in part, dependent upon the maintenance of the viscoelastic adsorbed layer. Proteins as a class of macromolecules are large, heterogeneous and have distributions of charge and hydrophobicity. The adsorbed conformation of the protein may not be thermodynamically ideal, i.e. hydrophobic regions of the protein may not be partitioned at the interface suitably, perhaps due to structural constraints in the protein. This surface denaturation process may begin seconds or hours after adsorption, depending on the flexibility and structure of the protein, and may continue to a lesser degree for days. Early observations of an insoluble coagulum formed by shaking a protein solution, support the presence of conformational changes or surface denaturation upon adsorption. More modern methods such as interfacial circular dichroism have allowed direct measurement of the change in structure of adsorbed protein (Smith and Clark, 1992). The basis for proteins forming a viscoelastic film is that of protein–protein interactions. Electrostatic processes probably play an important role in reducing interactions. Proteins have a distribution of both positive and negative charges throughout the molecule, and will therefore possess a specific net charge at a given pH. Thus, neighbouring molecules will have a similar net charge and will likely repel each other. This explains why protein foams are generally more stable around their isoelectric point, where their net charge is zero and repulsion is minimised (Buckingham, 1970; Graham and Phillips, 1976; Waniska and Kinsella, 1979; Kim and Kinsella, 1985; Le Meste *et al.*, 1990; Howell and Taylor, 1991). If acidic and basic proteins (opposite net charges at neutral pH) are foamed together, the corresponding charge interaction also increases foam stability (Poole *et al.*, 1984, 1986). Covalent bonds have been shown to develop between some proteins at the interface, increasing the surface viscosity even further (Dickinson and Matsumura, 1991).

Drainage of liquid from the foam also contributes to foam stability. The more liquid which remains in the foam lamellae, the less chance of the two interfaces coming together and rupturing. Liquid drainage from the thin film lamellae is affected by bulk viscosity, surface viscosity, resistance to flow and electrostatic repulsion of the two approaching interfaces. It has been shown that the bulk viscosity of protein solutions affects the foam stability, i.e. an increase in solution viscosity reduces drainage, which in turn extends the lifetime of the film and hence the foam stability (Townsend and Nakai, 1983; Kumagai *et al.*, 1991). The ability of a stiff viscoelastic surface to resist deformation impedes liquid drainage from the thin film. The interactions of proteins adsorbed in primary or secondary layers at the interface may cause local changes in bulk viscosity near the interface. Glycoproteins with large oligosaccharides protruding into the interlamellar liquid will have a similar effect, slowing the flow of liquid in the thin film and reducing drainage (Roberts, 1975). Electrostatic repulsion between the two interfaces of a thin film resists film thinning through drainage. The reduction in the rate of drainage, characteristic of protein-stabilised films, offers advantages over detergent films. As mentioned in section 5.1.1, detergents diffuse freely and rapidly at the surface, providing little resistance to rapid drainage, and so increasing the probability of film rupture (Lee and Tynan, 1988; Coke *et al.*, 1990).

Disproportionation is the transfer of gas from one bubble to another. The Laplace pressure, ΔP inside a foam bubble is given as:

$$\Delta P = \frac{2\gamma}{r}$$

where γ is the surface tension and r the radius of the bubble. Therefore, the gas in smaller bubbles is at a higher pressure than that in larger bubbles. This encourages the diffusion of gas from smaller bubbles into neighbouring, larger bubbles which reduces the number of bubbles, and increases drainage. There are two main ways to reduce disproportionation. One is to blow the foam with a gas which has poor solubility in the liquid phase, the other is to have a stiff interface, which resists compression and may reduce gas transport. Nitrogen, having a low solubility in aqueous systems, is a favoured gas for foams, as disproportionation is reduced significantly (Clark and Blackman, 1948; Bisperink *et al.*, 1992b). Protein foams also reduce disproportionation, as a stiff viscoelastic interface opposes the shrinkage of smaller bubbles, due to gas transfer.

Bulk solution conditions can also influence the stability of protein foams. They may alter either the characteristic properties of the protein, or the physical properties of the solution itself. As already mentioned, pH and ionic strength have a great effect on the electrostatic interactions which influence foam stability. Other conditions such as dielectric constant and temperature may influence the protein conformation (Clark and Smith, 1989; Yu and Damodaran, 1991a; Gabor, 1992). The density and viscosity of the solution may affect the film, and hence foam drainage. Reduced drainage also has an indirect effect in that it extends the timecourse of events, giving the proteins more time to adsorb, interact and unfold to form a more stable adsorbed layer at the interface.

The number of factors which affect, directly and indirectly, protein foam stability, makes studying the characteristics of the interface vital to understanding foam stability. Surface tension, surface viscosity, surface diffusion and thin film drainage help explain why certain individual proteins or mixtures of protein produce more stable foams than others.

5.1.4 *Improving protein foam production*

Improving the foaming properties of proteins has been the subject of much work over recent years, because it is of immense importance to industry.

There are three main methods of improving a protein foam: (i) modify the protein in some way; (ii) add protein functionality improvers; or (iii) modify the solution conditions.

Proteins have been modified in many different ways, including enzymatic modification or digestion, and chemical derivatisation. Chemical derivatisation is less favoured in the case of food applications because of the legislative controls which restrict the use and introduction of modified ingredients.

Digestion produces a range of peptides, which possess altered hydrophobicity, charge balance and conformation compared to the parent molecule. Their reduced molecular weight can increase the rate of diffusion to the interface, improving the foamability (Horiuchi *et al.*, 1978; Phillips and Beuchat, 1981; St. John Coglan *et al.*, 1992). The peptides alone or in combination may have enhanced functional properties over the native protein. Chemical modification of proteins has often been used to enhance protein functionality. Three types of modification have been employed, aimed at increasing hydrophobicity, altering surface charge or improving flexibility. Alkylation increases the hydrophobicity and surface activity of the protein, making adsorption more favourable, thus improving foamability (Watanabe *et al.*, 1981). C:6 and C:8 chains have been found most effective for foaming, whereas C:16 and C:18 enhance emulsifying properties. Amidation or succinylation alters the surface charge and isoelectric point of the protein (Groninger and Miller, 1975; Howell and

Taylor, 1991). This may shift the optimal foaming properties to a new pH region, or improve solubility in some cases.

Glycosylation influences protein flexibility. Electron spin resonance measurements on casein have shown that changes in flexibility increase the rate of adsorption, and hence improve foamability (Lorient *et al.*, 1989; Le Meste *et al.*, 1990). The effect on globular proteins is similar in that foamability is increased (Baniel *et al.*, 1992) but surface viscosity may be reduced (Waniska and Kinsella, 1988).

Modifying solution conditions is not always an acceptable option, since changing the pH, ionic strength or viscosity may have negative effects on other aspects of the product, particularly food. However, if changes are possible, significant enhancement of foaming properties is likely (Waniska and Kinsella, 1979; Townsend and Nakai, 1983). Certain additives can change solution conditions in an effective but subtle way. The modified polysaccharide, propylene glycol alginate (PGA) is such an example, and has been quite widely used in the food industry. It is a multifunctional additive because it is charged, and can therefore cross-link adsorbed proteins, increasing the surface viscosity (Ahmed and Dickinson, 1991). It is slightly surface active and therefore concentrates at the interface, where, being a moderately high molecular weight polysaccharide, it increases the local viscosity and reduces thin film drainage. It may also prevent extensive film thinning by steric effects between the two approaching interfaces. Hop acids in beer have also been shown to have a similar effect on protein foams by improving protein–protein interactions (Asano and Hashimoto, 1976; Clark *et al.*, 1991a).

5.2 Review

5.2.1 Methods of foam generation

Foams for industrial applications can be generated in different ways. Gas can be introduced into the product either mechanically, or by the release of dissolved gas due to a reduction in pressure. Examples of the latter include sparkling drinks, pressurised aerosol foams and instant 'cappuccino' style drinks. In the last of these, the gas is encapsulated in spray dried powder and released upon solubilisation in hot water. Mechanical methods offer more opportunity for control, and are generally the favoured methods for generating foam for detailed study. The main mechanical methods for foam formation can be split into four categories: (i) sparging; (ii) whipping; (iii) shaking; and (iv) pouring.

5.2.1.1 Sparging. This method involves forcing gas through the liquid via an aperture or frit to create bubbles and hence a foam. Applications include firefighting foam, 'draught' style cans of beer and fermenters. In fermenters the foam is unwanted. Sparging is not a particularly popular way of creating a foam in industry, but the release of pressure method is sometimes used to produce a foam *in situ*. For example, in firefighting foams (Corrie, 1976) a branch pipe introduces air into a flow of liquid and creates a foam through turbulence. Highly expanded firefighting foams are created by blowing air through a sieve into the liquid to create a large volume of foam. Such foams are most useful for pumping into inaccessible areas, compared with conventional firefighting foams which are sprayed over a fuel fire.

Sparging is much more widely used as a research tool (Gray and Stone, 1940; Miles *et al.*, 1945; Rudin, 1957; Wenn, 1972; Waniska and Kinsella, 1979; Halling, 1981; Wright and Hemmant, 1987; Hollemans and Tonies, 1989; Lorient *et al.*, 1989; Ahmed and Dickinson, 1990, 1991; Yu and Damodaran, 1991a,b; Bisperink *et al.*, 1992a,b; Guillerme *et al.*, 1993). It enables the delivery of controlled volumes and/or flow rates of a chosen gas to form a foam to a fixed volume or for a specific time. The gas may be delivered through either one or more fixed size orifices, or a glass sinter. The glass sinter creates a foam very quickly, and is therefore useful for detecting the very early stages of the drainage process. A single jet or orifice provides an alternative method, and applications have been reported over many years (Miles *et al.*, 1945; Wright and Hemmant, 1987; Coke *et al.*, 1990). This method is slower to generate the foam, but the initial bubble size is constant. This reduces disproportionation as the pressure difference between the bubbles is small. Sparging is a flexible method as the gas used can be varied. The choice of gas may depend on the particular application required. Alternatively, disproportion can be studied or reduced by choosing gases of differing solubilities.

5.2.1.2 Whipping. This method involves the high-speed movement of whisks or blades through a liquid, thus beating atmospheric air into the foam. Whipping is the most popular industrial method for forming a protein-stabilised foam, e.g. meringues, mousses, ice cream (release of pressure is also used in the final stages). Whipping generally produces the most stable foams, as it progressively decreases the bubble size and increases the liquid fraction of the foam with increasing time, giving a stiffer, more stable foam. There is great interest in the use of whipping as a research tool to study foams of the type produced in industry (Clark, 1948; Clark and Blackman, 1948; McKellar and Stadelman, 1955; Nakamura and Sato, 1964; Lawhon and Cater, 1971; Lawhon *et al.*, 1972; Halling, 1981; Watanabe *et al.*, 1981, 1982; Poole *et al.*, 1984, 1986; Phillips *et al.*, 1987, 1990; Kroll *et al.*, 1988; Maeda *et al.*, 1991). The devices used for foam generation range from domestic food whisks (Poole *et al.*, 1984) to precision six-bladed knives for high shear (Maeda *et al.*, 1991). Both approaches have produced sound results, demonstrating the sensitivity of protein foaming to the environment (Phillips *et al.*, 1987; Howell and Taylor, 1991; Maeda *et al.*, 1991; Gabor, 1992). Monitoring the foam volume can be more difficult with whipped samples, as the foams are stiffer, and tend not to form a flat top surface. Protocols usually involve whipping samples for a specific time, followed by measurement of the volume or density. One method examines the amount of energy required to create a specific volume of foam by measuring the torque required (Kroll *et al.*, 1988). It was demonstrated that although foamability and stability data may be similar for various systems, the energy required to generate the foam can be totally different. This is further confirmation that foamability and foam stability are distinct parameters. Much care needs to be taken while whipping foams as there are many sources of variability. An extensive study by workers in different laboratories (Phillips *et al.*, 1990) showed that the type and speed of beaters will give differing results. Using only one beater may not form uniform foams; bowl size and shape can also produce variability. Although whipping is not as controlled a method as sparging, if carefully controlled, the results can be extremely reproducible. Much higher initial concentrations of protein are usually required to form a stable whipped foam, as the expansion (foam volume and hence interfacial area) in whipped systems is usually much higher. Thus, the surfactant load per unit of interfacial area is an important factor when assessing foaming properties which involve whipping. Direct comparisons between whipping and a small-scale sparging apparatus have been made (Wright and Hemmant, 1987). It was found that the sparging experiments using 0.1 wt% protein concentrations correlated well with larger scale whipping experiments.

5.2.1.3 Shaking. This is used as a simple research method for generating foam in a closed system (Graham and Phillips, 1976; Wang and Kinsella, 1976; Horiuchi *et al.*, 1978; Halling, 1981). It has little industrial relevance, apart from the fact that a by-product from a shaking process may be an unwanted foam. In general, a sealed jar or graduated cylinder is shaken mechanically (or sometimes manually). Frequencies vary from 1 to 50 Hz, at a variety of amplitudes. The foaming properties have been reported to be dependent on the solution properties of the surfactant, the frequency and amplitude of agitation, and also on the shape and volume of the container. The rate of foam production is lower than whipping or sparging, because shaking is inefficient at producing gas bubbles. However, Graham and Phillips (1976) chose shaking to study the conformational changes due to foaming and their effect on foam stability. In their experiments, continuous shaking for up to 60 minutes ensured that conformational changes introduced into the protein did not derive from mechanical denaturation processes (e.g. whipping). Shaking ensured that the protein in solution was continually brought into contact with the interface. Shaking does have its uses in real systems: in aerosol generated foams, shaking is required to disperse the air roughly in the liquid before the final foam is created on dispensing. This shaking is essential, otherwise no foam would be created. In summary, shaking is an inefficient method of generating foam, but it is useful in specific studies or applications.

5.2.1.4 Pouring. This method is obviously of commercial interest as it is the one often used by consumers to generate foams (e.g. pouring a glass of beer). Some workers have developed research instruments for pouring in a standardised manner (Klopper, 1973; Gales, 1977, Jackson and Bamforth, 1982; Haugsted and Erdal, 1991; Hegarty *et al.*, 1991), to try and simulate the real system in a more scientifically reproducible way. Approaches adopted range from pouring from a fixed height (Klopper, 1973; Anon., 1976; Gales, 1977), to mechanical arrangements, where a glass and bottle are held in fixed positions and tipped at specific angles (Jackson and Bamforth, 1982). Again, this method of foam generation is not as controlled as sparging or whipping, so specific mechanisms may not always be investigated. However, in many cases the method allows generation of a foam which is more representative of the real situation. For example the American Society of Brewing Chemists use the Sigma method (Anon., 1976), which is a standardised pouring technique, to determine beer foam stability; this reference also describes the 'foam flashing' method, which is widely used in the brewing industry as a method of simulating the dispensing of a sample, by gas pressure, into a glass to form a foam.

5.2.2 Methods of protein foam measurement

Foams are difficult structures to study. They are dynamic systems, generally of only transient stability, and they do not respond well to invasive techniques. Direct measurement of the foam volume, bubble size distribution and lifetime provides a physical description of the foam, and the means of comparison with the properties of other foams. However, much higher resolution interfacial studies, at the microscopic and sometimes molecular level, are required to explain why one protein has better foaming properties than another. Such measurements are classified as indirect methods, as opposed to direct methods of bulk, physical characterisation. Consequently, this review of methods of foam measurement is divided into direct and indirect techniques; also included is a section on results analysis to enable more accurate interpretation of foaming experiments.

5.2.2.1 Direct methods. Methods for the direct measurement of foam properties can be split into three distinct areas: (a) determination of the volume of foam or drained liquid; (b) the physico-mechanical properties of the foam; and (c) the properties of the individual bubbles.

(a) Measurement of foam and drained liquid volume. This is the largest single area of study, and most methods yield a value for one of these parameters. The most common parameter used to define a foam is volume. This parameter is inextricably linked to the foamability of the system. The foamability of a system is often determined simply by measuring the foam volume. More correctly, it should be expressed as a volume of foam obtained for a given input of work. This involves generating a foam under fixed conditions (e.g. time of foaming, volume of gas used, etc.) These fixed values must be chosen very carefully, because a highly efficient system may overflow the sample chamber, or a poor foamer may reach its maximum sustainable volume (i.e. where the rate of drainage/collapse matches the rate of foam creation) before the foam generation period is terminated. Monitoring foam volume during production is the best way to determine whether the foam has reached its maximum sustainable volume. Another way to assess the foamability is to foam the system to a predetermined volume (Halling, 1981), carefully chosen to be within the capabilities of the sample, and measure the amount of time taken, the energy input or the volume of gas sparged to create a given volume of foam. Again, the foamability can be expressed as an efficiency of foaming for comparison with other techniques or samples. This method is easiest to use with the sparging technique. Crude estimation of the foam volume can be achieved by eye. The foam is simply generated in, or transferred to, a suitable graduated container, where the volume is read. If the foam is particularly stiff, it may not have a uniform surface, which makes reading the volume difficult. A plunger style device has been used to flatten the surface of whipped foams (McKellar and Stadelman, 1955), overcoming the problem. Depending upon experimental design and operator variability, the measurements are usually reliable, but they can be subjective. Image analysis has been used as a less subjective (but more expensive) approach. The systems range from a low-resolution photodiode array (Gales, 1977) to high-quality video cameras linked to computers where the image analysis is performed, and the foam volumes are calculated (Haugsted *et al.*, 1990; Loisel *et al.*, 1992). Here there is no need to flatten uneven surfaces, as these systems can calculate volumes of any irregular shape.

Once a given volume of foam has been generated, and the foamability determined, the stability of the foam needs to be measured. This involves the measurement of the foam volume and/or drainage volume with time. Figure 5.4 shows a typical profile of foam volume with time. Once foaming has ceased, the shape of the V_f curve may vary as a result of stability and composition of the foam. The foam volume can be constantly monitored, or the time taken for half the liquid to drain or half the foam to collapse can

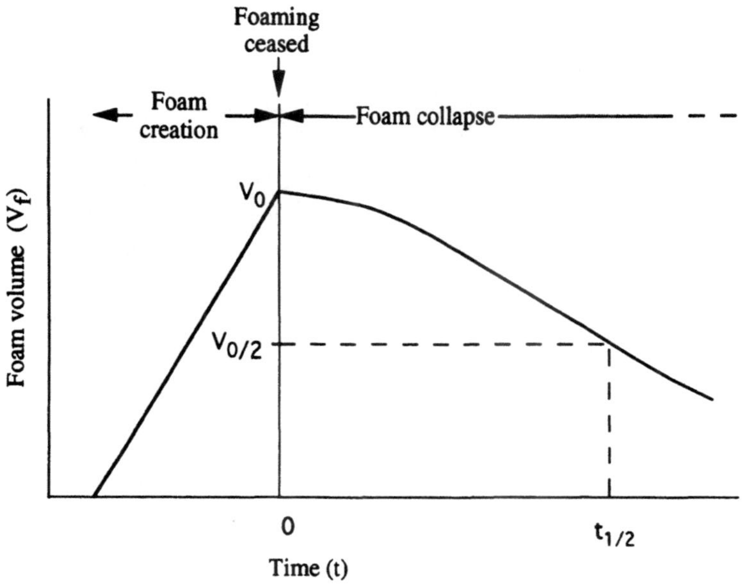

Figure 5.4 Time profile of foam volume. An initial linear increase in volume as the foam is created may become non-linear after prolonged foaming and is followed by decay once foaming has ceased.

be recorded. Foam stability can then be expressed as a half-life term, or a foam/drained liquid volume ratio after a certain time has elapsed. Generally speaking the longer it takes for a foam to collapse, the more stable the foam. Numerous methods have been employed for continuous monitoring of foam volumes. Foam volume can be monitored by the same methods described above for measuring the foamability (i.e. simple observation or image analysis). German and McCarthy (1989) used a different approach with an image analysis technique which qualitatively studied foam stability by viewing the foam on a glass of beer from above. The foam appeared white, but as it collapsed the dark background of the beer below the foam showed through and image analysis quantified this darkness into a grey scale number (0–255) and continually monitored the foam with time. The half-life of the beer foam was then calculated in the real environment of a glass following pouring or dispensing in other ways. Another direct volume measurement of stability is to monitor the volume of liquid as it drains from the foam. The simplest way to measure the volume is, again, by eye. Rudin (1957) developed a simple method which has been adopted as one of the standards in the beer industry. In this method, shown diagramatically in Figure 5.5, a beer is sparged to 3.25–3.5 times its liquid volume, and the rise of the foam–liquid boundary due to drainage of the foam is timed between two fixed markers. This corresponds

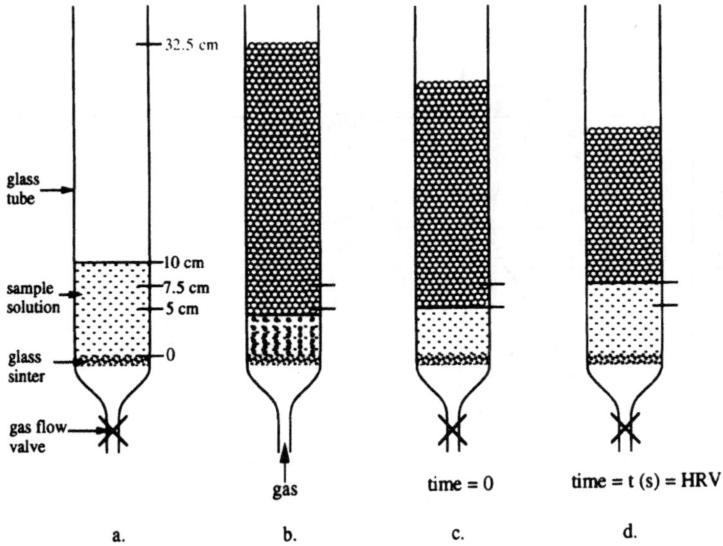

Figure 5.5 Rudin method. 10 cm of sample (a) is foamed to 32.5 cm (b). The time of the foam/liquid boundary to pass between 5 cm (c) and 7.5 cm (d), is recorded as the head retention value (HRV).

to the half-life of the foam and is referred to as the head retention value (HRV). This value has become a standard measure of beer foam stability amongst brewers. Other methods usually measure the foam volume or drained volume after a fixed period, say 1, 5 and 30 minutes (Nakamura and Sato, 1964; Groninger and Miller, 1975; Wang and Kinsella, 1976; Watanabe *et al.*, 1981; Poole *et al.*, 1984, 1986; Mitchell, 1986). These methods can distinguish between foams with large differences in stability, but much information is not utilised, therefore many workers have looked to more automated forms of monitoring foam stability. A collaborative project involving several groups (Phillips *et al.*, 1990) attempted to design a standard method for assessing the quality of whipped foams. The sample was whipped for a fixed time, and the entrapped liquid was allowed to drain through a hole in the base of the bowl onto a balance (Figure 5.6). The weight of the drained liquid was continuously monitored, and the time taken for half the liquid to drain away determined and used as a measure of foam stability. Firefighting foams have been characterised in a similar fashion. Corrie (1976) reported a method whereby the foam was generated in a large container which was mounted on scales. As the liquid drained away through a hole in the base of the container, the weight of the remaining foam was monitored. Loisel *et al.* (1992) used conductivity to monitor the volume of the drained liquid. A schematic diagram of the apparatus is shown in Figure 5.7. A pair of electrodes is positioned across

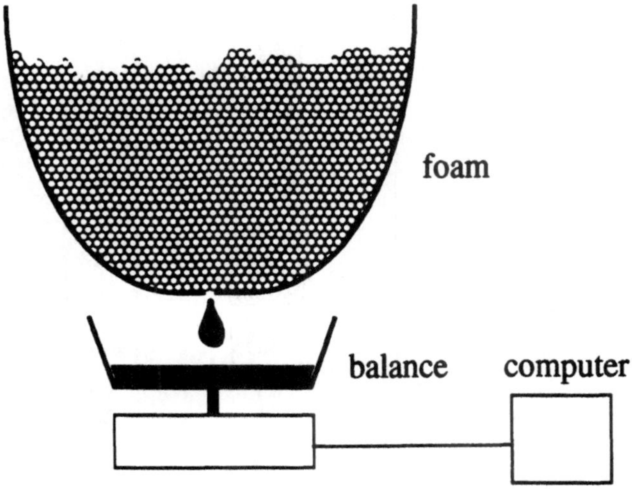

Figure 5.6 Monitoring drainage from whipped foam. Once the foam has been created, drained liquid is allowed to drain through a hole onto a balance. The mass of the fluid may be recorded via a computer for constant accurate monitoring.

the meniscus of the sample to be foamed, giving an instant readout corresponding to the initial sample volume. As the foam is generated, the liquid level reduces, and the conductivity decreases respectively. In most instances the conductivity due to the foam will be negligible compared to the bulk liquid. Once foaming is terminated, the electrodes monitor the rise in the liquid level as it drains from the foam. Conductivity has also been used to measure the foam density to determine drainage and stability. This will be discussed in more detail in the next section. The volume of liquid in the foam can be determined by measuring the drained volume, and the rate of drainage can be expressed as a proportion of the liquid in the foam. In certain circumstances, accurate measurement of the liquid volume in the foam is not possible, so some methods, such as the Sigma method and the foam flashing method (Anon., 1976), collapse the foam with a known volume (2 ml) of isopropanol after they have measured the drainage of fluid from an unknown quantity of foam. The foam liquid volume can then be determined.

Other, novel methods have been developed to monitor foam or drainage volume. Klopper (1973) used an ingenious approach, known as the Nibem method, for detecting the top of a foam (Figure 5.8). A conducting needle is lowered onto the surface of the foam; when contact is made, the motors controlling the vertical position of the needle stop. As the foam collapses, the needle follows the upper surface of the foam, recording foam height with time. Although this may not provide an accurate indication of the foam height as the top surface of a foam is rarely flat, it does give very

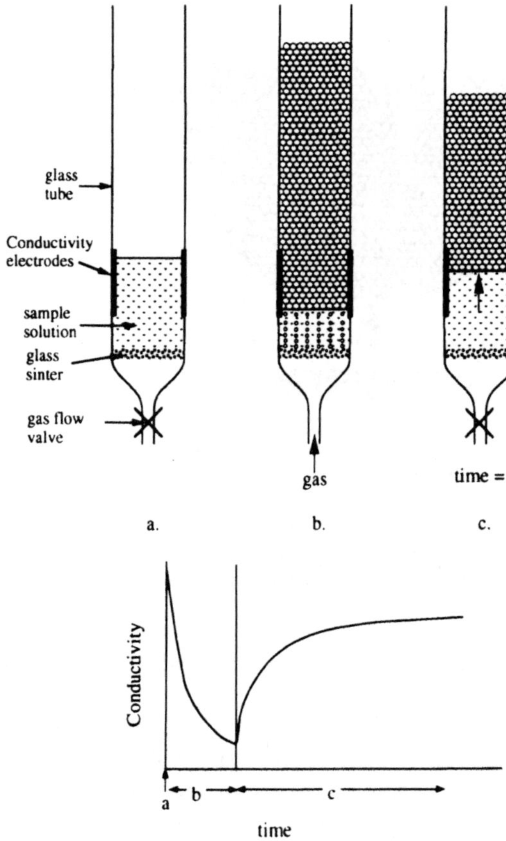

Figure 5.7 Monitoring drainage volume by conductivity. (a) The electrode records the initial sample volume; (b) liquid volume, hence conductivity, drops during foaming; (c) once foaming ceases, conductivity recovers with drainage.

useful information about the stages and rates of foam collapse. Measuring the change in surface area of a foam is a very useful tool for studying foam collapse and disproportionation. One such method measures the change in pressure of a foam in a closed system as the bubbles collapse and release their pressure (Nishioka and Ross, 1981; Yu and Damodaran, 1991a,b; Dickinson *et al.*, 1993) (Figure 5.9). Theoretically the change in pressure of the system is directly proportional to the change in surface area of the foam. At the end of the experiment, an antifoaming agent is added to collapse the foam, and determine the final pressure at zero surface area. The change in relative surface area with time can then be calculated and is related to foam stability. More recently, magnetic resonance imaging has been used purely as a non-invasive research tool to look at cross-sections of

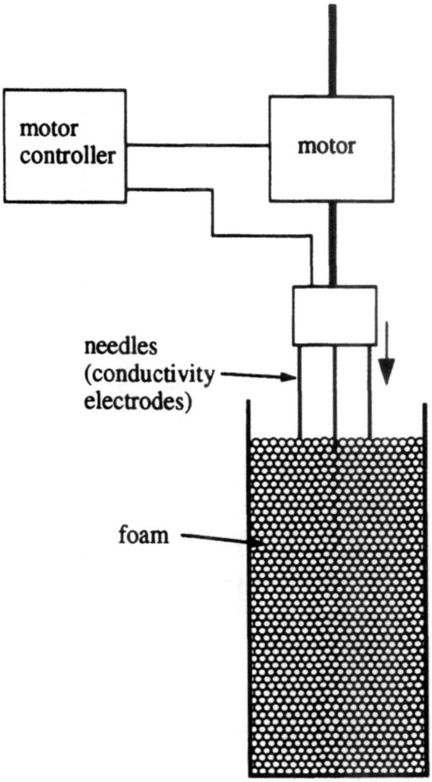

Figure 5.8 The Nibem technique. Conducting needles fed back to controlling motors are lowered onto a foam. Once the needles come into contact with the foam, the motors stop. As the foam collapses, the needles are lowered again until they contact the foam once again.

foam (German and McCarthy, 1989; McCarthy, 1990). This enables the calculation of drainage and foam volumes from the same image, although only at great expense. More importantly, the NMR signal may be able to give some structural or environmental information about the proteins in the foam (McCarthy, 1990).

(b) Foam properties. The physical properties of a foam are often determined as a measure of foamability or stability. Measuring the density of a foam is a simple method of determining the fraction of gas to liquid phase of a foam. This is generally only possible with relatively stable foams. In such cases foams can be transferred carefully to a container of known mass and volume, and their density determined by weighing. The foam expansion can then be calculated and used as a measure of foamability. Robust, whipped foams offer greatest resistance to mechanical

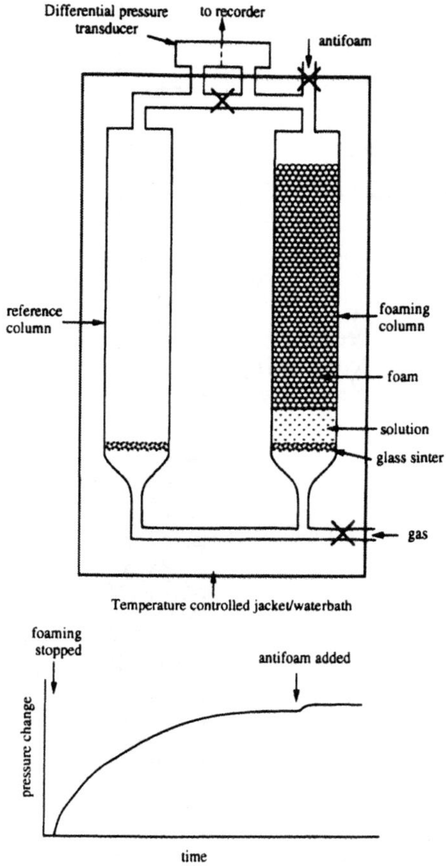

Figure 5.9 Schematic diagram of the differential pressure technique of Nishioka and Ross (1981), showing a typical pressure response with time. The antifoam is added to collapse the foam and obtain a pressure reading for zero surface area of the foam.

disturbances that such transfers can generate. For less robust foams, measurement of the foam density *in situ* is preferable. Clark (1948) showed that the conductivity of a foam is proportional to its density; more recently, it was shown that the conductivity also relates to the amount of liquid in the foam (Guillerme *et al.*, 1993). Conductivity is a very popular method of measuring foam density, as it is non-invasive and can be constantly monitored, enabling the study of rapidly draining samples. Figure 5.10 shows a typical foam conductivity apparatus, where the foam is formed, and electrodes in the wall of the foaming tube monitor the conductivity with time. This technique was used as far back as 1945 (Miles *et al.*, 1945) when a single conductivity measurement was used to measure foam

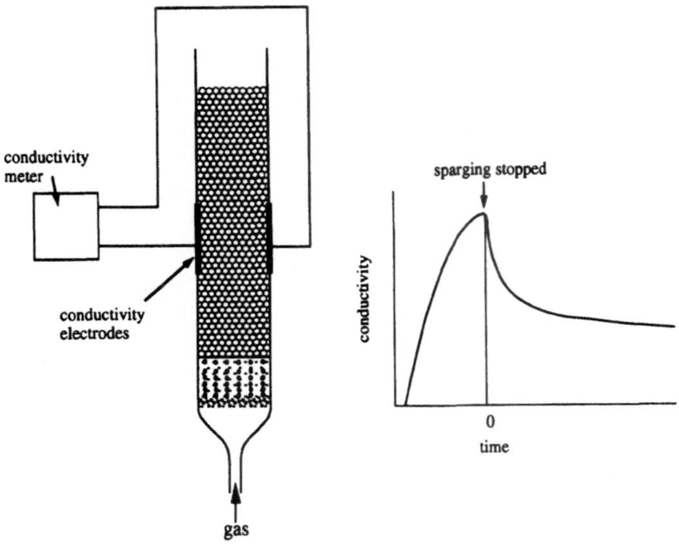

Figure 5.10 Diagram of foam conductivity apparatus. The conductivity increases as the foam is created. Once foaming ceases, the conductivity decays with the foam density as liquid drains from the foam.

density. More recently, conductivity has been used to study the dynamics of foam drainage and collapse (Kato *et al.*, 1983a,b; Wright and Hemmant, 1987; Coke *et al.*, 1990; Guillerme *et al.*, 1993). It is sensitive to very small changes in foam properties. In isolation, the conductivity of the foam cannot distinguish between drainage and film rupture. However, if the foam volume and drainage volume are monitored simultaneously, further information about the amount of liquid in the foam can be obtained (Guillerme *et al.*, 1993). A combination of image analysis and conductivity methods allows development of a powerful technique for accurately monitoring the foam volume, density and liquid drainage, as shown schematically in Figure 5.11. Separate pairs of electrodes monitor the foam conductivity and drainage level, and a video camera monitors the foam volume; the signals, all analysed by computer, give the complete information required (density, volume and drainage data) on a single foam. The shear viscosity of a foam may be a physical property for certain applications. The total viscosity of a foam is a combination of the bubble size, bulk liquid viscosity and surface viscosity. As proteins tend to generate foams with a high surface viscosity, their foams also tend to be quite stiff. The viscosity of foams has often been determined by standard rotational, or torsion wire viscometers (Corrie, 1976; Morr, 1985). Alternatively, foam strength can be measured. The foam strength is related to the foam viscosity by the fact that both properties depend on the

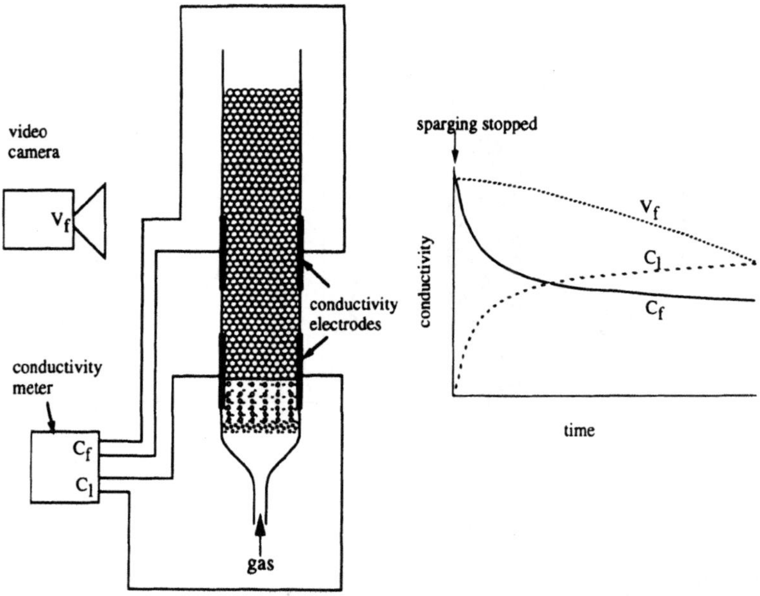

Figure 5.11 Multi-parameter foaming technique (Loisel *et al.*, 1992; Guillerme *et al.*, 1993). Relative foam conductivity (C_f), foam/liquid boundary level (C_l), and foam volume (V_f) are measured simultaneously.

rigidity of the adsorbed layers in the foam. Most methods for measuring foam strength involve dropping a weight through the foam (Mangan, 1958; Buckingham, 1970; Waniska and Kinsella, 1979). Generally this involves generation of a foam in a column. Once a preset time has elapsed, to allow for drainage of surplus liquid, a brass weight, perforated with a series of holes (to allow passage through the foam), is lowered onto the upper surface of the foam and released to fall through the foam. The time taken for the weight to traverse a known distance is measured. Although this technique is qualitative, it has been shown to be an effective relative measurement of protein foam strength, and is sensitive enough to detect pH and ionic-induced changes in foam strength (Waniska and Kinsella, 1979). The stability of foams to reduced pressure has also been examined (Lalchev *et al.*, 1979; Kruglyakov *et al.*, 1992). A technique has been described where a foam formed by sparging is subjected to a negative pressure applied through the sinter used for foam generation. The time taken for total foam collapse is recorded as a function of increasing pressure until a pressure is reached where the foam instantly collapses. This technique was initially designed to study the effect of thin film stability on foam stability. However it also gives a good indication of the stiffness of the interfacial protein layer, and its resistance to collapse under extreme conditions.

A property of beer foam which has wide consumer appeal is the ability of the foam to cling to the side of the glass as the beer is consumed. This is known as lacing, and is measured by a clingmeter (Klopper, 1973). In a typical experiment, a glass of beer is poured and the liquid in the foam is allowed to drain. Subsequently, a probe is lowered into the glass to measure the reflectivity of light from the inside of the glass. The reflectivity is proportional to the amount of foam clinging to the side of the glass, and can be monitored with time, allowing different beers and processes to be tested. Jackson and Bamforth (1982) used a more indirect approach to measuring foam lacing. Beer was poured into a container and allowed to drain slowly, allowing the foam to cling to the side of the glass. The UV (230 nm) absorbance of the drained beer, which is an indication of protein content, was measured. The UV absorbance for the original beer and the drained beer are compared and used to calculate the lacing index, which is related to the amount of protein remaining in the foam. This index correlates closely with the visually estimated degree of lacing. An alternative approach to quantification of foamability involves measurement of the work required to create a foam. Kroll *et al.* (1988) developed a whipping instrument where the torque on the rotating shaft is monitored, allowing calculation of the energy input required to create a unit volume of foam. They observed that the work required to maintain the beaters at constant speed increased as the foam was generated, reaching a plateau when the maximum foaming capacity was attained. The foaming properties can be measured from the rate at which the torque reaches its maximum, and also the maximum value of the torque. The wider implications of this technique involve the testing of the efficiency of whipping equipment.

(c) Bubble properties. Bubbles constitute a fundamental component of foams. Consequently, many workers have studied the properties of bubbles *en masse* or individually to identify the mechanisms of protein foam stability. The most common measurement is that of bubble size distribution. Monitoring bubble size distribution with time provides information about bubble rupture and disproportionation. Unfortunately, accurate measurement of the bubble size distribution is not simple. Many workers have observed the bubbles that are positioned against a glass wall of the foam cylinder (Clark and Blackman, 1948; Nishioka and Ross, 1981; Guillerme *et al.*, 1993). Although this makes for easy observation, small bubbles tend to be excluded giving a bias to the measurement. Statistical treatment, involving a relatively small correction of the overall distribution, can be employed to overcome this (Jashnani and Lemlich, 1974). Clark and Blackman (1948) showed very clearly the effects of disproportionation using this technique. By photographing the same set of bubbles with time, they showed how the smaller bubbles became smaller and the larger bubbles increased in size, and how the rates varied with gases of different

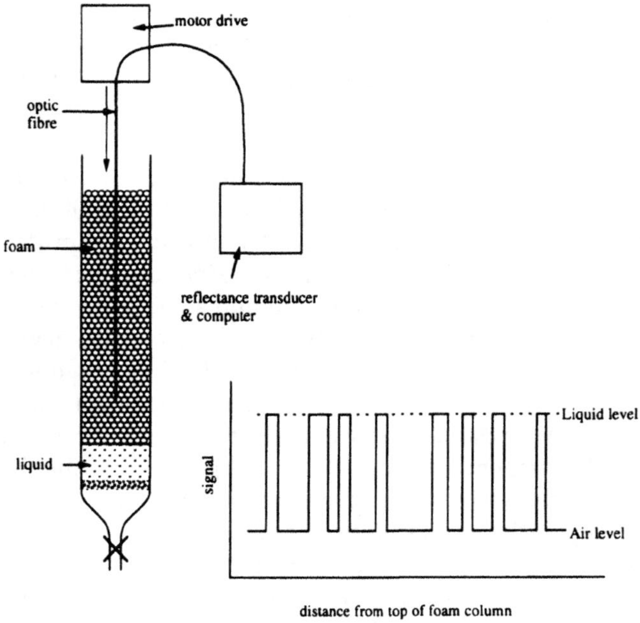

Figure 5.12 Schematic diagram of the foam analyser (Bisperink *et al.*, 1992a,b). The optic fibre passes quickly through the foam. When a thin film is encountered the signal increases, and when the tip is within a bubble the signal drops.

solubilities. Kitabatake *et al.* (1982) measured the bubble size distribution of freeze-dried foams by scanning electron microscopy. They found that changes in bubble and lamellar structure corresponded with the effectiveness of different proteins at producing stable foams. What influence freeze drying had upon these observations is unclear. More recently, an unusual method of measuring the bubble size distribution, which involves lowering a fibre optic probe through the foam at a constant rate, has been developed (Bisperink *et al.*, 1992a,b). Figure 5.12 demonstrates that the instrument measures changes in reflectance detected by a probe tip as it passes through a foam lamella (or thin film) while being lowered into a foam. The distance between each lamella is measured and can be mathematically transformed to give the bubble size distribution. This is necessary since the probability that the probe will pass through the centre of each bubble is small. The data collection rate is high as each run takes only one second. This allows the bubble size distribution to be followed with time; the change in foam volume and the drainage rate can also be measured accurately.

Bubble stability can be monitored using a simple technique to study individual bubbles. Two bubbles are formed and brought together (Foulk and Miller, 1931), or a bubble is formed in a solution and allowed to rise to the air/water interface (Cumper, 1953). The behaviour of the bubbles can be studied in more detail, and the times of coalescence measured to determine the stability. Such experiments do not truly reflect the environment of a bubble in a foam, because of buoyancy effects and the large difference in pressure between the bubble and the adjacent atmosphere. Despite this, they are useful for studying stability and the results show good correlation with surface behaviour and composition.

5.2.2.2 Indirect methods. Indirect measurements are those made on a model system, rather than directly on a foam, to give information about foam behaviour. They include: (a) surface techniques on macroscopic, planar interfaces, (b) thin film studies; and (c) measurement of the intrinsic properties of proteins.

(a) Interfacial properties. The surface properties of proteins have a major influence on the stability of protein foams. Surface tension, surface viscosity and surface elasticity have all been linked with protein foam formation. Surface tension can be determined by the Wilhelmy plate (Paulsson and Dejmek, 1992), or Du Nouy ring method (Weser, 1980). Both methods involve translating the intermolecular force (surface tension) to a force balance or transducer. Water alone has a very high surface tension (>72 mN/m), due to the forces between water molecules at the surface. When a surface active molecule adsorbs to the surface, these forces are broken down and the surface tension is reduced. This reduction is often expressed as a surface pressure π:

$$\pi = \gamma_{\text{solvent}} - \gamma$$

where γ is the surface tension, and γ_{solvent} is the surface tension of the solvent alone. The rate of increase in surface pressure is related to foamability (Graham and Phillips, 1976; Kalischewski and Schurgeri, 1979), i.e. the ability to form a stable interface around the bubbles as they appear. The concentrations of the protein in the surface tension experiments need to be very low (less than 0.01 mg/ml) in order to slow down the adsorption process sufficiently to allow accurate monitoring. A large reduction in surface tension does not necessarily result in good foaming properties; surface viscosity/elasticity is generally a stronger determinant (Phillips, 1981; Mitchell, 1986). Surface shear viscosity is often measured by a torsion wire technique (Boyd and Sherman, 1970; Lee *et al.*, 1991), whereby a plate is aligned with the surface, and the sample dish is revolved at a constant rate. A detector measures the twist in the wire due to the surface shear viscosity. Surface elasticity relates to the ability of the interface to stretch and deform, and is a property associated with stable protein foams. It is usually measured by changing the interfacial area sinusoidally whilst measuring the surface tension. There is usually a delay or phase lag between the area change and the change in surface tension. The magnitude of the surface tension change, the increase in surface area and the extent of the phase lag are used to calculate the surface dilational elasticity. The measurements can be performed on a Langmuir trough with moving barriers, or by dipping and withdrawing a ground glass ring through the interface and measuring the surface tension change inside the ring (Kokelaar *et al.*, 1991). The latter technique has the advantage that there is no chance of leakage of surfactant through the points of contact of the barriers, and if the plate is close to the middle of the ring, the expansion of the surface is almost purely dilational, with very little shear.

(b) Thin films. Air-suspended thin liquid films are a good model of the foam lamellae which separate gas bubbles in a foam. The properties of these films cannot be observed easily in the foam itself; however, microscopic observation by epi-illumination of isolated model thin films formed on a support provides vital information about drainage and stability behaviour (Lee and Tynan, 1988; Coke *et al.*, 1990). A typical thin film apparatus is shown in Figure 5.13. The films can be mounted vertically or horizontally, depending on whether the effect of gravity on drainage is

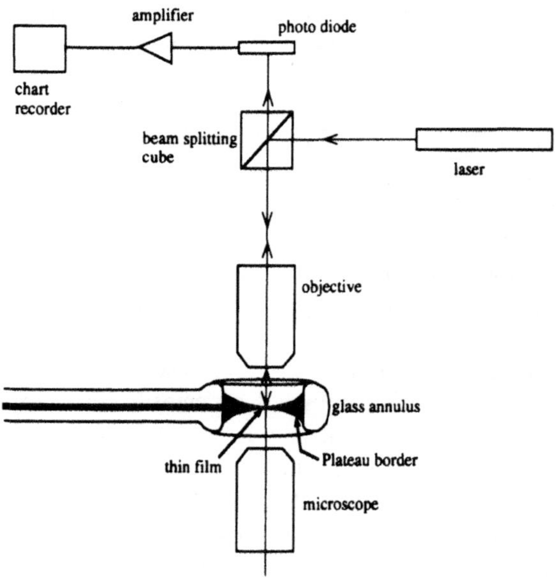

Figure 5.13 Diagram of the model thin film apparatus. The film can be viewed from below by epi-illumination. Alternatively, the thickness can be determined from above by measuring the reflected intensity of a laser beam.

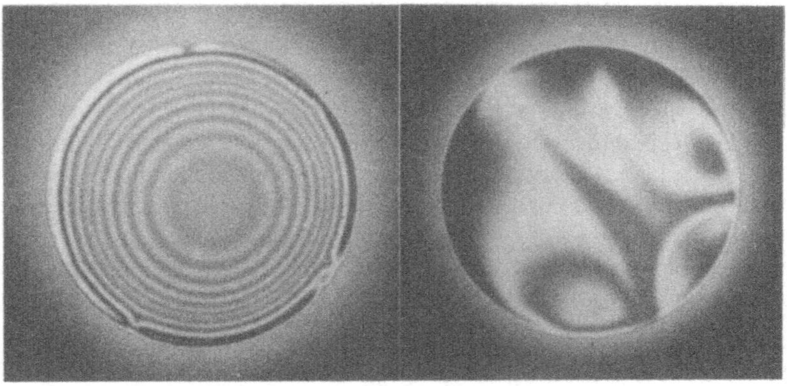

Figure 5.14 Photographs of model thin films. (a) A typical protein-stabilised film, slowly draining with concentric Newton's rings; (b) a surfactant-stabilised film showing rapid, chaotic drainage with no Newton's rings.

of interest. Observation of film drainage gives an immediate indication of the stiffness of the adsorbed layers, and hence the most likely mechanism of stabilisation (Figure 5.2). Figure 5.14a shows thin films stabilised by protein (i.e. the mechanism described by Figure 5.2b); Figure 5.14b shows the typical behaviour of a thin film stabilised by a surfactant (Figure 5.2a), or a protein film which has been destabilised by breakdown of protein–protein interactions (Figure 5.2c). As expected, the protein film is very slow to thin, because of the stiff adsorbed protein layer which resists mechanical deformation and impedes liquid flow towards the Plateau borders. It has been shown that films formed from solutions containing protein and anti-foaming agents or surfactants, have their drainage properties transformed from protein-like, to those resembling a surfactant film. When this change occurs, a marked decrease in stability is observed (Lee and Tynan, 1988; Coke *et al.*, 1990; Wilde *et al.*, 1993).

The thickness of the film can be measured by interferometry (Clark *et al.*, 1990), where the interference of the reflected light from the lower and upper interfaces of the film allows calculation of the film thickness. Thicker films obviously hold more liquid, are less likely to rupture, and produce more stable foams. The time taken to drain to equilibrium thickness is also important. The longer the film remains thick, the longer the protein has to adsorb and form a stiff interface. The breakdown of the interactions in the adsorbed protein layer correlates with the onset of lateral surface diffusion of protein, which can be measured by fluorescence recovery after photobleaching (FRAP) (Coke *et al.*, 1990; Wilde *et al.*, 1993). This sophisticated technique measures directly the rate of redistribution of molecules in and out of a localised, illuminated circular spot on the film surface. Normally with a protein film, no diffusion takes place in the adsorbed layer, as the protein forms a highly interacting, stiff, viscoelastic film. However if the protein–protein interactions are reduced by competitive adsorption of surfactants or lipids, which are present in many real systems, surface diffusion can occur. The strength of the interactions is paramount for a stable protein film, and clear correlations have been reported between protein surface diffusion and foam stability (Coke *et al.*, 1990; Clark *et al.*, 1991a,b; Wilde *et al.*, 1993).

(c) Protein properties. Differences in the foaming behaviour of proteins can often be explained by their intrinsic properties. The structure, surface hydrophobicity, charge, pI and flexibility all contribute to the properties of a protein-stabilised foam. Knowledge of these properties and the effect of environmental conditions or treatments on them can often explain changes in foaming behaviour. The general structure of a protein may give clues to its surface properties. Globular proteins tend to form highly elastic adsorbed layers which contribute to enhanced foam stability, whereas more flexible proteins form a stable interfacial layer more rapidly, which provides enhanced foamability properties. The usefulness of detailed analysis of protein structure, in solution or crystal, using NMR or X-ray crystallography, in understanding foaming properties is presently unclear. However, some efforts to correlate data from these techniques, and electron spin resonance, with increased foamability have been attempted (Le Meste *et al.*, 1990). The surface hydrophobicity of a protein has been shown to be related to foam formation (Horiuchi *et al.*, 1978; Kato *et al.*, 1983a; Townsend and Nakai, 1983). Fluorescence titrations employing hydrophobicity probes such as ANS or parinaric acid (see Chapter 8) have successfully been used to determine this property (Horiuchi *et al.*, 1978). Surface hydrophobicity influences the initial surface activity of a protein, and its capability for rapid formation of a stable interface (i.e. foamability).

One of the most important features of a protein-stabilised interface is the extent to which intermolecular interactions are formed in the adsorbed layer. Electrostatic interactions have an important role to play in this. At the isoelectric point (pI) of a protein, electrostatic repulsion is minimised and foam stability is usually maximal (Graham and Phillips, 1976; Waniska and Kinsella, 1979; Kim and Kinsella, 1985). Determination of the pI of a protein using isoelectric focusing is fairly simple. Changing the charge distribution by chemical modification can also affect the protein interactions, as discussed earlier. It is possible to move the pI of a protein to the working pH of a system so that foam stability can be markedly increased (Howell and Taylor, 1991).

5.2.3 *Analysis*

Treatment of foaming data is not usually very complex, but care must be taken with expression and interpretation. The results are generally extracted by two methods, either direct measurement, or by fitting the data to a foam collapse/drainage model. In the following sections we will review the various methods by which authors have expressed foamability and stability data.

5.2.3.1 Foamability. The simplest and most popular expression of foamability is merely the foam volume created during a defined period of foam generation, Vf_0 (McKellar and Stadelman, 1955; Nakamura and Sato, 1964; Grunden *et al.*, 1974; Groninger and Miller, 1975; Wang and Kinsella, 1976; Townsend and Nakai, 1983; Ahmed and Dickinson, 1991; Gabor, 1992). This value is dependent on a number of parameters, mainly the initial sample volume. Therefore, to facilitate comparison of results between workers, it is preferable to express Vf_0 as a function of the initial liquid volume Vl_i. Foam expansion, FE%, is Vf_0 expressed as a percentage of Vl_i (Lawhon and Cater, 1971; Lawhon *et al.*, 1972; Halling, 1981; Poole *et al.*, 1984, 1986; Mitchell, 1986; Closs *et al.*, 1990; Howell and Taylor, 1991) and is normally calculated thus:

$$\text{FE\%} = \frac{Vf_0}{Vl_i} \cdot 100\%$$

For whipped systems, it is often more difficult to measure the foam volume accurately. For such systems, a sample of liquid is weighed, Wl, the same volume of the foam is weighed, Wf, and the foam expansion is calculated as %overrun (Richert *et al.*, 1974; Phillips *et al.*, 1987, 1990):

$$\text{\%overrun} = \frac{Wl - Wf}{Wl} \cdot 100\%$$

Both FE% and %overrun are used in systems where the volume of the foam is most important. However, the %overrun provides no information about foam volume, it merely infers it because it is mainly used with whipped systems, where normally all of the sample is incorporated into the foam. The %overrun value is therefore misleading in a system where not all of the liquid is incorporated in the foam. To correct this, the %overrun should be normalised to the percentage of Vl_i in the foam after foaming. Therefore %overrun becomes:

$$\text{\%overrun} = \frac{Wl - Wf}{Wl} \cdot \frac{Vl_i - Vl_0}{Vl_i} \cdot 100\%$$

where Vl_0 is the liquid volume remaining after foaming.

The sparging technique for foam generation not only allows simple measurement of Vf_0, but also allows control of the volume of gas sparged Vg. The efficiency or foaming capacity (FC) can then be determined (Loisel *et al.*, 1992):

$$FC = \frac{Vf_0}{Vg}$$

Foam volume is not always the most important feature of a foam. The ability of a foam to take up and hold liquids, i.e. foam density, is also of value. Conductivity is an indirect method of measuring the foam density, the relationship between foam conductivity and density having been shown theoretically (Prager, 1963) and experimentally (Kumagai *et al.*, 1991; Torikata *et al.*, 1991). Therefore, the foam conductivity when the foam has been created, C_i, can be regarded as a measure of foamability (Kato *et al.*, 1983a,b; Bacon *et al.*, 1988; Baniel *et al.*, 1992). The usefulness of C_i alone is limited, as it is dependent on the conductivity of the bulk liquid C_l. A more useful expression is the relative foam conductivity C_f:

$$C_f = \frac{C_i}{C_l} \cdot 100\%$$

This is only an estimate of the relative density of the foam. As a result of adsorption at the interface, the charge distribution through a foam lamella may not be the same as that in the bulk solution. Therefore C_f is not a true value of the liquid content of the foam, but is proportional to it; if the relationship is known, the liquid content of the foam can easily be calculated.

In summary, it is possible to measure two independent properties which relate to the foamability of a system: (i) the foam volume; and (ii) the foam density. For whipped foams, which incorporate all of the liquid into the foam, the foam density is inversely proportional to the foam volume. However, in the case of other methods of foam generation, particularly sparging, the values are independent of each other.

5.2.3.2 Foam stability. Measuring the change in foam volume with time is a popular and generally highly relevant direct method for assessing foam stability. The stability, expressed in its simplest state, is the foam volume after a certain time t has passed, Vf_t (Nakamura and Sato, 1964; Lawhon and Cator, 1971; Wang and Kinsella, 1976). This value on its own has little comparative value. Therefore the stability is more commonly expressed as the foam volume stability $FVS\%_t$ at time t (Poole *et al.*, 1986; Howell and Taylor, 1991):

$$FVS\%_t = \frac{Vf_t}{Vf_0} \cdot 100\%$$

The problem with FVS% is the arbitrary nature by which the time is chosen to assess the stability. The use of time itself as a determinant of stability removes this problem. To this end, some authors have used the half-life ($t_{1/2}$) of the foam volume (Horiuchi *et al.*, 1978; Loisel *et al.*, 1992), where $t_{1/2}$ is the time taken for the foam to collapse to half of its initial volume Vf_0. It is also possible to calculate the rate of foam collapse, and express this as a stability parameter (Horiuchi *et al.*, 1978). The time taken for total foam collapse can be used (Ahmed and Dickinson, 1990; Closs *et al.*, 1990), but is only suitable for foams with relatively short lifetimes. A more specialised method which uses time as a parameter is the Nibem method (Klopper, 1973). As described previously (section 5.2.2.1(a)), this technique mechanically follows the upper surface of the foam and measures the times taken for the foam to collapse by 10, 20 and 30 mm. This is a very specific assessment and can only really be compared with other measurements using identical devices and the same method of foam generation, sample sizes etc. When these precautions are taken, the method is reproducible and is used as a standard measurement.

The other main measure of stability is the volume of liquid which drains from the foam. The ability of a foam to hold liquid and reduce drainage is very important in many applications. Again, the liquid volume at time t, Vl_t, is not a particularly useful value (McKellar and Stadelman, 1955; Graham and Phillips, 1976), but is normally expressed as a proportion of the initial sample volume, Vl_i, and the liquid volume immediately after foaming ($t = 0$), Vl_0. This is referred to as the foam liquid stability (FLS; Watanabe *et al.*, 1981; Poole *et al.*, 1984; Howell and Taylor, 1991; Mitchell, 1986), where:

$$FLS\% = \frac{Vl_i - Vl_t}{Vl_i - Vl_0} \cdot 100\%$$

Thus FLS% is a direct measure of the relative volume of liquid still remaining in the foam at time t. Still, there is the arbitrary time t to be chosen. Again, to overcome this, the $t_{1/2}$ value for the FLS can be presented as a suitable indication of stability (Phillips *et al.*, 1987; Lorient *et al.*, 1989). Two of the standard methods for the brewing industry are designed to give time values for foam liquid drainage. The Rudin method (as described in section 5.2.2.1(a)) (Rudin, 1957; Slack and Bamforth, 1983; Yokoi *et al.*, 1989) gives the $t_{1/2}$ value, referred to as the head retention value (HRV), and the Sigma (Σ) method (Anon., 1976) calculates the time for the liquid in the foam to drain to $1/e$ of its initial value $t_{1/e}$. The latter method is a little more complicated than the Rudin technique. The sample is foamed to a fixed height. After 45 seconds, the drained liquid is removed; time $t = 0$. After a further 200 seconds, slowly remove the drained liquid (Vd); time $= t$ (s). The remaining foam is collapsed with 2 ml of isopropanol and the volume of liquid remaining in the foam Vl_f is determined. The Σ value is calculated as:

$$\Sigma = \frac{t}{2.303 \cdot \log \dfrac{Vd + Vl_f}{Vl_f}}$$

where Σ is actually the $t_{1/e}$ value. This is calculated from the rate of drainage over time t. The approach ensures that all the samples are treated consistently and, for very stable foams, the length of the experiment is no longer than that of an unstable foam.

The capillary model, derived by Haas and Johnson (1967), has been used to model drainage of protein foams (Kumagai *et al.*, 1991; Torikata *et al.*, 1991). The volume fraction $\varepsilon_{t,z}$ at time t and height z above the foam liquid boundary is given by:

$$\varepsilon_{t,z} = \frac{\varepsilon_{0,0}\,(1 - z/L)}{1 + \varepsilon_{0,0}\,(t/T)}$$

Where L is the height of the foam and T is the time constant for liquid drainage given by:

$$T = \frac{4\pi\mu k^2 L}{\varrho g d^2}$$

Where μ and ϱ are the liquid viscosity and density respectively, g is gravitational acceleration, d is the average bubble diameter and k is a constant. The change in height Δh of the foam liquid boundary is measured as a function of time and fitted to the equation:

$$\Delta h = \frac{t}{a(a + bt)}$$

Where the constants a and b are defined as:

$$a = (TL/2)^{-\frac{1}{2}}\, T/\varepsilon_{0,0} \text{ and } b = (TL/2)^{-\frac{1}{2}}$$

T is calculated from the fitted parameters, and has been shown to be an indicator of foam stability. However, the protein solutions tested have been of high concentration, where the foam is highly stable, and the viscosity of the solution may influence the drainage behaviour. There are no reports of the model being tested on a system where the stability is limited by the level of adsorption and the interfacial properties of the adsorbed layer.

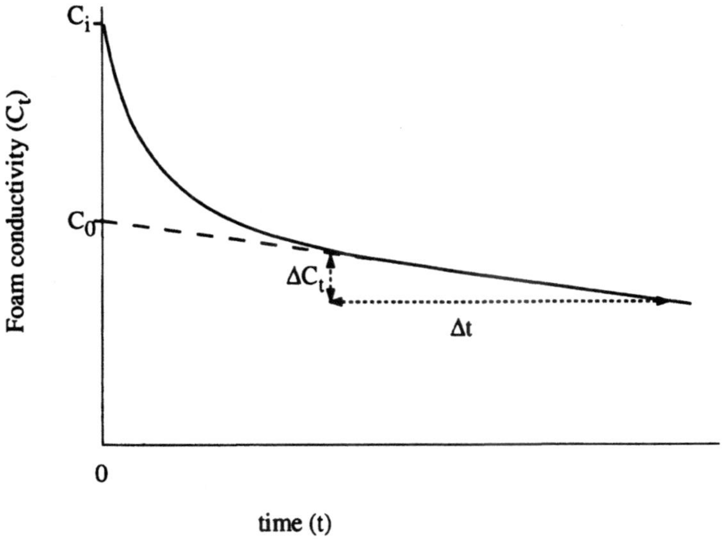

Figure 5.15 Foam conductivity decay curve, illustrating the parameters used for calculating foamability and foam stability values.

Foam conductivity is a convenient method for monitoring the foam density with time, as the values can be easily logged by computer. There are several ways of analysing conductivity decay curves. Figure 5.15 shows theoretical conductivity data, illustrating the reference conductivity values. The simplest way to express the stability is to take the foam conductivity at time t (C_t) as a percentage of C_i (Bacon *et al.*, 1988). This can be described as the foam density stability FDS% where it can be equated to the percentage of the original foam density, as the conductivity is proportional to the foam density (Clark, 1948; Torikata *et al.*, 1991; Loisel *et al.*, 1992):

$$\text{FDS\%} = \frac{C_t}{C_i} \cdot 100\%$$

Again there is the problem of choosing a suitable time for expressing the stability, so some authors quote the $t_{1/2}$ of the FDS% as an indicator of stability (Lorient *et al.*, 1989). Others have used a different approach involving measurement of the slope of the second phase of the conductivity decay curve (Figure 5.15) (Kato *et al.*, 1983b; Baniel *et al.*, 1992). These authors define the foam stability index (FSI) as:

$$\text{FSI} = C_0 \cdot \frac{\Delta t}{\Delta C_t}$$

There is a reasonable correlation between the $t_{1/2}$ values and FSI for these samples (Baniel *et al.*, 1992). However, the C_0 term biases the FSI towards the initial conductivity value. Therefore, the foamability of the sample is contributing to the stability value. In addition, differences in the bulk solution conductivity C_1 will also contribute to the FSI. A similar treatment used by Wright and Hemmant (1987) avoids this problem. Here the data are linearised by means of the following equation:

$$F(t) = \frac{C_0 \cdot t}{(C_0 - C_t)}$$

$F(t)$ plotted against t gives a straight line. Linear regression of these data gives a slope M. The slope shows a good correlation with FDS% and $t_{1/2}$, and is independent of time and sample conductivity C_1. Although the parameter M is not directly related to a foaming property, it can be converted into an equivalent FDS% over the timescale of the experiment, without having to choose an arbitrary value of t:

$$\text{FDS\%} = (1 - \frac{1}{M}) \cdot 100\%$$

Some workers have attempted to fit a model to the foam conductivity data. Wright and Hemmant (1987) tried to fit a double exponential equation to the data such that:

$$C_t = A_1 \cdot e^{-t/\tau_1} + A_2 \cdot e^{-t/\tau_2}$$

where A_1, A_2 are scaling constants and τ_1 and τ_2 are time constants for the foam conductivity decay. τ_1 is believed to be related to gravitational drainage, and τ_2 to disproportionation. The equation was originally derived by Monsalve and Schechter (1984) to describe the change in interfacial area of a foam. Unfortunately, Wright demonstrated that the results for foam conductivity give poor reproducibility, and generally do not correlate well with other accepted stability parameters.

Another means of fitting a model to foam decay has been the very specific method of Nishioka and Ross (Nishioka and Ross, 1981; Nishioka *et al.*, 1983; Yu and Damodaran, 1991a,b; Dickinson *et al.*, 1993). As described in section 5.2.2.1(a), this method monitors the pressure change of a foam in a closed system during collapse. It has been shown that the pressure change ΔP is proportional to the change in interfacial area of the foam ΔA as it collapses, so that:

$$\Delta A = \frac{-3V\Delta P}{2\gamma}$$

where V is the total volume of the closed system, and γ is the surface tension. Yu and Damodaran (1991a) fitted the decay in surface area to the biphasic first-order equation:

$$A_t/A_0 = Q_g e^{-k_g t} + Q_d e^{-k_d t}$$

Where A_0 and A_t are the interfacial areas at time zero and t, respectively, Q_g and Q_d are the amplitude parameters for gravitational drainage and disproportionation respectively, and k_g and k_d are the alleged time constants for foam collapse due to gravitational drainage and disproportionation, respectively. A_0 is also used as an indicator of foamability. k_g appeared to be a good indicator of stability, but again the values obtained were not tested on a wide range of systems, nor were they compared with more conventional means of expressing the stability. The values of k_d were not very reproducible but, as the system under test was sparged with nitrogen, the rates of disproportionation would be very low, due to low solubility. No data have been published showing a change in k_d by using a gas of higher solubility. The most recent paper using this technique, Dickinson *et al.* (1993), in fact uses only a single exponential term to obtain a rate constant for collapse. This value compares well with measured foam volume stability.

In summary, the analysis of foaming properties is generally fairly straightforward. However, a careful decision needs to be made as to which parameters are important (e.g. volume, drainage or density). Once this is done, the necessary foamability and stability parameters can be calculated. As discussed above, more detailed analysis attempting to extrapolate more information about specific processes of destabilisation, has met with only limited success. If such information is required, it is best achieved using alternative, more appropriate techniques.

The study and measurement of the properties of protein-stabilised foams has progressed significantly over recent years. The complex behaviour of proteins at interfaces, and how this relates to foam formation and stability, is now more widely understood, mainly through the detailed study of pure protein systems. Armed with this underpinning knowledge, the investigation of more complex, multi-component systems, more closely related to real systems, is a more practical proposition and is meeting with increasing success. However, our ability to predict the foaming properties of a system merely by knowing the list of ingredients, the degree of processing and the environmental conditions, is still some time away.

5.3 Protocols

The choice of method to study the foaming properties of proteins is dependent upon two main factors: (i) the objectives of the work or application; and (ii) the protein system under test. Protein systems which have been studied can be divided into three general groups:

(a) *Pure proteins* – single-component or controlled multi-component model systems comprising purified proteins. Common proteins which have been tested include bovine serum albumin, β-lactoglobulin, β-casein, ovalbumin, seed legumins and α-lactalbumin.
(b) *Crude protein extracts* – extracted from a variety of sources, these are often used in commercial foam-based products and can therefore also be classed as real systems. Common preparations include caseinates, whey protein, egg white, seed and bean proteins, and meat and fish protein extracts.
(c) *Real systems* – the final product. Very few commercial foams have been studied scientifically in any real detail, apart from bread and cake products which have not been covered in this chapter. The main systems that have been studied are egg white and beer foam.

We will now look at these three groups in more detail. This will include a description of the methods which are frequently employed to study each group, some considerations to be taken into account, potential pitfalls to be aware of, together with the relevant references that contain further information.

5.3.1 Pure proteins

These systems are usually studied when a more extensive knowledge of the foaming and interfacial behaviour of proteins is required. All parameters must be carefully controlled, namely sample composition, pH, ionic strength, etc. Consequently, the foaming conditions also tend to be well controlled. This is probably why most workers studying pure protein foams use the sparging method (section 5.2.1.1), so that flow rate, pressure, foam volume and even bubble size may be carefully controlled. Whipping of these systems has also been studied, when comparison with real whipped systems has been required. However, in general, when studying pure proteins, it is the functional and interfacial properties of the proteins that are of interest. If total control is required for more accurate interpretation of the results, then the bubble size distribution must be considered. Sparging through a sinter will produce a wide range of bubble sizes. For a thorough understanding of the foaming behaviour, the bubble size distribution must be measured, as smaller bubbles result in a denser, and mechanically more stable foam, irrespective of the properties of the protein itself. Bubble sizing methods have been described in section 5.2.2.1(c). A simple way to avoid discrepancies in bubble size is to control the bubble size as the foam is being generated. Sparging through a single orifice (Miles *et al.*, 1945; Wright and Hemmant, 1987; Coke *et al.*, 1990) can successfully produce a foam of a very narrow bubble size distribution. This effectively eliminates any effects of initial bubble size on the foam stability. Measurement of the foamability and foam stability is of course largely determined by which property of the foam is of interest. Whether the researcher measures the foam volume, density or drainage, depends upon the application, if any, or what property of the foam the system is supposed to affect. For example, if the system under study is designed to reduce interlamellar drainage, then the rate of drainage from the foam (Kumagai *et al.*, 1991; Torikata *et al.*, 1991), or foam density (Torikata *et al.*, 1991; Baniel *et al.*, 1992; Loisel *et al.*, 1992) should be measured. If the system is sensitive to coalescence, then the foam volume may be important (Kalischewski and Schurgeri, 1979; Townsend and Wakai, 1983; Ahmed and Dickinson, 1991; Loisel *et al.*, 1992; Guillerme *et al.*, 1993). However, from a fundamental research point of view, it is recommended that as much information as possible is collected, to give more accurate interpretation of the results. A very important consideration when dealing with pure proteins is experimental scale. Pure protein systems can be expensive, so low sample volume is an important criterion. However, low sample volume will lead to low foam and drainage volumes. Measurement of these parameters therefore becomes less accurate. This may necessitate increasing the sophistication of the measuring technique. This is why image analysis or video monitoring of such foams is becoming more popular (Loisel *et al.*, 1992; Guillerme *et al.*, 1993). Another sensitive method used in pure protein systems is conductivity (Kato *et al.*, 1983a,b; Bacon *et al.*, 1988; Lorient *et al.*, 1989; Kumagai *et al.*, 1991; Torikata *et al.* 1991; Baniel *et al.*, 1992), which is very sensitive to small changes in foam density, or drainage volume level. Another technique is to measure the pressure change of the foam in a closed system, to give information about the interfacial area of the foam (Nishioka and Ross, 1981, Yu and Damodaran, 1991a,b; Dickinson *et al.*, 1993). This is a useful academic research tool, but is very expensive to set up because of the accuracy of the pressure monitoring and thermal control.

As already mentioned, the study of pure protein systems is often associated with the desire for a more detailed understanding of the interfacial properties of the protein. To correctly interpret how protein foams are stabilised, therefore, the interfacial properties of the proteins should also be studied (section 5.2.2.2(a)). Surface activity of proteins can be measured by measuring the surface tension (Graham and Phillips, 1976; Kalischewski and Schurgeri, 1979; Kim and Kinsella, 1985; Mitchell, 1986; Closs *et al.*, 1990; Paulsson and Dejmek, 1992). Surface viscosity of protein interfaces has frequently been correlated with foam stability (Phillips, 1981; Kim and Kinsella, 1985; Mitchell, 1986; Maeda *et al.*, 1991). The study of thin films is a good method for determining the method of stabilisation of the foam lamellae (section 5.2.2.2(b)) (Lee and Tynan,

1988; Coke *et al.*, 1990), the surface lateral mobility (Coke *et al.*, 1990; Clark *et al.*, 1991a,b; Wilde *et al.*, 1993) and, in particular, the effect of low molecular weight surfactants on protein foams. Looking in finer detail, the structural properties of the proteins themselves are related to interfacial and foam stability (section 5.2.2.2(c)). Protein structure (Le Meste *et al.*, 1990), surface hydrophobicity (Horiuchi *et al.*, 1978; Kato *et al.*, 1983a; Townsend and Nakai, 1983), isoelectric point (Graham and Phillips, 1976; Waniska and Kinsella, 1979; Kim and Kinsella, 1985; Howell and Taylor, 1991) and flexibility (Townsend and Nakai, 1983; Mitchell, 1986; Closs *et al.*, 1990; Le Meste *et al.*, 1990) are all aspects which have been linked with particular foaming properties. If a profound understanding of the foaming properties of pure proteins is to be developed, then these interfacial and structural parameters need to be determined. Great care needs to be taken over the purity of alleged pure protein fractions. Cruder, commercial protein extracts often contain a significant proportion (up to 5% w/w) of lipid material. These systems should be characterised with this in mind. If allegedly pure samples are contaminated, this can seriously affect the results, and their interpretation. Recent work has shown that minute quantities of surface active contaminants can significantly affect the surface behaviour of pure protein systems. Surface rheological measurements are particularly sensitive to this effect.

5.3.2 Crude protein extracts

These are a step nearer to real, commercial systems. In fact, some crude protein extracts are used in manufacture of commercial products, so in some senses, they are real systems. Assuming that these protein extracts are under test with a specific application in mind, then the methods used for assessing their foaming properties should reflect how they are likely to be processed. By far the most common method to generate foams when studying these systems is the whipping method (section 5.2.1.2). This is usually more vigorous than sparging, as the crude proteins generally form poorer foams than their purified counterparts. Again, the method of assessing the foamability and foam stability depends upon the application. Overrun or foam density are the most common assessments of foamability. The foam stability is expressed either as the rate and extent of foam collapse (Watanabe *et al.*, 1981; Poole *et al.*, 1986; Closs *et al.*, 1990; Gabor, 1992) or the rate and extent of liquid drainage from the foam (Grunden *et al.*, 1974; Richert *et al.*, 1974; Groninger and Miller, 1975; McKellar and Stadelman, 1975; Poole *et al.*, 1984; Mitchell, 1986; Phillips *et al.*, 1987; Phillips *et al.*, 1990; Gabor, 1992). The method of Phillips *et al.* (1990) is recommended for those who wish to study this aspect of foam measurement in more detail. With this method the rate of drainage is determined by allowing the liquid draining from the foam to collect onto a balance (section 5.2.2.1(a) and Figure 5.6). The paper by Phillips *et al.* (1990) is particularly thorough, as the authors give consideration to the experimental design of a whipping experiment, imparting information about bowl shape, mixer type, beater frequency, protein dispersion technique and other considerations that should be taken into account when designing such a method. However, not all applications rely on whipping for foam generation – some other forms of foam dispensing do occur. Consequently, if testing a particular application, the real methods used to generate the foams should be simulated in controlled laboratory conditions, if this is feasible and cost-effective.

5.3.3 Real systems

Simulating the foaming mechanisms of the real situation applies even more so to the measurement of foaming properties of real systems. Egg white protein is a widely used protein foaming product; it can be regarded as a real system as it is used both as a raw material, commercially and domestically, and as a final product. Most applications which generate egg white protein foams use some form of whipping technique. It is no surprise therefore that the vast majority of research into the foaming properties of egg white also uses the whipping method. This method has been detailed in section 5.3.2, therefore no further discussion is necessary here. Another common final product which has been extensively studied is beer foam. A range of techniques has been employed throughout the literature to investigate the foaming properties of beer. Sparging and foam drainage data are often measured by the Rudin method (Rudin, 1957). This is a simple standard method used by the brewing industry, but does not reflect strongly the real dispensing situation. The simplest form of dispensing beer is pouring, therefore this method has often been utilised to generate beer foam (Anon., 1976; Klopper, 1973; Gales, 1977; Haugsted and Erdal, 1991; Hegarty *et al.*, 1991; Jackson and Bamforth, 1982) (section 5.2.1.4). Pressurised gas dispensing is also used (Anon., 1976). The foams generated can be assessed, either by foam volume change (Klopper, 1973; Haugsted and Erdal, 1991), or by the rate of drainage (Gray and Stone, 1940; Rudin, 1957; Anon., 1976; St. John Coglan *et al.*, 1992). Ideally, information both on foam volume stability and on liquid drainage is required to assess the stability and quality of the created foam.

In summary, it is essential to reiterate certain critical points concerning the methods of protein foam measurement. A variety of methods has been used in the literature over many years. When choosing a particular technique, it is important to consider certain criteria, and to ask the following questions. What information is required from the measurements? What is the basis of the work – is it fundamental research or does it relate to a commercial product? What type of protein systems will be studied? All these factors must be considered when selecting a foaming technique. In addition, care must be taken in the analysis of the results (5.2.3), to ensure accurate interpretation of what will hopefully be accurate and reliable data.

References

Ahmed, M. and Dickinson, E. (1990) Effect of ethanol content on the foaming of alcoholic beverages. *Food Hydrocolloids*, **4** (5), 77–82.

Ahmed, M. and Dickinson, E. (1991) Foaming of aqueous solutions of protein and propylene glycol alginate. *Food Hydrocolloids*, **4** (1), 395–402.

Anon. (1976) Foam collapse rate: A – Sigma value method. B – Foam flashing method in *American Society of Brewers and Chemists Methods of Analysis 7th edn.* BEER-22, 1–6.

Asano, K. and Hashimoto, N. (1976) Contribution of hop bittering substances to head formation of beer. *Report of the Research Lab of Kirin Brew.*, **19**, 9–16.

Bacon, J.R., Hemmant, J.W., Lambert, N., Moore, R. and Wright, D.J. (1988) Characterisation of the foaming properties of lysozymes and lactalbumins: a structural evaluation. *Food Hydrocolloids*, **2** (3), 225–45.

Badwan, A.A., James, K.C. and Pugh, W.J. (1977) A relationship between foaming and HLB. *J. Pharm. Pharmacol.*, **29** (suppl., Br. Pharm. Conf. 1977), 22P.

Baniel, A., Caer, D., Colas, B. and Gueguen, J. (1992) Functional properties of glycosylated derivatives of the 11S storage protein from pea (*Pisum sativum* L.) *J. Agric. Food Chem.*, **40** (2), 200–5.

Bisperink, C.G.J., Akkerman, J.C., Prins, A. and Ronteltap, A.D. (1992a) A moving optical fibre technique for structure analysis of heterogenous products: application to the determination of the bubble size distribution in liquid foams. *Food Structure*, **11** (2), 101–8.

Bisperink, C.G.J., Ronteltap, A.D. and Prins, A. (1992b) Bubble size distribution in foams. *Adv. Colloid Interface Sci.*, **38**, 13–32.

Boyd, J. and Sherman, P. (1970) Two-dimensional rheological studies on surfactant films at interfaces. *J. Colloid Interface Sci.*, **34** (1), 76–80.

Buckingham, J.H. (1970) Effect of pH, concentration and temperature on the strength of cytoplasmic protein foams. *J. Sci. Food Agric.*, **21**, 441.

Clark, N.O. (1948) The electrical conductivity of foam. *Trans. Faraday Soc.*, **44**, 13–15.

Clark, N.O. and Blackman, M. (1948) The degree of dispersion of the gas phase in foam. *Trans. Faraday Soc.*, **44**, 1–7.

Clark, D.C. and Smith, L.J. (1989) Influence of alcohol-containing spreading solvents on the secondary structure of proteins: A circular dichroism investigation. *J. Agric. Food Chem.*, **37** (3), 627–33.

Clark, D.C., Dann, R., Mackie, A.R. *et al.* (1990) Surface diffusion in sodium dodecyl sulphate-stabilised thin liquid films. *J. Coll. Interface Sci.*, **138** (1), 195–206.

Clark, D.C., Wilde, P.J. and Wilson, D.R. (1991a) The effect of pre-isomerised hop extract on the properties of model protein stabilised foams. *J. Inst. Brew.*, **97** (3), 169–72.

Clark, D.C., Wilde, P.J. and Wilson, D.R. (1991b) Destabilisation of α-lactalbumin foams by competitive adsorption of the surfactant Tween 20. *Colloids and Surfaces*, **59**, 209–23.

Closs, B., Courthaudon, J.L. and Lorient, D. (1990) Effect of glycosylation on the surface properties of the soluble fraction of casein. *J. Food Sci.*, **55** (2), 437–9.

Coke, M., Wilde, P.J., Russell, E.J. and Clark, D.C. (1990) The influence of surface composition and molecular diffusion on the stability of foams formed from protein/surfactant mixtures. *J. Coll. Interface Sci.*, **138** (2), 489–503.

Corrie, J.G. (1976) Experimental methods for the study of fire-fighting foams, in *Foams* (ed. Akers, R.J.), Academic Press, London, pp. 195–215.

Cumper, C.W.N. (1953) The stabilisation of foams by proteins. *Trans. Faraday Soc.*, **49**, 1360–9.

Damodaran, S. (1990) Interfaces, protein films and foams. *Adv. Food Nutr. Res.*, **34**, 1–79.

DeFeijter, J.A., Benjamins, J. and Veer, F.A. (1978) Ellipsometry as a tool to study the adsorption behaviour of synthetic and biopolymers at the air–water interface. *Biopolymers* **17**, 1759–72.

Dickinson, E. and Matsumura, Y. (1991) Time dependent polymerisation of β-lactoglobulin through disulphide bonds at the oil–water interface in emulsions. *Int. J. Biol. Macromol.*, **13**, 26–30.

Dickinson, E., Izgi, E. and Nelson, V. (1993) Determination of protein foaming stability from time dependent pressure monitoring. *Food Hydrocolloids*, **7** (4), 307–18.

Evans, J.I. (1972) Control of beer foam. *Process Biochem.*, **Apr.**, 29–31.

Foulk, C.W. and Miller, J.N. (1931) Experimental evidence in support of the balanced-layer theory of liquid film formation. *Ind. Eng. Chem.*, **23** (11), 1283–8.

Gabor, E. (1992) Change of foam properties of gelatin solutions under different conditions. *Acta Alimentaria*, **21** (1), 77–86.

Gales, P.W. (1977) Gales foam analyser – a novel approach to beer foam measurement. *ASBC J.*, **35**, 197–200.

German, J.B. and McCarthy, M.J. (1989) Stability of aqueous foams: analysis using magnetic resonance imaging. *J. Agric. Food Chem.*, **37** (5), 1321–4.

Graham, D.E. and Phillips, M.C. (1976) The conformation of proteins at the air–water interface and their role in stabilising foams, in *Foams* (ed. Akers, R.J.), Academic Press, London, pp. 195–215.

Graham, D.E. and Phillips, M.C. (1979a) Proteins at liquid interfaces I. Kinetics of adsorption and surface denaturation. *J. Colloid Interface Sci.*, **70** (3), 403–14.

Graham, D.E. and Phillips, M.C. (1979b) Proteins at liquid interfaces II. Adsorption isotherms. *J. Colloid Interface Sci.*, **70** (3), 415–26.

Graham, D.E. and Phillips, M.C. (1979c) Proteins at liquid interfaces III. Molecular structures of adsorbed films. *J. Colloid Interface Sci.*, **70** (3), 427–39.

Gray, P.P. and Stone, I. (1940) The measurement and study of beer foam. *Proc. Am. Soc. Brew. Chemists*, **11**, 11–23.

Groninger, H.S. Jr. and Miller, R. (1975) Preparation and aeration properties of an enzyme-modified succinylated fish protein. *J. Food Sci.*, **40** (2), 327–30.

Grunden, L.P., Vadehra, D.V. and Baker, R.C. (1974) Effects of proteolytic enzymes on the functionality of chicken egg albumin. *J. Food Sci.*, **39** (4), 841–3.

Guillerme, C., Loisel, W., Bertrand, D. and Popineau, Y. (1993) Study of foam stability by video image analysis. Relationship with the quantity of liquid in the foams. *J. Texture Studies*, **24** (3), 287–302.

Haas, P.A. and Johnson, H.F. (1967) A model and experimental results for drainage of solution between foam bubbles. *Ind. Eng. Chem. Fundam.*, **6** (2), 225–33.

Halling, P.J. (1981) Protein-stabilized foams and emulsions. *CRC Critical Reviews in Food Science and Nutrition*, **Oct.**, 155–203.

Haugsted, C. and Erdal, K. (1991) Head hunting, *Proc. EBC Congress, 22nd Symposium (Lisbon)*, pp. 449–56.

Haugsted, C., Pederson, M.B. and Erdal, K. (1990) An opto-electrical foam assay system. *Monatsschrift für Brauwissenschaft*, **43** (10), 336–9.

Hegarty, P.K., Cope, R., Crompton, I.E. and Wallach, E.R. (1991) An image analysis system for measuring beer foam stability. *Proc. EBC Congress, 22nd Symposium (Lisbon)*, pp. 465–72.

Hollemans, M. and Tonies, A.R.J.M. (1989) The role of specific proteins in beer foam. *Proc. EBC Congress, 21st Symposium (Zurich)*, pp. 561–8.

Horiuchi, T., Fukishima, D., Sugimoto, H. and Hattori, T. (1978) Studies on enzyme-modified proteins as foaming agents: effect of structure on foam stability. *Food Chem.*, **3**, 35–42.

Howell, N.K. and Taylor, C. (1991) Effect of amidation on the foaming and physicochemical properties of bovine serum albumin. *Int. J. Food Sci. Technol.*, **26**, 385–95.

Jackson, G. and Bamforth, C.W. (1982) The measurement of foam lacing. *J. Inst. Brew.*, **88**, 378–81.

Jashnani, I.L. and Lemlich, R. (1974) Foam drainage, surface viscosity and bubble size bias. *J. Coll. Interface Sci.*, **46** (3), 13–16.

Kalischewski, K. and Schurgerl, K. (1979) Investigation of protein foams obtained by bubbling. *Coll. Polym. Sci.*, **257**, 1099–110.

Kato, A., Osako, Y., Matsudomi, N. and Kobayashi, K. (1983a) Changes in the emulsifying and foaming properties of proteins during heat denaturation. *Agric. Biol. Chem.*, **47** (1), 33–7.

Kato, A., Takhashi, A., Matsudomi, N. and Kobayashi, K. (1983b) Determination of foaming properties by conductivity measurements. *J. Food Sci.*, **48**, 62–5.

Kim, S.H. and Kinsella, J.E. (1985) Surface activity of food proteins: relationships between surface pressure development, viscoelasticity of interfacial films and foam stability of bovine serum albumin. *J. Food Sci.*, **50**, 1526–30.

Kitabatake, N., Sasaki, H. and Doi, E. (1982) Scanning electron microscopy of freeze-dried protein foams. *Agric. Biol. Chem.*, **46** (11), 2881–3.

Klopper, W.J. (1973) Foam stability and foam cling. *Proc. EBC, 14th Congress (Salzburg)*, **14**, pp. 363–71.

Kokelaar, J.J., Prins, A. and De Gee, M. (1991) A new method for measuring the surface dilational modulus of a liquid. *J. Colloid Interface Sci.*, **146** (2), 507–11.

Kroll, J., Gassmann, B., Dahme, A. and Hahnemann, J. (1988) Foaming properties of protein based whipping agents: new characteristics. *Die Nahrung*, **32**, 79–81.

Kruglyakov, P.M., Exerowa, D. and Khristov, Khr. (1992) New possibilities for foam investigation: creating a pressure difference in the foam liquid phase. *Adv. Colloid Interface Sci.*, **40**, 257–81.

Kumagai, H., Torikata, Y., Yoshimura, H., Kato, M. and Yano, T. (1991) Estimation of the stability of foam containing hydrophobic particles by parameters in the capillary model. *Agric. Biol. Chem.*, **55** (7), 1823–9.

Lalchev, Z. and Exerowa, D. (1981) Concentration of proteins by foaming. *Biotechnol. Bioeng.*, **23**, 669–76.

Lalchev, Z., Khristov, K. and Exerowa, D. (1979) Drainage and stability of foams from aqueous protein solutions when a pressure is applied in their Gibbs–Plateau borders. *Coll. Polym. Sci.*, **257**, 1248–50.

Lalchev, Z., Dimitrova, L. and Tzvetkova, P. (1982) Foam separation of DNA and proteins from solutions. *Biotechnol. Bioeng.*, **24**, 2253–62.

Lawhon, J.T. and Cater, C.M. (1971) Effect of processing method and pH of precipitation on the yields and functional properties of protein isolates from glandless cottonseed. *J. Food Sci.*, **36**, 372–7.

Lawhon, J.T., Cater, C.M. and Mattil, K.F. (1972) A comparative study of the whipping potential of an extract from several oilseed flours. *Cereal Science Today*, **17** (9), 240–6; 294.

Lee, J.C. and Tynan, K.J. (1988) Antifoams and their effects on coalescence between protein-stabilised bubbles. *Proc. 2nd Int. Conf. on Bioreactor Fluid Dynamics*, Elsevier Applied Science, Cambridge, pp. 353–77.

Lee, H.O., Jiang, T.-S. and Avramidis, K.S. (1991) Measurements of interfacial shear viscoelasticity with an oscillatory torsional viscometer. *J. Colloid Interface Sci.*, **146** (1), 90–122.

Le Meste, M., Colas, B., Simatos, D., Closs, B., Courthaudon, J.L. and Lorient, D. (1990) Contribution of protein flexibility to the foaming properties of casein. *J. Food Sci.*, **55** (5), 1445–7.

Loisel, W., Gueguen, J. and Popineau, Y. (1992) A new apparatus for analysing foaming properties of proteins. Poster from: *4th Symposium on Food Proteins 'Structure and functionality relationships', Reinhardsbrunn, October 1992.*

Lorient, D., Closs, B. and Courthaudon, J.-L. (1989) Surface properties of the bovine casein components: relationships between structure and foaming properties. *J. Dairy Res.*, **56**, 495–502.

McCarthy, M.J. (1990) Interpretation of the magnetic resonance imaging signal from a foam. *AIChE J.*, **36** (2), 287–90.

McKellar, D.M.B. and Stadelman, W.J. (1955) A method for measuring volume and drainage of egg white foams. *Poultry Science*, **34**, 455–8.

MacRitchie, F. (1978) Proteins as interfaces. *Advances in Protein Chemistry*, **32**, 283–326.

MacRitchie, F. and Alexander, A.E. (1963) Kinetics of adsorption of proteins at interfaces. Part II. The role of pressure barriers in adsorption. *J. Colloid Sci.*, **18**, 458–63.

Maeda, K., Yokoi, S., Kamada, K. and Kamimura, M. (1991) Foam stability and physicochemical properties of beer. *Am. Soc. Brew. Chem. J.*, **49** (1), 14–18.

Mangan, J.L. (1958) Bloat in cattle. *N.Z. J. Agric. Res.*, **1**, 140–7.

Miles, G.D., Shedlovsky, L. and Ross, J. (1945) Foam drainage. *J. Physical Chem.*, **49**, 93–107.

Mitchell, J.R. (1986) Foaming and emulsifying properties of proteins, in *Developments in Food Proteins – 4* (ed. Hudson, B.J.F.), Elsevier, London, pp. 291–338.

Monsalve, A. and Schechter, R.S. (1984) The stability of foams: dependence of observation on the bubble size distribution. *J. Colloid Interface Sci.*, **97**, 327–35.

Morr, C.V. (1985) Composition, physicochemical and functional properties of reference whey protein concentrates. *J. Food Sci.*, **50**, 1406–11, 1421.

Nakamura, R. and Sato, Y. (1964) Studies on the foaming properties of the chicken egg white. Part x: on the role of ovomucin (B) on the egg white foaminess. *Agr. Biol. Chem.*, **28** (8), 530–4.

Neurath, H. and Bull, H.B. (1938) The surface activity of proteins. *Chemical Reviews*, **23**, 391–435.

Nishioka, G.M. and Ross, S. (1981) A new method and apparatus for measuring foam stability. *J. Coll. Interface Sci.*, **81** (1), 1–7.

Nishioka, G.M., Ross, S. and Whitworth, M. (1983) The stability of foam: comparison of experimental data and computed results. *J. Coll. Interface Sci.*, **95**, (1), 435–42.

Paulsson, M. and Dejmek, P. (1992) Surface film pressure of β-lactoglobulin, α-lactalbumin and bovine serum albumin at the air/water interface studied by Wilhelmy plate and drop volume. *J. Colloid Interface Sci.*, **150** (2), 394–403.

Pearce, R.J., Marshall, S.C. and Dunkerley, J.A. (1991) Reduction of lipids in whey protein concentrates by microfiltration – effect on functional properties. *IDF special issue* **9201** – *New applications of membrane processing*, pp. 118–29.

Phillips, L.G., Hague, Z. and Kinsella, J.E. (1987) A method for the measurement of foam formation and stability. *J. Food Sci.*, **52** (4), 1074–77.

Phillips, L.G., German, J.B., O'Neill, T.E., Foegeding, E.A., Harwalkar, V.R., Kilara, A., Lewis, B.A., Mangino, M.E., Morr, C.V., Regenstein J.M., Smith, D.M. and Kinsella, J.E. (1990) Standardised procedure for measuring foaming properties of three proteins, a collaborative study. *J. Food Sci.*, **55** (5), 1441–4; 1453.

Phillips, M.C. (1981) Protein conformation at liquid interfaces and its role in stabilising emulsions and foams. *Food Technol.*, **35** (1), 50–7.

Phillips, R.D. and Beuchat, L.R. (1981) Enzyme modification of proteins. *ACS Symposium Series*, **147**, 275–98.

Poole, S., West, S.I. and Walters, C.L. (1984) Protein–protein interactions: their importance in the foaming of heterogeneous protein systems. *J. Sci. Food Agric.*, **35**, 701–11.

Poole, S., West, S.I. and Fry, J.C. (1986) Lipid-tolerant protein foaming systems. *Food Hydrocolloids*, **1** (1), 45–55.

Prager, S. (1963) Diffusion and viscous flow in concentrated suspensions. *Physica*, **29**, 129–39.

Richert, S.H., Morr, C.V. and Cooney, C.M. (1974) Effect of heat and other factors upon foaming properties of whey protein concentrates. *J. Food Sci.*, **39**, 42–8.

Roberts, R.T. (1975) Glycoproteins and beer foam. *Proc. EBC Congress, 15th Congress (Nice)*, **15**, pp. 453–64.

Rudin, A.D. (1957) Measurement of the foam stability of beers. *J. Inst. Brew.*, **63**, 506–9.

Saraga, L.T.-M. (1981) Protein denaturation on adsorption and water activity at interfaces: an analysis and suggestion. *J. Colloid Interface Sci.*, **80**, 393–401.

Slack, P.T. and Bamforth, C.W. (1983) The fractionation of polypeptides from barley and beer by hydrophobic interaction chromatography: the influence of their hydrophobicity on foam stability. *J. Inst. Brew.*, **89**, 397–401.

Smith, L.J. and Clark, D.C. (1992) Measurement of the secondary structure of adsorbed protein by circular dichroism. 1. Measurements of the helix content of adsorbed melittin. *Biochim. Biophys. Acta*, **1121**, 111–18.

St. John Coglan, D., Woodrow, J., Bamforth, C.W. and Hinchcliffe, E. (1992) Polypeptides with enhanced foam potential. *J. Inst. Brew.*, **98** (3), 207–13.

Torikata, Y., Kato, M., Kumagai, H. and Yano, T. (1991) Estimation of foam stability by parameters in the capillary model. *Agric. Biol. Chem.*, **55** (5), 1307–12.

Townsend, A. and Nakai, S. (1983) Relationships between hydrophobicity and foaming characteristics of food proteins. *J. Food Sci.*, **48**, 588–94.

Wang, J.C. and Kinsella, J.E. (1976) Functional properties of alfalfa leaf protein: foaming. *J. Food Sci.*, **41**, 498.

Waniska, R.D. and Kinsella, J.E. (1979) Foaming properties of proteins: evaluation of a column aeration apparatus using ovalbumin. *J. Food Sci.*, **44** (5), 1398–402; 1411.

Waniska, R.D. and Kinsella, J.E. (1988) Surface viscosity of glycosylated derivatives of β-lactoglobulin at the air–water interface. *Food Hydrocolloids*, **2** (1), 59–67.

Ward, A.F.H. and Tordai, L. (1946) Time dependence of boundary tensions of solutions. I. The role of diffusion in time-effects. *J. Chem. Phys.*, **14**, 453–61.

Watanabe, M., Toyokawa, H., Shimada, A. and Arai, S. (1981) Proteinaceous surfactants produced from gelatin by enzymatic modification: evaluation of their functionality. *J. Food Sci.*, **46**, 1467.

Watanabe, M., Fujii, N. and Arai, S. (1982) Characterization of foam and emulsion-stabilizing functions of enzymically modified proteins with surfactancy. *Agric. Biol. Chem.*, **46** (6), 1587–92.

Wenn, R.V. (1972) Foam loving complex nitrogen materials in beer. *J. Inst. Brew.*, **78**, 404–6.

Weser, C. (1980) Measurement of interfacial tension and surface tension – General review for practical man. *GIT Fachzeitschrift für das Laboratorium*, **24**, 642–8; 734–42.

Wilde, P.J., Clark, D.C. and Marion, D. (1993) Influence of competitive adsorption of a lysopalmitoylphosphatidyl choline on the functional properties of puroindoline, a lipid-binding protein from wheat flour. *J. Agric. Food Chem.*, **41** (10), 1570–6.

Wright, D.J. and Hemmant, J. (1987) Foaming properties of protein solutions: comparison of large scale whipping and conductimetric methods. *J. Sci. Food Agric.*, **41**, 361–71.

Yokoi, S., Maeda, K., Xiao, R., Kamada, K. and Kamimura, M. (1989) Characterisation of beer proteins responsible for foam of beer. *Proc. EBC Congress, 21st Symposium (Zurich)*, pp. 593–600.

Yu, M.-A. and Damodaran, S. (1991a) Kinetics of protein foam destabilisation: evaluation of a method using bovine serum albumin. *J. Agric. Food Chem.*, **39** (9), 1555–62.

Yu, M.-A. and Damodaran, S. (1991b) Kinetics of destabilisation of soy protein foams. *J. Agric. Food Chem.*, **39** (9), 1563–7.

6 Emulsions

S.E. HILL

6.1 Introduction

Emulsification is a complex process and many different approaches have been taken to gathering more information about this important phenomenon. Major advances have been made in the understanding of the molecular forces that stabilize or destabilize the emulsion structure. However, much of the methodology used to elucidate the principles of emulsion formation and stability is not applicable to real food systems, or even to models that could relate to food applications. Of more relevance are the tests described as '*ad hoc*' by Parker (1987). It is these tests which have been extensively used for comparative purposes and not withstanding all their faults and lack of standardization, they are probably the best indication of the potential of proteins as emulsifiers.

This chapter is designed for those who must take the pragmatic approach. It will not deal with methods to measure the interfacial qualities of purified proteins, nor will much discussion be given to methods utilizing specialist equipment, although on occasions references pointing the reader to texts that could be appropriate will be given. It is the intention to introduce the '*ad hoc*' test procedures often quoted. Much of the terminology associated with emulsions is also given. However, it is the author's belief that Lewis Carroll's Humpty Dumpty was a 'functional property' scientist, as does he not say "When I use a word, it means just what I choose it to mean – neither more nor less"? It is essential when using the functional tests that it is realized that the numerical values also have a 'looking glass' quality. Values quoted by different authors can rarely be compared, as slight changes in the method used to generate the results can have marked differences in the values obtained. This being the case, lists of values indicating emulsion performance for different materials will not be given except to illustrate specific points or if the values have been obtained using the detailed methods described in the text. Although much of the methodology used to assess proteins as emulsifiers is not fundamental, understanding of the reasons for emulsion formation and destabilization will help in the application of the most appropriate test regime and the interpretation of the results, and therefore some explanation about forces important to emulsions is given.

6.2 Emulsion definition

One of the primary functional requirements of several food systems is the ability to form emulsions. An emulsion is defined as a dispersion or suspension of two immiscible liquids but the structure of most food emulsions is much more complex (Dickinson and Stainsby, 1982). Food emulsions can be products such as margarines and butter, which are water-in-oil (W/O) emulsions, or milk where the water is the continuous phase, i.e. an oil-in-water (O/W) emulsion. In many food systems solids are an important part of the matrix (for example in meat sausages) and often a foam as well as an emulsion is present (whipping dessert toppings). In some meat emulsions the fat droplets can be bigger than 100 μm while in milk dispersions some of the droplets will only be 1 μm.

An emulsion is intrinsically unstable and with time the droplets of the dispersed phase will be attracted to each other and the emulsion will undergo processes which will result in the separation of oil and water phases. The physical stability for food W/O emulsions is primarily governed by the size and position of the solid particles; for butter these are the fat crystals (Walstra, 1987a). In O/W emulsions destabilization results in flocculation, coalescence, creaming and oiling-off, which may occur singly or in combination.

Flocculation is said to have occurred when the emulsified droplets become associated in flocs without destruction of the individual droplets. When the walls of the droplets are destroyed and larger drops are thus formed, the process is known as coalescence. When the dispersed phase is at a lower density than the continuous phase, creaming takes place, but this is not coalescence (Figure 6.1).

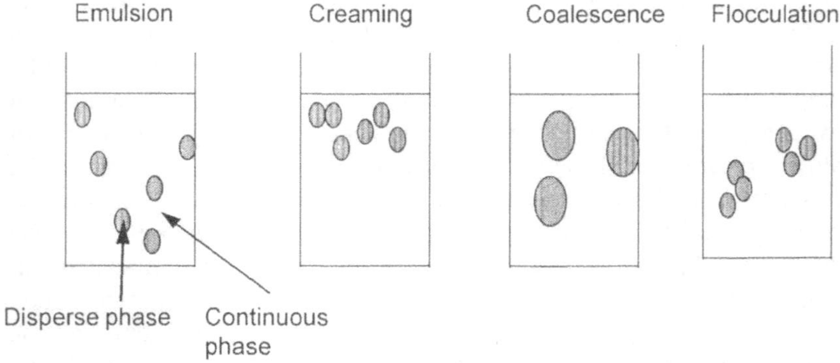

Figure 6.1 Schematic diagram of the states of an emulsion during destabilization.

6.3 Forces involved with emulsions

The stability of an emulsion depends on the balance of the forces associated with the interface of the oil and water.

6.3.1 Attractive forces

As the droplets of the disperse phase approach each other they are attracted by van der Waals forces. The range of these forces is in the order of 20 nm and at greater distances they would be countered by the thermal energy of the droplets. When the distance between the droplets (h) is small compared to the radius of the droplets (a) the attractive potential (V_a) can be represented by:

$$V_a = A^* \, a/12h$$

A^* is not a constant and depends on the distance between the particles and the temperature of the matrix.

As the droplets approach, the liquid film between them drains away and the rate of approach becomes dependent on the viscosity of the liquid draining, i.e. the continuous phase. As the droplets come together, extremely short-range repulsive Born forces will become relevant. Emulsion droplets are normally deformable and as they approach, the sides closest to each other flatten and can even 'dimple' (Parker, 1987). If the attractive forces at the region of closest approach are not balanced by repulsive forces the droplets fuse and coalescence occurs. The critical thickness of the thin liquid lamella in a model single thin film was thought by Manev *et al.* (1974) to range from 30 to 100 mm but Parker (1987) reported that values of lower than 30 nm were observed before rupture.

6.3.2 Repulsive forces

If an emulsion is to be stable then the attractive forces need to be balanced by repulsive forces. In the DVLO theory of stabilization the attractive van der Waals forces are balanced by the electrostatic repulsive forces (Derjaguin and Landau, 1941; Verwey and Overbeek, 1948). Electric charges on the droplets in emulsions can arise by ionization, adsorption or frictional contact. In the absence of emulsifiers only frictional contact appears capable of charging droplets. For an ionic emulsifier (such as a protein) adsorbed at the liquid/liquid interface, counter ions must be present in the continuous phase to stabilize the system. These counter ions will surround the charged droplets and create electrical double layers, which will stop the droplets from coming sufficiently close to one another to coalesce (Figure 6.2).

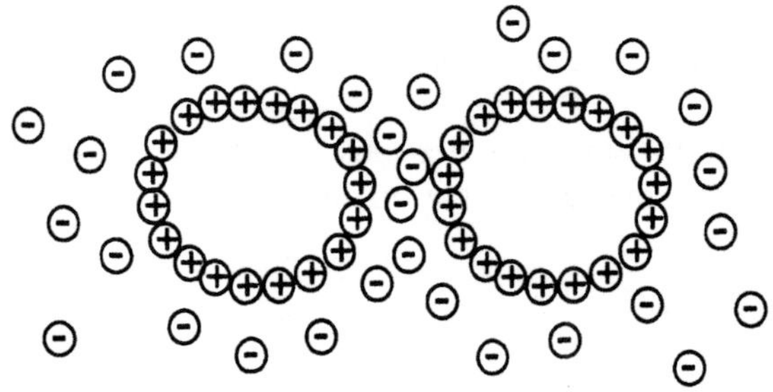

Figure 6.2 Schematic diagram of electrostatic stabilization.

The thickness of the electrical double layer (known as the Debye–Hückel length) is the important factor in the quantity of repulsive force. This force falls off approximately exponentially with distance. The thickness of the double layer is dependent on the square root of the electrolyte concentration in the continuous phase. Typically the Debye–Hückel length would fall from 100 nm at 10^{-5} M to 1 nm at 10^{-1} M for a univalent electrolyte (Parker, 1987). In many food products low ionic strengths cannot be utilized as higher salt concentrations may be necessary for flavour enhancement, pH control or microbiological stability. The high forces are not only favoured by low ionic strength of the continuous phase, but also by a pH well away from the pI of the protein being used as the stabilizer.

It should be noted that some disperse phases possess internal double charged layers. This effect is small in O/W emulsions as the electrical permittivity of the oil phase is low. However, the effect is relevant to W/O emulsions, and hence in these emulsions electrostatic repulsion is not effective as a stabilizing factor.

6.3.3 Steric forces

Although not fully understood, it is recognized that protein will align at an oil/water interface. The protein regions which are predominately hydrophobic will adopt a position in the oil phase, while those with mostly hydrophilic regions will tend to protrude into the aqueous phase. As shown diagrammatically in Figure 6.3 this leads to a certain thickness at the interface and an area of protein protruding from the interface. As droplets approach, the protein 'loops' and 'tails' at the interface will start to interact and steric effects come into play (see Figure 6.4).

The forces which disfavour the approach of the droplets because of the overlap of the proteins have been described by Napper (1983) as 'volume restriction' and 'osmotic effects'. The 'volume restriction' occurs as the portion of protein in the continuous phase loses some of its possible

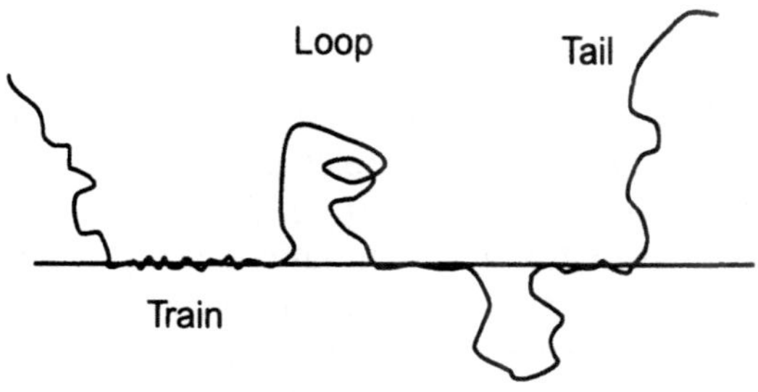

Figure 6.3 Representation of a protein molecule aligning at an oil/water interface.

Figure 6.4 Representation of the steric effects due to protein adsorption at an oil droplet interface.

conformation, while the reduction of the concentration of the continuous phase due to the presence of the protein gives rise to the 'osmotic effects'. These events are normally thermodynamically disfavoured and hence give rise to repulsive forces. These steric repulsive forces will depend on the length of the protein segment in the continuous phase and they need to be of greater length than the range of the attractive van der Waals forces. Unlike protein stabilization due to electrostatic effects, the steric effects are not directly influenced by salt concentration.

On occasions, rather than adsorbed protein leading to repulsive forces, the net effect can be attractive. This is most likely to occur if the protein molecule is adsorbed by more than one droplet, when 'bridging' can occur, or if the continuous phase is a poor solvent for the adsorbed molecules.

6.3.4 *Depletion forces*

Long-range forces, which can be attractive or repulsive, can occur due to depletion factors. Contributions to the droplet interactions can be influenced by polymers which are not at the interface. This is often the case in food emulsions where high polymer concentrations are present, for example due to the presence of starches. Polymers in solution can restrict the movement of particles towards each other due to geometric factors; this gives rise to repulsive forces. If these forces are overcome and polymer is removed from between the droplets then, in a good solvent, there is a net increase in the free energy of the system. This increase continues until there is essentially pure solvent between the droplets. As the droplets approach the pure solvent is displaced to once again freely mix with the polymer and therefore the forces between the droplets are then attractive.

6.4 Emulsion stability

In most cases, microemulsions being the exception, the free energy of an emulsion is higher than the energy of the separated liquid phases because of the energy associated with the interface between the liquids. Hence, an energy input is required to form emulsions and when they are formed they are unstable.

The structure of an emulsion can change in several ways, all of which have been used as quantifiers of the suitability of a protein as an emulsifying or stabilizing agent. Creaming, flocculation, coalescence and phase inversion are all processes undergone during destabilization.

6.4.1 *Creaming and drainage*

Creaming denotes a movement of the disperse phase due to the density differences of the two phases, the oil phase normally being less dense than the aqueous phase. At low disperse phase volume (ϕ) the process is called creaming, but at higher ϕ the process is termed drainage with the continuous phase flowing past the droplets.

An indication of the velocity of creaming can be obtained from Stokes' law:

$$\text{velocity} = \frac{g \Delta \varrho d^2}{18 \eta_o}$$

where g is the gravitational force, $\Delta\varrho$ is the phase density difference, d is the diameter of the droplet and η_o is the continuous phase viscosity. If creaming can be predicted from Stokes' law and the electrical charge on the droplets does not have a major influence on the rate, then creaming can be reduced by the following actions:

1. matching the densities of the continuous and disperse phase;
2. increasing the viscosity of the dispersion medium;
3. reducing the droplet size.

From the above equation it can be calculated that droplets of 1 μm in an oil-in-water emulsion would rise to the surface at a rate of 1 mm/h. Droplets smaller than 1 μm would not cream as the Brownian movement would be greater than the movement due to the gravitational separations. High-pressure homogenization can be used to reduce the average droplet size and, although a portion of the droplets may still show creaming behaviour, the overall stability will be increased. For commercial cream liqueurs the shelf-life is required to be more than a year and to achieve this over 95 wt% of the cream droplets must be smaller than 0.8 μm. Even so the product is packaged in coloured glass bottles so that the creaming which still does take place is not obvious to the consumer. The creaming is reversible and uniform redistribution of the creamed droplets can be achieved by gentle agitation, such as occurs during pouring.

Matching the densities of the two phases is often very difficult in food emulsions. The density difference between water and oil is in the region of 50 kg m^{-3}. The density of the continuous phase is often moved even further away from that of the disperse oil phase due to the incorporation of sugars or proteins into the aqueous phase. However, the density of the droplets will be affected by the amount of material adsorbed at their interface. Adsorbed polymer typically has a density higher than that of the phases, therefore its adsorbance will have a marked affect on the density of the smaller droplets, but large droplets will retain a density close to that of the disperse phase. This is an additional reason for small droplet size giving rise to emulsions that are stable against creaming. Increasing the viscosity of the dispersion phase also leads to a decrease in the creaming rate. The addition of gums and starches to food emulsions can often achieve this.

There are difficulties to the application of Stokes' law to many real food emulsions as it is only strictly applicable to rigid unaggregated spheres at infinite dilution. For further details on the theoretical basis of creaming the reader is referred to Dickinson (1992).

6.4.2 *Flocculation*

Flocculation is normally a reversible loose aggregation of droplets held together by weak forces. The actual size of the force depends on the structure of the adsorbed layer and the thermodynamic quality of the solvent between the droplets. The droplets in a stable emulsion are

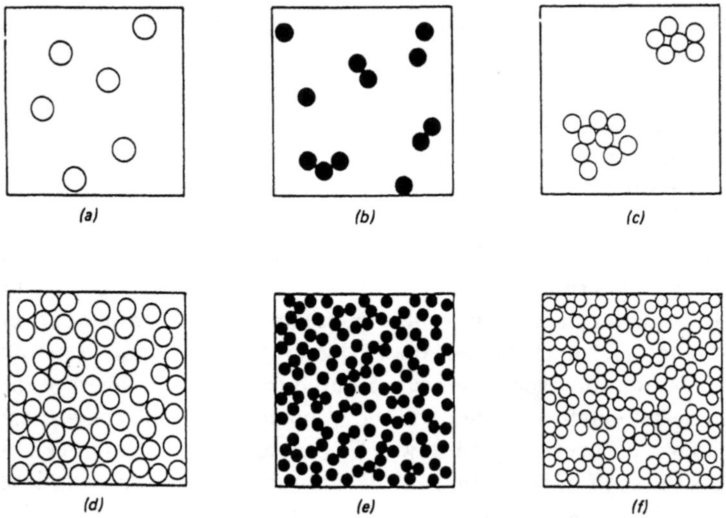

Figure 6.5 Schematic diagram of flocculation at low (a, b and c) and high (d, e and f) oil volume fractions. a and d are stable, b and e are weakly flocculated, and c and f are strongly flocculated.

prevented from flocculating by electrostatic and/or steric stabilization. Summaries of the theoretical equation governing flocculation have been given by Tadros and Vincent (1983) and are explained well by Dickinson (1992).

A floc will depend on the volume fraction and the strength of the interactions between the droplets. Figure 6.5 shows a schematic representation of strongly and weakly flocculated emulsions at different phase volumes. If large aggregates are formed in dilute emulsions the system is very unstable and creaming is likely to occur. At high phase volume weak aggregates are likely to give thixotropic rheological properties as the aggregates break down with shear. Strongly flocculated emulsions will have definite elastic properties and will be stable in respect to creaming, but the gel-like structure may have high syneresis.

If a protein is to be used in an emulsified product it must be added in sufficient quantity to give complete coverage of the interface. If the protein levels are low it is possible that the droplets will share protein molecules and this will lead to flocculation. Displacement of adsorbed protein from the interface by a low molecular weight amphiphile may lead to rapid flocculation as the steric stabilization layer is lost. Systems with a specific binding affinity for materials may also floc rapidly in the presence of that material; for example, on addition of calcium to a casein-stabilized emulsion (Dickinson *et al.*, 1987a).

Depletion forces will also lead to flocculation. It has also been shown that mixtures of biopolymers tend to phase separate and this will alter the relative conditions in each phase. More details of depletion flocculation and the effects on emulsions are given by Williams (1995).

6.4.3 Coalescence

If the attractive forces are great enough to overcome the energy barrier to bring two droplets into close proximity with one another and the stabilizers fail the droplets will coalesce. Much has still to be elucidated about the mechanism and reasons for the thinning and rupture of the stabilizing film. During the coalescence procedure the aqueous layer between the droplets has to drain away. As it does so surface active material is swept from the contact region and the resulting interfacial tension gradient causes the Gibbs–Marangoni effect, which opposes the film-thinning process. The geometry of the droplets changes as they approach each other and therefore there must be rearrangement of the materials at the surface. Protein, adsorbed at the interface, may not be able to migrate quickly to a thinning part of the film, yet most proteins will provide a viscoelastic film. It has been reported (Dickinson and Stainsby, 1988) that a positive correlation exists between the viscoelasticity of the adsorbed film and the stability against coalescence. It is thought that spontaneous film breakage is very low and coalescence usually occurs only when droplets have been together for an extended period of time, as they are in a creamed layer or a floc.

The likelihood of coalescence can be enhanced by the presence in the disperse phase of solid particles such as fat crystals, ice crystals, water droplets or protein aggregates. Storage temperature may induce solids (e.g. solidifying of vegetable oil) and hence accelerate destabilization. The solids, especially if needle-like crystals are present, pierce the thin film around the droplets to form a liquid bridge and hence induce coalescence. The process is increased by stirring the emulsion. Even in samples not containing solids, stirring is an important parameter in the stability of emulsions. The work of Lips *et al.* (1993) indicates that if a concentrated emulsion is stirred for long periods (hours) little change in droplet size occurs initially, but after a certain period there is a very rapid coalescence. This 'critical' kinetic behaviour should be appreciated in the production and use of food emulsions.

In some emulsions stabilization is facilitated by the presence of solids. An example is mayonnaise, which contains egg yolk particles. To give this 'Pickering stabilization' for an O/W emulsion it is essential that the particulate material accumulates at the interface. For an O/W emulsion the particles also need to be predominantly hydrophillic.

6.4.4 Phase inversion

Phase inversion occurs when an O/W emulsion becomes a W/O emulsion, or W/O becomes O/W. Churning cream into butter is the changing of the O/W cream emulsion into an emulsion where the fat becomes the continuous phase (W/O), although the details of this change are exceedingly complicated. Large amounts of energy are usually required for the transition and it is possible that multiple emulsions are formed during the process. Phase inversion is complex and probably involves creaming, flocculation and coalescence. The volume fractions in the two phases are a very important factor in the change from one phase to another. As the amount of disperse phase (ϕ) increases, the viscosity of the system tends to increase, but above a certain level (ϕ_i) the viscosity suddenly decreases and indicates that inversion has taken place. The high quantity of disperse phase becomes the continuous phase and the disperse phase will equal ($1 - \phi_i$). The marked change from one type of emulsion to another indicates that for a large range of volume fractions stable states, i.e. W/O or O/W, can exist, but something intermediate is not favoured. The samples show a hysteresis, the (ϕ_i) to change O/W to W/O will not be the same as changing the same system from W/O back to O/W. The history of the samples is also important and minor differences in the sample or its preparation can have a marked effect on the phase inversion. Table 6.1 shows the physical factors that may affect the quantity and stability of an emulsion.

Table 6.1 Twelve key factors affecting the stability and rheology of emulsions. 0 = not important, 1 = sometimes important, 2 = often important, 3 = generally important. Taken from Dickinson (1992)

Factor	Creaming	Flocculation	Coalescence	Rheology
Droplet size	3	2	1	1
Droplet size distribution	3	2	0	2
Droplet volume fraction	3	3	3	3
Density difference between phases	3	0	0	0
Rheology of continuous phase	3	3	2	3
Rheology of dispersed phase	0	0	0	1
Rheology of adsorbed layer	0	0	3	2
Thickness of adsorbed layer	1	2	3	2
Electrostatic interactions	1	3	2	1
Steric (polymer interactions)	0	3	2	2
Fat crystallization	0	0	3	3
Liquid crystalline phases	1	1	2	2

6.5 Stabilizers and emulsifiers

The terms 'stabilizer' and 'emulsifier' are often used interchangeably, but by some definitions their role is different. Dickinson (1988) defined an 'emulsifier' as a single or mixture of chemical components having the capacity for promoting emulsion formation and stabilization by interfacial action, while a 'stabilizer' is a single or mixture of chemical components able to confer long-term stability on an emulsion. Stabilizers may be adsorbed at the interface, but this is not necessarily the case. The agents used to produce food emulsions can be low molecular weight emulsifiers or macromolecular materials. Typical stabilizers used in food emulsions include monoglycerides, phospholipids (e.g. lecithin), fatty acids and their esters, as well as the topic of this chapter, proteins.

To produce an emulsion, an emulsifier initially has to facilitate the production of a new interface by lowering the internal free energy, and then has to provide an adsorbed layer at the oil/water interface. It is important that the emulsifier is adsorbed rapidly at the oil/water interface as it is being formed. The interfacial properties of the surfactants are therefore crucial in the production and stability of emulsions, as is the case for foam production (see Chapter 5). The role of the stabilizer is to retard the collision frequency of the droplets. Proteins can act as emulsifiers and as stabilizers, but polysaccharides are often also used to provide improved stabilization.

6.5.1 Interfacial properties

The unfolding and structure of the protein at the interface must be important in the stability of the film formed. Phillips (1981) summarized the possible behaviour of proteins at interfaces. The thickness, cohesiveness and charge are just some of the properties of the layer formed by the adsorbed protein that are important in emulsification.

Emulsifiers have to be amphiphiles, with polar and non-polar portions in each molecule. On migration to the interface they align so that free energy is at a minimum and the interfacial tension between the oil and the water is reduced. This is why food emulsifiers are often classed according to their hydrophile–lipophile balance (HLB) (see Dickinson, 1992, p. 47 for details). Proteins contain both hydrophobic and hydrophilic side chains and regions. However, the folding of the proteins often results in the hydrophobic regions of the molecule being buried inside it. The average hydrophobicity of proteins is calculated by the method of Bigelow (1967) and shows some correlation with foaming potential, but is not a very good indicator of emulsification properties. However, estimation of the hydrophobicity at the surface of the protein molecules is a much better indicator of emulsification properties. Correlation for surface hydrophobicity and emulsifying activity or interfacial tension is found for a wide range of proteins, both in the native and denatured state (Kato and Nakai, 1980). It is reasoned that these differences are because foam formation takes more time than does forming the interface for an emulsion, and therefore the protein has more time to unfold at the foam interface and hence expose the internal hydrophobic regions (Mitchell, 1982).

6.5.2 *Types of protein*

A whole range of proteins is used in food emulsions. Table 6.2 shows some proteins that have been assessed for emulsifying properties. The major components of caseinate, α_{s1} and β-casein, are still the most extensively used proteins for the stabilization of food emulsions. These proteins show the features thought to be of major importance for emulsifying properties: adequate solubility of the protein, flexibility of the backbone of the molecule and sufficient hydrophobicity. The ability of the protein to lose tertiary structure is normally considered a positive attribute for an emulsifier. It is therefore not surprising that heat denaturation often improves emulsifier performance, as long as solubility is not lost. Breakage of disulphide bonds also tends to enhance the functionality.

A surface of more than 1 mg m^{-2} corresponds to a monolayer coverage of unfolded peptide chains and can only be formed if the protein concentration is small compared to the interfacial surface area and if time is available for the proteins to unfold when adsorbed. Globular proteins generally cover the surface at three times this amount and if aggregates are formed then levels can be greater than 5 mg m^{-2}. The protein lysozyme is rigid and remains in a globular form at the interface while the casein is flexible and forms random coils. Hence the films of lysozyme are much more condensed than those of casein. To stabilize the films more lysozyme will be required.

The local polymer concentration at an interface can be sufficiently high for the protein to adopt gel-like tendencies, with yield values in the order of 10^{-4}–10^{-3} N m^{-1}. The layer will become substantially stronger if additional protein accumulates beyond the interfacial layer. For β-casein, when the bulk protein concentration is sufficiently high, thick films are formed. At the interface an irreversible adsorbed layer of about 3 mg m^{-2} is formed, with secondary layers of reversible adsorbed layers. Gelatin in bulk solution will form gels which with time become stronger. This behaviour is also shown in emulsions where gelatin is adsorbed at the interface and in the additional reversible adsorbed protein layers.

6.5.3 *Protein modification*

If it is accepted that surface activity is required to produce useful emulsifiers a strategy can be adopted to modify proteins so that their

Table 6.2 Emulsion data on a range of proteins using the techniques described in detail in this chapter. Taken from Hill (1986)

Protein powders	Type	Emulsion capacity g oil/g powder	Oil phase volume (%)	Emulsion stability (24 hours)	Emulsion stability (14 days)
Soya products	Soya isolate	277	82.9	94.0	88.6
	Soya flour	184	79.8	100	100
Egg powders	Albumen powder	226	82.9	11.8	3.3
	Albumen flakes	240	83.7	3.5	1.1
	Liquid egg white	215	82.2	0.0	2.5
Dairy products	Casein	336	87.8	5.2	41.0
	Whey powder	190	80.3	100	100
	Lactose	160	77.4	100	100
Fish products	Hydrolysed *Haplochromis* fish	160	77.4	97.1	99.0
Animal products	Soluble collagen	167	78.1	92.2	83.1
	Beef marrow (dried)	281	85.7	70.9	100
	Meat protein (dried)	216	82.2	4.2	6.0
Blood powders	Red blood cells (dried)	271	85.3	10.9	12.8
	Blood cells (hydrolysed)	47	49.6	100	100
	Bovine albumin (dried)	270	85.2	0.4	0.9
	Plasma (dried)	275	85.5	0.5	13.4
	Bovine plasma (thawed)	296	86.4	0.7	11.1
	Bovine sera (fresh)	316	87.1	1.1	13.2
	Bovine sera (dried)	287	83.7	1.1	1.3
	Porcine plasma (fresh)	247	84.1	10.4	14.5
	Porcine plasma (dried)	263	84.1	2.0	3.8

molecular configurations have suitable amphiphilic structures. Table 6.3 indicates that increasing the level of succinylation on a protein increases its emulsifying activity. This can be done in a number of ways and some interesting examples are given by Arai and Watanabe (1988). However, it must be recognized that many food producers would rather use emulsifiers that can be considered natural (Dickinson, 1993).

6.5.4 *Mixed biopolymers*

Commercial food emulsifiers are generally a mixture of components. The use of mixed polymers can affect the system in ways that are difficult to predict as they will depend on the polymer interactions at the surface of the disperse phase, the interactions with the aqueous phase, and polymer/ polymer interactions.

When two types of protein are present during emulsification, the more surface-active component predominates at the interface. However, the longer one protein has been adsorbed at the interface the more resistant it is to displacement by a more surface-active protein (Dickinson *et al.*, 1987b). In a mixture of protein and polysaccharide there may be direct interaction between the two. This interaction could be due to covalent linkage of the protein to the polysaccharide as would seem to occur in the experiments of Kato and co-workers (Nakamura *et al.*, 1991; Kato *et al.*, 1993, Kato *et al.*, 1995) and Dickinson and Semenova (1992). In these studies incubation of dextrans with protein, produced compounds with enhanced surface active properties and hence good emulsion stabilization properties. Even if the polymers are not formed via covalent linkage, changes in the properties can be observed. Work by Dickinson and Euston

Table 6.3 Emulsifying activity index of a range of proteins at two pHs (concentration 0.5%, phosphate buffer ionic strength (I) = 0.1 molar). Taken from Pearce and Kinsella (1978)

Protein		Emulsifying activity index ($m^2\ g^{-1}$)	
		pH 6.5	pH 8.0
Bovine serum albumin			197
Sodium caseinate		149	166
Whey protein		102	101
Soy protein A		41	92
Soy protein B		26	66
Egg albumin			49
Yeast protein, succinylated	(0%)	40	
	(24%)	110	204
	(62%)	262	332
	(88%)	322	341

(1991) indicated that the interfacial properties of a protein could be enhanced by the interactions between the adsorbed protein and the polysaccharide, which may have led to the formation of a secondary macromolecular stabilizing layer.

As well as considerations due to the chemical interaction of the different polymers, thought must also be given to the general thermodynamics of the system. Mixed biopolymers are difficult to maintain in an homogeneous state and some sort of phase separation could occur (Dickinson and Stainsby, 1988). This may be very important when considering material adsorbed at an interface where the density of the material is high compared to the surrounding continuous phase (Dickinson, 1995). Surfactant may partially displace the surface protein and controlled destabilization of interfacial films by movement of other materials to the interface is essential for the production of some foods such as ice cream.

6.6 Forming emulsions

Oil and water are immiscible, therefore they require mechanical agitation for dispersion. The interfacial tension between the two phases is in the region of 14 to 20 mN m^{-1} and for formation of a stable emulsion the emulsifier should reduce the interfacial tension to below 10 mN m^{-1}. The more effective the emulsifier, the less the intensity of agitation required (Nakai and Li-Chan, 1993). The method of emulsion formation is of utmost importance as it will affect the droplet size of the emulsion (Table 6.4). Tornberg (1980) stated that the emulsifying conditions were the dominant factors in emulsion formation and that the protein was of only minor importance.

When preparing an emulsion large droplets are formed initially and are subsequently disrupted. The mechanical stress applied has to be sufficient

Table 6.4 Turbidity measurements of emulsions made using two different blenders, a range of oils and whey powders (concentration 2%, pH 8.0, $\phi = 0.25$ and diluted 1/2500). Taken from Pearce and Kinsella (1978)

Oil	Absorbance at 500 nm	
	Janke–Kunkel blender	Waring blender
Peanut	0.312	0.152
Safflower	0.301	0.146
Soybean	0.297	0.148
Corn	0.273	0.160
Paraffin	0.250	0.089

to disrupt the pressure inside the droplet (ΔP), which is given by the Laplace equation:

$$\Delta P = \frac{4\gamma}{d}$$

where d is the droplet diameter and γ interfacial tension. The smaller the droplet, the greater the mechanical energy that has to be supplied to disrupt it. The stress acting on the droplet can come from a shearing stress. In water, shearing stresses of greater than 10 Pa are difficult to produce, and correspond to a droplet size no less than about 1 mm (Walstra, 1987b). To make good emulsions, therefore, it is usually necessary to produce turbulent flows. Hand-shaking an emulsion gives a coarse emulsion with a wide range of droplet sizes, with the smallest droplets probably in excess of 20 μm. High-speed mixers, e.g. Ultra Turrax or Silverson mixers, fitted with suitable baffles can produce droplets in the range of 5–10 μm (Figure 6.6). Ultrasonic homogenizers, which induce cavitation, are often used in laboratory studies to prepare emulsions and can produce droplets smaller than 1 μm. This method of emulsion formation has the advantage that small samples can be produced with ease.

For the manufacture of food emulsions high-pressure valve homogenizers are widely used. These work by forcing a coarse mixture of oil and aqueous phase through a valve which, at high pressure, is lifted to produce a narrow gap (Figure 6.6). The pressure drop across the valve disrupts the droplets by a combination of turbulence and intense laminar shear flow, with

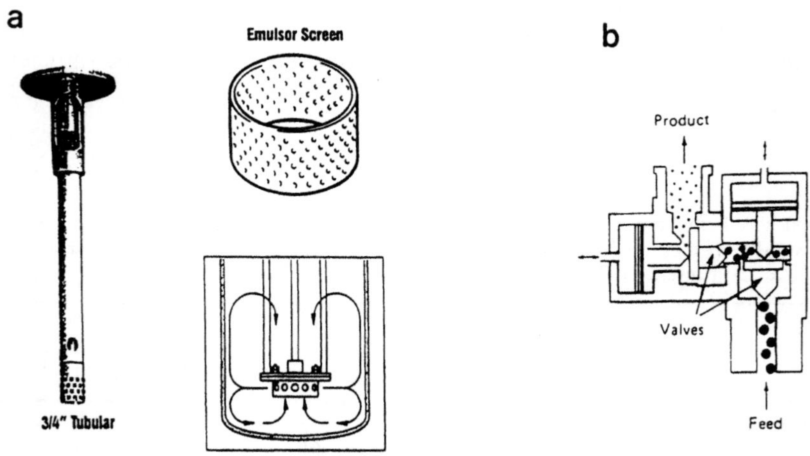

Figure 6.6 Diagrams of different methods for emulsion preparation: (a) equipment and method of high-shear mixing using Silverson equipment; (b) emulsion production by means of a two-valve homogenizer.

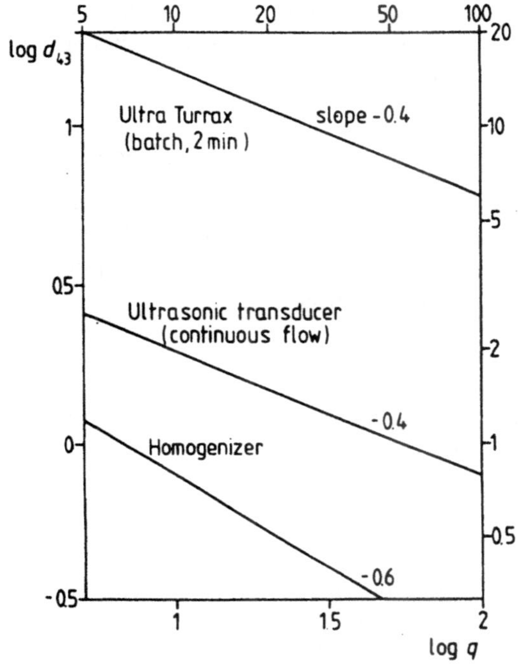

Figure 6.7 Average droplet size (d_{43}, μm) as a function of the net energy input in the emulsion (q, MJ m^{-3}) for emulsions in various machines. Taken from Walstra (1983)

turbulence on the low-pressure side of the valve probably the most important factor leading to fine droplet size. Commercial homogenizers usually have a second valve operating at a pressure of 10–20% of that of the first valve. Samples are often warmed by several degrees after a single pass through the homogenizer. The average droplet size produced as a function of the net energy input to the emulsion is shown in Figure 6.7.

A small-scale, high-pressure homogenizer developed by Dickinson *et al.* (1987b) produces emulsions of similar particle size to normal laboratory scale machines. Microfluidization has also been developed as a means of producing emulsions of small scale (US Patent 4 533 254).

A major problem in comparing any published emulsion results is that the method of their preparation varies. The initial droplet size will influence the way the emulsion behaves; for example, the dependence of creaming on droplet size. Surface area requiring coverage by the emulsifying agent is dependent on the droplet size. What may have been high protein concentration at the interface for large droplets may be an insufficient amount of protein if the droplets are much smaller. The emulsifiers themselves may also be affected by the high force fields and some denaturation and loss of molecular weight could occur. Incorporation of air into the emulsion during its preparation also changes the emulsifying properties (Pearce and Kinsella, 1978).

6.6.1 Procedure for the laboratory scale production of an emulsion

Although it would be beneficial if all laboratories had access to small-scale homogenizers, this is often not the case and high-shear mixing is employed to produce an emulsion. The following procedure can be used to produce emulsions for further investigation.

Apparatus

Silverson Mixer model L2R serial no. L2R 4297, with ¾ inch (19 mm) disintegrating or emulsifying head;
250 ml plastic beakers.

Method

1. Prepare a 2% (w/v) solution of protein solution (this level of protein addition is suitable for a powder of high (>80%) solubility and good emulsification properties). The protein solution can be prepared in buffer to control the pH and the ionic strength. Note the specific gravity of the solution.
2. To a 25 ml aliquot of the protein solution add 80 ml of oil (most types of vegetable oil can be used satisfactorily. Note the type, as the quality of emulsion may vary, especially if storage temperature is to be investigated). It is often easier to measure the weight of oil rather than the volume.
3. Mix the oil and protein solution together for 2 minutes using the Silverson at half maximum speed. The mixing probe of the Silverson must be moved about within the emulsion to ensure homogeneity.

Results. For a series of protein powders formed from dried blood plasma the resulting emulsion was thick and shiny with a viscosity of about 4500 Pa.s. No significant differences were observed between emulsions produced with the disintegrating and the emulsifying head, although it is recommended that only one type of head is used for any batch of experiments. The average size of the droplets was about 30 μm as assessed by Coulter counting (Table 6.5). This is greater than the value that could be expected from homogenization. The coefficient of variation for this average droplet size was about 8%. The method of particle estimation may have given falsely high values, as will be explained in section 6.7.1. Altering the mixing time from 2 to 4 minutes tended to reduce the size of the droplets by about 10%.

Table 6.5 Estimates of droplet size using blood plasma powders, a mixing time of 2 minutes, a protein concentration of 2%, a pH of 8.1 and an oil volume of 80% w/w. Details of estimates are given in the text, all samples were diluted with 0.1% sodium dodecyl sulphate

Method	Coulter counter	Malvern	Turbidity (*R*)
Droplet size (μ) mean ± S.D.	33.6 ± 3.7[a]	26.2 ± 2.6[a]	15.6 ± 2.2[b]

[a]Diameter estimates.
[b]Radius estimates.

6.6.1.1 Assessing emulsion qualities. It must be remembered that real food products may have to be stable over a wide range of conditions. An ice cream emulsion must be stable at very low temperatures while an emulsified meat sauce must give the expected quality at elevated temperatures. Salt levels and low pH also have to be considered when the probable stability of a stabilized emulsion is being reviewed for certain food products.

The details given in the previous sections should indicate that a number of features influence the stability of emulsions:

1. Interfacial tension between the two phases;
2. Characteristics of the adsorbed film in the interface;
3. Magnitude of the electrical charge on the globules;
4. Size and surface/volume ratio of the globules;
5. Weight/volume ratio of dispersed and dispersion phases;
6. Viscosity of the dispersion phase.

All these areas could be studied to indicate the quality of the emulsion. Factors which affect the behaviour of the protein emulsifier also need to be taken into consideration. Temperature, pH, ionic strength, concentration, prehistory of the sample and the presence of other materials are all important factors in emulsion behaviour. The quantities of protein or the phase volume used in many functional tests for emulsion quality can make an enormous difference to quoted values for assessed emulsion characteristics.

An example of the influence of ionic strength on emulsion capacity was shown by Krasnowska *et al.*, (1995) for offal proteins, where high salt concentrations were better at stabilizing emulsions than were lower salt levels. Table 6.3 indicates the changes that occur in the emulsifying activity index when a range of samples are examined at two different pHs. A commonly found feature is that proteins show their lowest ability to stabilize proteins near their isoelectric point, where it can also be expected that the solubility of the protein will be minimum and the electrostatic repulsive forces between the droplets and the adsorbed protein will be least

(Crenwelge *et al.*, 1974; Huffman *et al.*, 1975; McWatters and Holmes, 1979; Morr and Ha, 1993). Using the same emulsion formation regime as for Table 6.2, the amounts of protein added to the system were altered and the emulsion capacity (EC) and oil phase volume (OPV) assessed (details of the methods and terminology are given in sections 6.7.4 and 6.7.5). On 0.1 g of protein being added the EC (g oil/g powder) was assessed as 2800 while the OPV was 0.557, but on addition of 1 g of powder the EC and OPV were 250 and 0.895, respectively (Hill, 1986). This clearly demonstrates that the amount of protein added does markedly affect the quoted values for emulsifying properties. Altering the amount of aqueous phase in a test regime also results in very different values for EC and OPV (Figure 6.8). The oil used and the method of mixing are also reflected in the quoted values for emulsifying properties and this is reflected in Table 6.4.

6.7 Methods of measuring emulsification properties

A variety of tests has been applied to indicate the value of a protein in an emulsion. Most of these tests are not applicable to W/O emulsions or to the

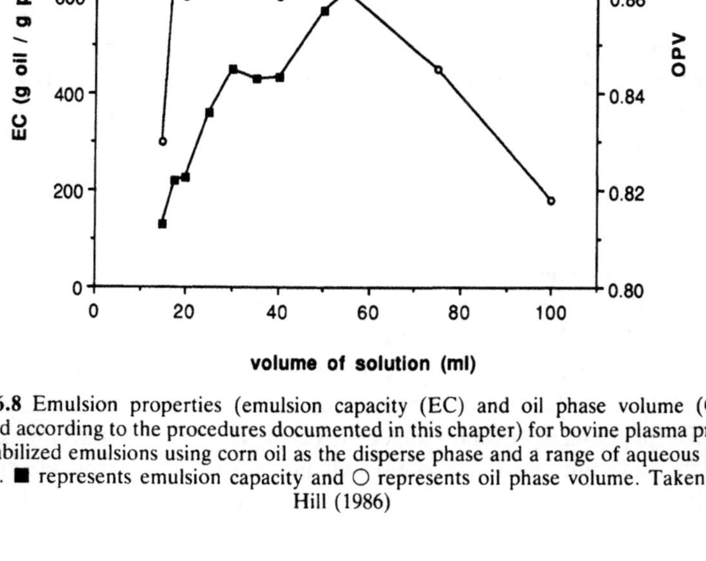

Figure 6.8 Emulsion properties (emulsion capacity (EC) and oil phase volume (OPV) estimated according to the procedures documented in this chapter) for bovine plasma protein (2%) stabilized emulsions using corn oil as the disperse phase and a range of aqueous phase volumes. ■ represents emulsion capacity and ○ represents oil phase volume. Taken from Hill (1986)

study of microemulsions which have properties very different from ordinary food emulsions.

As indicated in the previous section a range of experimentally controllable parameters will alter the measured emulsification properties of a protein being evaluated. These include:

type of equipment used to produce the emulsion;
energy input into the system;
amount of protein used;
phase volumes used;
ionic strength;
pH;
type of oil used.

With this range of variables some preliminary screening is normally required so that the correct method is chosen for the evaluation of the emulsion properties of a particular protein. The following sections give details of the functional evaluation tests most commonly applied. Some details about the tests are given in relation to what physical properties or forces the test is evaluating. Details of a method that works for the evaluation of protein powders of good functionality are then described. Some typical, or range of, values given for the tests in the literature will also be outlined to give some indication of values and/or the problems that occur when interpreting reported data.

It is now considered that estimation of droplet size is the most appropriate measure of emulsion quality and some methods, including calculation of emulsion activity index, are given to show how this can be carried out. The empirical tests of emulsion capacity, oil phase volume and emulsion stability are also given. However, before estimates on the emulsion can be attempted, the emulsion must be formed and, as discussed previously, this is not trivial (Table 6.4 and Figure 6.7).

6.7.1 Measurement of droplet size

The sizes of the droplets forming the non-continuous phase of most emulsions are highly polydispersed and range from 0.1 micron up to 1 mm. A direct approach for evaluation of the emulsifying properties of a system is to measure the individual droplets. Microscopy can be used to follow the variation in droplet size with time due to coalescence in a protein-stabilized emulsion.

A common method for protein size estimation by microscopy is to dilute the emulsion with a solution that will gel. A thin film is formed on a slide and covered. A haemocytometer cell can be used (Walstra, 1968; Howell, 1981) and the size of the globules can then be estimated. Walstra (1968) concluded that microscopic counting and measurement was difficult (mainly due to the poor detection of small globules), tedious and time consuming, and the results showed poor reproducibility. Fluorescence microscopy was considered slightly preferable and a Coulter counter, used for a limited size range, could determine globule size and size distributions with good replication and little bias.

However, proper characterization of the globule size polydispersity is a major problem. The techniques used tend only to estimate a certain range of droplet size. Optical microscopy or Coulter counting is useful for globules over 10 μm, but the small particles, <0.1 μm, must be estimated by a technique such as dynamic light scattering, which cannot measure the large globules.

The method most favoured by workers to estimate globule size distribution of oil in water emulsions is spectroturbidity. The emulsion can be regarded, in its light scattering behaviour, as a collection of spheres that fulfils the conditions of the Mie theory. In the method of Walstra (1968) the optical density of the emulsions is measured at several wavelengths and, using data on concentration and reflective indices, a specific turbidity spectrum is calculated and plotted. This is compared with theoretical spectra that are computed for assumed globule size distributions. An example of the particle size distributions for emulsions at pH 7 and 5.5 when freshly made and after standing for a period of time is shown in

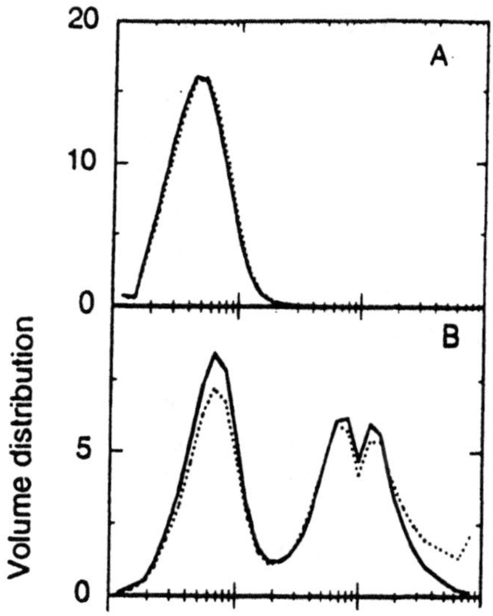

Figure 6.9 Particle size distribution (vol% versus size) as measured on a Malvern Mastersize of emulsions (20% soya oil with 1 wt% whey protein isolate). Solid lines are fresh emulsions and dotted lines are emulsions aged for (A) 3 weeks and (B) 2 days. Samples for (A) were prepared at pH 7.0, while samples for (B) were prepared at pH 5.5. Taken from Hunt and Dalgleish (1994)

Figure 6.9 (Hunt and Dalgleish, 1994). For further explanation on how to estimate the number and size of particles the reader is referred to the volume by Washington (1992), which deals with a range of methods including light scattering, Coulter counting and light microscopy.

6.7.2 *Emulsifying activity (EA)*

The terms emulsifying activity (EA) and emulsifying activity index (EAI) are widely used in the functional studies of proteins. However, the terms are used for many different types of analysis, the results of which cannot be compared. All the estimates are meant to reflect the ability of the protein to aid formation and stabilization of newly created emulsions (Ivey *et al.*, 1970).

6.7.2.1 Emulsifying activity as assessed by particle size. Emulsifying activity can be measured by determining the particle size distribution of the dispersed phase. This can be done by microscopy, Coulter counting or spectroturbidity estimations (Walstra *et al.*, 1969) or by using laser scattering techniques as employed by the Malvern particle sizer or equivalent. This work has been taken further by Pearce and Kinsella (1978), who defined an emulsifying activity index (EAI), which had units of area of interface stabilized per unit weight of protein and was estimated from the turbidity of the emulsion at a wavelength of 500 nm. As well as using turbidimetric measurements, droplet size estimates by laser diffraction can be used to calculate EAI (Relkin *et al.*, 1993).

Other quoted values based on turbidity/particle size estimate readings include IFT and Index R. Ivey *et al.* (1970) used the term IFT, which is an approximation for the hydrated interfacial film thickness and is dependent on the diameter of the droplets. They found that the stability of emulsions was related to the IFT. The term Index-R has been used as an estimate of emulsion quality. It is derived from diluted emulsion turbidity measurements at 800 and 400 nm, the ratio of which gives the index (Kaufman and Garti, 1984). The value has been used to describe the emulsion properties of soya as it is denatured (Nir *et al.*, 1994).

6.7.2.2 Emulsifying activity as assessed by various methods. Emulsifying activity has also been used in other contexts. Kato *et al.* (1985) defined emulsion activity of proteins as the difference between the conductivity of the protein solution and the minimum conductivity of the emulsion during homogenization for one minute. The idea of emulsifying activity as an indication of emulsion capacity of protein was developed by Yasumatsu *et al.* (1972); the method is the same as for emulsion stability except that no heat is involved.

6.7.2.3 Procedure for the estimation of emulsion turbidity and emulsion activity index (EAI), R *(an indication of globular size) and* N *(an indication of number of globules).* The method given is based on that of Pearce and Kinsella (1978).

Apparatus.

Spectrophotometer
Hot air oven

Method.

1. Follow the standard procedure to produce an emulsion.
2. Prepare a 0.1% (w/v) solution of sodium dodecyl sulphate (SDS) in distilled water.
3. Weigh 1.00 g of emulsion into a 100 ml volumetric flask, make up to volume with the 0.1% SDS and mix thoroughly.
4. Take 1.00 ml of the thoroughly mixed 1/100 dilution and add to 4 ml of 0.1% SDS to form a dilution of 1/500.
5. Determine the absorbance of the 1/500 diluted emulsion in a 1 cm pathlength cuvette at a wavelength of 500 nm. If the absorbance reading exceeds 0.8 dilute the sample further and allow for this dilution factor in the calculations.
6. The turbidity of the sample (T) is given by the following equation:

$$T = \frac{2.303 \times \text{absorbance}}{\text{pathlength of cuvette}}$$

Estimation of the oil volume fraction of the emulsion.

7. The oil volume fraction (ϕ) is estimated by drying 5 g of emulsion to constant weight at 105°C.
8. Parameters required:

   D_o density of oil
   D_m density of protein solution
   E concentration of solutes; mass per unit mass solvent
   C weight of protein per unit volume of aqueous phase
   W_l (loss of weight of emulsion on heating/weight of emulsion)
   W_d (dried weight/weight of emulsion)

9. Calculation of the oil volume fraction (ϕ):

$$\phi = \frac{W_d - (E \times W_l)}{W_d + W_l\,[((1 + E) \times D_o)/D_m) - E]}$$

Estimation of emulsifying activity index (EAI).

10. Calculate the EAI

$$\text{EAI} = \frac{2 \times \text{turbidity}}{\phi \times C}$$

EAI has units of area of interface stabilized per unit weight of protein, i.e. $m^2\ g^{-1}$. Some values calculated for a range of proteins are given in Table 6.3.

Estimates of globular size (R).

11. R is the volume/area mean radius of the dispersed particles and can be calculated from:

$$R = \frac{3 \times \phi}{2 \times \text{turbidity}}$$

Equivalent number density (N).

12. N is the equivalent number density of particles (the hypothetical number density of particles, all of radius R, which have the same T as that observed for the polydisperse system) and can be calculated from:

$$N = \frac{2 \times (\text{turbidity}^3)}{(9\,\Pi\,\phi^2)}$$

The samples must be diluted for estimation by spectroscopy, and this is also the case for the other estimations of droplet size. In Table 6.5 three different methods of estimating particle size are given for some blood plasma powders produced by the standard methods. The methods were turbidity measurements to calculate R values, the use of a Coulter Counter TA11 with population accessory utilizing a 200 μ orifice tube (particle range 4–80 μ) (1 g of emulsion made up to 100 ml with 0.1% SDS), and the use of a Malvern Particle sizer fitted with a 100 μ lens (particle range 2–190 μ) (diluted with 0.1% SDS to an appropriate turbidity).

6.7.3 Emulsion capacity (EC)

Emulsion capacity (EC) is the parameter most commonly estimated in the various studies on oil-in-water emulsions. When an aqueous solution containing a protein is vigorously stirred while oil is added, an emulsion is formed. After a certain volume of oil has been added, the emulsion undergoes a sudden change referred to as inversion or breaking. The EC gives the maximum amount of oil that is emulsified before inversion, under specific conditions, by a standard amount of protein and is normally expressed as the amount of oil emulsified by unit weight of protein. The system introduced by Swift *et al.*, (1961), mixing oil with a protein until the emulsion formed breaks, is the method most commonly adopted to estimate EC.

There are three indicators of the inversion of the emulsion: (i) the change in visual appearance; (ii) the sudden drop in viscosity; and (iii) the sudden increase in electrical resistance. The sudden drop in viscosity which occurs as the emulsion collapses has been used as the end-point by many workers. However, the method requires an experienced operator and is not applicable to emulsions of low viscosity (Swift *et al.*, 1961). To aid in the visual detection of breaking, Marshall *et al.* (1975) added oil-red-O to the oil before mixing.

Webb *et al.* (1970), Satterlee *et al.*, (1973), Kato *et al.* (1985) and Tejada *et al.* (1987) investigated the sudden increase in electrical resistance as the oil phase became continuous, for the objective determination of end-points in emulsification. The perception of the drop in amperage required to drive the mixing apparatus which occurs at the breaking point has also been investigated (Crenwelge *et al.*, 1974).

6.7.3.1 Procedure for the estimation of emulsion capacity. The method given is based on that of Swift *et al.* (1961).

Apparatus

Peristaltic pump
100 ml plastic beaker
as listed for formation of an emulsion (section 6.6.1)

Method

1. Place a 25 g sample of an emulsion, as formed by the standard procedure, in a 100 ml plastic beaker.
2. Set the peristaltic pump to deliver oil at the mixing head of the Silverson at a rate of 5 ml per minute. Fill all tubing to deliver oil to the emulsion with oil. Fill the container which was to be used as the pump feed with oil and weigh.
3. Mix the emulsion using the Silverson at half maximum speed for half a minute then start the pump. Move the mixing probe within the emulsion and keep a close visual examination of the emulsion.
4. When the nature of the emulsion changes, as marked by decreased homogeneity and loss of viscosity, stop the pump and weigh the oil feed container. Measure the weight of oil required to break the 25 g of emulsion and calculate the weight of oil required to break the emulsion formed by 1 g of powder under the above conditions.

Emulsion capacity has units of g oil/g of powder or g oil/g of protein. These values can be very different if the powder contains other materials beside protein, e.g. lactose in whey powders. As stated previously, if different amounts of protein are added or, as shown in Figure 6.8, the volume of the aqueous phase is altered for the assessment of emulsion capacity the final calculated figures are markedly affected. Some values for a range of proteins using the above method are shown in Table 6.2.

6.7.4 *Oil phase volume (OPV)*

When the maximum amount of oil is incorporated into an oil-in-water emulsion, theoretically the addition of more oil would form a water-in-oil system. According to Ostwald's geometric phase theory the critical value for inversion, if the droplets are spherical, is 0.74 (dispersed phase/total volume) (Halling, 1981). Different techniques and stabilizers for emulsifying can induce deformation of the spheres, producing polyhedra droplets and/or heterogeneous sizes for the droplets. Emulsions have been prepared containing up to 99% internal phase (Becher, 1966).

Acton and Saffle (1972) reported a narrow range of oil phase volume values when the protein concentrations differed widely and suggested that emulsifying capacity expressed in 'oil phase volume' would minimize the variations caused by protein concentration and allow results to be compared.

6.7.4.1 *Procedure for the estimation of oil phase volume*

Method

1. Find the emulsion capacity as estimated in the standard procedure. Note the specific gravity of the oil and the amount of protein and water in each emulsion at the break point.
2. Calculate the oil phase volume from the ratio of oil at the point where the emulsion breaks compared to the total volume of emulsion.

As shown in Figure 6.8 the OPV still shows a wide range of values depending on the amount of aqueous phase originally used to form the emulsion.

6.7.5 *Emulsion stability (ES)*

Different procedures have been used to measure emulsion stability (ES), but it is commonly measured in terms of the oil and/or cream separating from an emulsion during a certain period of time at a stated temperature and gravitational field. Some studies measure emulsion destabilization by coalescence and by creaming (Petruccelli and Añón, 1994). In dilute systems it is possible to follow the formation of layers by eye or by light scattering techniques (Dagorn-Scaviner *et al.*, 1987). In dilute and more concentrated systems visualization of the process may be possible by application of confocal scanning light microscopy (Blonk and Vanaalst, 1993).

The destructive technique of sampling layers, diluting them and measuring the turbidity is also carried out. Freezing and then sectioning the sample for analysis by electron microscopy is a further method for monitoring destabilization. More recently, creaming has been measured by ultrasound (McClements, 1995) and this can detect the occurrence of creaming before it is visible to the naked eye (Dickinson, 1992). Ultrasound is a non-invasive method and the kinetics of the creaming process can be followed as long as the sample does not contain air (Povey, 1988). A clear demonstration of the usefulness of this method is given in Figure 6.10.

Basic methods for measuring coalescence include that developed by Tornberg and Hermansson (1977). Having realized that to be able to measure oil release from an emulsion, only weak or centrifuged emulsions

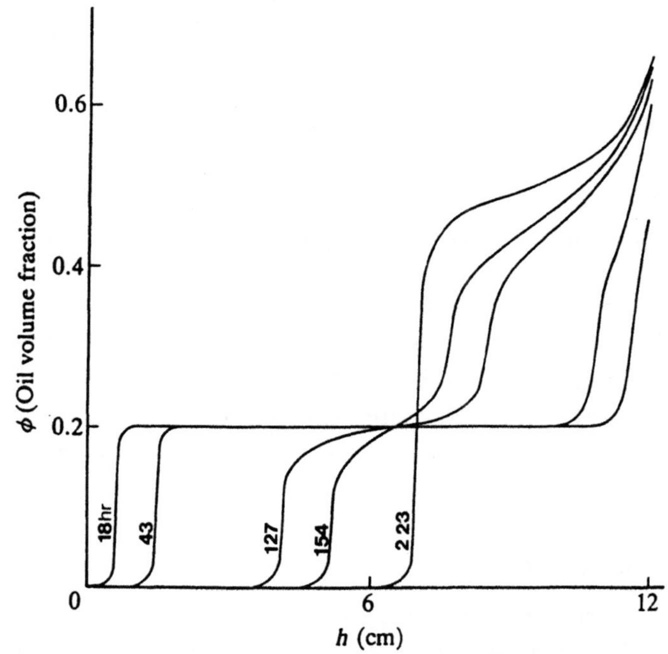

Figure 6.10 Estimates of the rate of creaming using ultrasound detection for an emulsion (15 wt% oil, 0.75 wt% protein, 0.05 wt% xanthan gum, oil volume fraction over whole sample = 0.2). The oil volume fraction ϕ is plotted against height (h) of sample. Values are hours of storage. Taken from Dickinson (1992)

could be studied, these workers developed a creaming stability test based on the percentage of fat in the lower portion of an emulsion after mild centrifugation. Only low-viscosity emulsions were used. Other workers have also used this method (Hayes *et al.*, 1979; Yamauchi *et al.*, 1980). Some methods (Inklaar and Fortuin, 1969; Yasumatsu *et al.*, 1972; Hutton and Campbell, 1977; Betschart *et al.*, 1979; Schachtel, 1981; Xu and Diosady, 1994) are based on taking an aliquot of emulsion before it breaks, heating it (up to 90°C) and centrifuging the sample. Kato *et al.* (1985) maintained that a value for emulsion stability could be extrapolated from the reciprocal of the initial slope of the electrical conductivity curve for an emulsion after it is mixed. Pearce and Kinsella (1978) proposed a turbimetric method for the determination of emulsion stability based on the methods developed for emulsifying activity indices.

6.7.5.1 Procedure for the estimation of emulsion stability

Apparatus

Water bath
Glass Universal bottles (internal diameter 21 mm) with metal screw caps and rubber liners
Bench centrifuge

Method

1. Form an emulsion by the standard procedure and then weigh two separate 10 g aliquots of the emulsion into glass universals.
2. Place the universals, well capped, into a water bath at 80°C for 1 hour and then allow to stand at ambient temperature. Leave one sample for 24 hours and the other for 14 days. At the end of the standing period centrifuge the universals at 1400 *g* for 15 minutes.
3. Remove the free oil using a pasteur pipette, or filter paper if the amount of oil is low, and weigh. The weight of oil released, divided by the original weight of oil in the sample can then be expressed as a % oil released from the emulsion.

Some results found for a range of proteins using the above method are given in Table 6.2. As can be seen the method only measures the gross changes and if more detailed stability measurements are required dilute emulsions must be used.

6.8 Other useful parameters when assessing emulsion properties

Rather than expressing results for emulsion behaviour in terms of 'g of powder' the values can usefully be given as 'g of protein' or 'g of soluble protein'. This requires the assessment of protein content and solubility. The measurable parameter that would seem to correlate well with emulsion behaviour is surface hydrophobicity.

6.8.1 Surface hydrophobicity

Chapter 8 deals with the measurement of hydrophobicity. The most common method used to estimate surface hydrophobicity and to correlate it with emulsion activity is quantification of the amount of *cis*-parinaric acid bound to the hydrophobic regions of the protein (Nakai, 1983). The method of Kato and Nakai (1980) adapted from Sklar *et al.* (1977) is suitable for most proteins.

6.8.2 Viscosity

The viscosity of many emulsions has been measured. McWatters and Cherry (1977) gave descriptions to a range of materials varying in their viscosity, from thick mayonnaise-like, to pourable suspensions. Many authors quote values based on Brookfield measurements of viscosity. Values of viscosity, like so many other of the assessments of emulsions, need to be carefully considered in light of the method used to form the emulsion and the equipment and method used in the measurement.

6.9 Conclusions

This chapter should indicate that there is *no* ideal way to assess the ability of a protein to form or stabilize an emulsion. All the methods have faults. Lack of standard procedures for emulsion preparation, lack of consistent terminology and no available control material (so that at least comparisons can be made) are amongst some of obvious weaknesses which workers in the area could address.

References

Acton, J.C. and Saffle, R.L. (1972) Stability of oil-in-water emulsions. *J. Food Sci.*, **37**, 904–7.

Arai, S. and Watanabe, M. (1988) Emulsifying and foaming properties of enzymatically modified proteins, in *Adv. in Food Emulsions and Foams* (eds E. Dickinson and G. Stainsby), Elsevier Applied Science, London, pp. 189–220.

Becher, P. (1966) *Emulsions: Theory and Practice*, Reinhold, New York.

Betschart, A.A., Fong, R.Y. and Hanamoto, M.M. (1979) Safflower protein isolates: functional properties in simple systems and breads. *J. Food Sci.*, **44**, 1022–6; 1035.

Bigelow, C.C. (1967) On the average hydrophobicity of proteins and the relationship between it and protein structure. *J. Theoretical Biology*, **16**, 187–211.

Blonk, J.C.G. and Vanaalst, H. (1993) Confocal scanning light-microscopy in food research. *Food Research International*, **26**, 297–311.
Crenwelge, D.D., Dill, C.W., Tybor, P.T and Landman, W.A. (1974) A comparison of the emulsification capacities of some protein concentrates. *J. Food Sci.*, **39**, 175–7.
Dagorn-Scaviner, C., Gueguen, J. and Lefebvre, J. (1987) Emulsifying properties of pea globulins as related to their adsorption behaviours. *J. Food Sci.*, **52**, 335–41.
Derjaguin, B.V. and Landau, L.D. (1941) Theory of the stability of strongly charged lyophobic sols and of the adhesion of strongly charged particles in solutions of electrolytes. *Acta Physicochimica URRS*, **14**, 633.
Dickinson, E. (1988) The structure and stability of emulsions, in *Food Structure: Its Creation and Evaluation* (eds J.M.V. Blanshard and J.R. Mitchell), Butterworths, Sevenoaks, UK, pp. 41–57.
Dickinson, E. (1992) *An Introduction to Food Colloids*, Oxford Univeristy Press, Oxford.
Dickinson, E. (1993) Towards more natural emulsifiers. *Trends in Food Science and Technology*, **4** (10), 330–4.
Dickinson, E. (1995) Mixed biopolymers at interfaces, in *Mixed Biopolymers* (eds S.E. Harding, S.E. Hill and J.R. Mitchell), Nottingham University Press, Nottingham, pp. 349–72.
Dickinson, E. and Euston, S.R. (1991) Stability of food emulsions containing both protein and polysaccharide, in *Food Polymers, Gels and Colloids* (ed. E. Dickinson), Royal Society of Chemistry, London, pp. 132–46.
Dickinson, E. and Semenova, M.G. (1992) Emulsifying properties of covalent protein/dextran hybrids. *Colloids and Surfactants*, **64**, 299–310.
Dickinson, E. and Stainsby, G. (1982) *Colloids in Food*, Applied Science, London.
Dickinson, E. and Stainsby, G. (1988) Emulsion stability, in *Advances in Food Emulsions and Foams* (eds E. Dickinson and G. Stainsby), Elsevier Applied Science, London, pp. 1–44.
Dickinson, E., Whyman, R.H. and Dalgleish, D.G. (1987a) Colloidal properties of model oil-in-water food emulsions stabilized separately by α_{s1}-casein, β-casein and κ-casein, in *Food Emulsions and Foams* (ed. E. Dickinson), Royal Society of Chemistry, London, pp. 40–51.
Dickinson, E., Murrey, A., Murray, B.S. and Stainsby, G (1987b) Properties of adsorbed layers in emulsions containing a mixture of caseinate and gelatin, in *Food Emulsions and Foods* (ed. E. Dickinson), Royal Society of Chemistry, London, pp. 86–99.
Halling, P.J. (1981) Protein stabilized foams and emulsions. *Critical Reviews in Food Science and Nutrition*, **15**, 155–203.
Hayes, J.F., Stranaghan, B. and Dunkerley, J.A. (1979) The emulsifying properties of whey protein concentrates in a model system. *N.Z. J. Dairy Sci. Technol.*, **14**, 195–6.
Hill, S.E. (1986) *Processing and Functional Properties of Bovine Plasma*, Ph.D. Thesis, Loughborough University.
Howell, N. (1981) *Separation and Functional Properties of Blood Plasma Proteins*, Ph.D. Thesis, University of Nottingham.
Huffman, V.L., Lee, C.K. and Burns, E.E. (1975) Selected functional properties of sunflower meal (*Helianthus annus*). *J. Food Sci.*, **40**, 70–4.
Hunt, J.A. and Dalgleish, D.G. (1994) Effect of pH on the stability and surface composition of emulsions made with whey protein isolate. *J. Agric. Food and Chem.*, **42**, 2131–5.
Hutton, C.W. and Campbell, A.M. (1977) Functional properties of a soy concentrate and a soy isolate in simple systems and in a food system. *J. Food Sci.*, **42**, 457–60.
Inklaar, P.A. and Fortuin, J. (1969) Determining the emulsifying and emulsion stabilizing capacity of protein meat additives. *Food Technol.*, **23**, 103–7.
Ivey, F.J., Webb, N.B. and Jones, V.A. (1970) The effect of disperse phase droplet size and interfacial film thickness on the emulsifying capacity and stability of meat emulsions. *Food Technol.*, **24**, 1279–81.
Kato, A. and Nakai, S. (1980) Hydrophobicity determined by a fluorescence probe method and its correlation with surface properties of proteins. *Biochimica et Biophysica Acta*, **624**, 13–20.
Kato, A., Fujishige, T., Matsudomi, N. and Kobayashi, K. (1985) Determination of emulsifying properties of some proteins by conductivity measurements. *J. Food Sci.*, 50, 56–8; 62.

Kato, A., Minaki, K. and Kobayashi, K. (1993) Improvement of emulsifying properties of egg white proteins by the attachment of polysaccharide through Maillard reaction in a dry state. *J. Agric. Food Chem.*, **41**, 540–3.

Kato, Y., Aoki, T., Kato, N., Nakamura, R. and Matsuda, T. (1995) Modification of ovalbumin and glucose 6-phosphate by amino-carbonyl reaction. Improvement of protein heat stability and emulsifying activity. *J. Agric. Food and Chem.*, **43**, 301–5.

Kaufman, V.R. and Garti, N. (1984) Effect of cloudy agents on the stability and opacity of cloudy emulsions for soft drinks. *J. Food Technol.*, **19**, 255–61.

Krasnowska, G., Gorska, I. and Gergont, J. (1995) Evaluation of functional properties of offal proteins. *Nahrung-Food*, **39**, 149–55.

Lips, A., Westbury, T., Hart, P.M., Evans, I.D. and Campbell, IJ. (1993) On the physics and shear-induced aggregation in concentrated food emulsions, in *Food Colloids and Polymers: Stability and Mechanical Properties* (eds E. Dickinson and P. Walstra), Royal Society of Chemistry, Cambridge, pp. 31–44.

McClements, D.J. (1995) Advances in the ultrasound in food analysis and processing. *Trends in Food Sci. and Technol.*, **6**, 293–9.

McWatters, K.H. and Cherry, J.P. (1977) Emulsification, foaming and protein solubility proteins of defatted soybean, peanut, field pea and pecan flours. *J. Food Sci.*, **42**, 1444–8.

McWatters, K.H. and Holmes, M.R. (1979) Salt concentration, pH and flour concentration effects on nitrogen solubility and emulsifying properties of peanut flour. *J. Food Sci.*, **44**, 765–9.

Manev, N.E., Scheludko, A. and Exerowa, D. (1974) Effect of surfactant concentrations on the critical thickness of liquid films. *Colloid and Polymer Science*, **252**, 586.

Marshall, W.H., Dutson, T.R. and Carpenter, Z.L. (1975) A simple method for emulsion end-point determinations. *J. Food Sci.*, **40**, 896–7.

Mitchell, J.R. (1982) Foaming and emulsifying properties of protein, in *Development in Food Proteins* 4, (ed. B.J.F. Hudson), Elsevier Applied Science, London, pp. 291–338.

Morr, C.V. and Ha, E.Y.W. (1993) Whey protein concentrates and isolates: processing and functional properties. *Critical Reviews in Food Science and Nutrition*, **33**, 431–76.

Nakai, S. (1983) Structure–function relationships of food proteins with an emphasis on the importance of protein hydrophobicity. *J. Agric. Food Chem.*, **31**, 676–83.

Nakai, S. and Li-Chan, E. (1993) Recent advances in structure and function of food proteins, QSAR approach. *Critical Reviews in Food Science and Nutrition*, **33**, 477–500.

Nakamura, S., Kato, A. and Kobayashi, K. (1991) New antimicrobial characteristics of lysozyme dextran conjugate. *J. Agric. Food Chem.*, **39**, 647–50.

Napper, D.J. (1983) *Polymeric Stabilization of Colloid Dispersions*, Academic Press, London.

Nir, I., Feldman, Y., Aserin, A. and Garti, N. (1994) Surface properties and emulsification behaviour of denatured soy proteins. *J. Food Sci.*, **59**, 606–10.

Parker, N.S. (1987) Properties and functions of stabilizing agents in food emulsions. *Critical Reviews in Food Science*, **25** (4), 285–316.

Pearce, K.N. and Kinsella, J.E. (1978) Emulsifying properties of proteins. Evaluation of a turbidimetric technique. *J. Agric, Food Chem.*, **26**, 716–23.

Petruccelli, S. and Añón, M.C. (1994) Relationship between the method of obtention of the structure and functional properties of soy protein isolates. 2. Surface properties. *J. Agric. Food Chem.*, **42**, 2170–6.

Phillips, M.C. (1981) Protein conformation at liquid interfaces and its role in stabilizing emulsions and foams. *Food Technol.*, **35** (1), 50–7.

Povey, M.J.W. (1988) Ultrasonics as a probe of food emulsions and dispersions, in *Advances in Food Emulsions and Foams* (eds E. Dickinson and G. Stainsby), Elsevier Applied Science, London, pp. 285–327.

Relkin, R., Liu, T. and Launay, B. (1993) Effects of pH on gelation and emulsification of a β-lactoglobulin concentrate. *J. Dispersion Sci. and Technol.*, **14**, 335–54.

Satterlee, L.D., Zachariah, B. and Levin, E. (1973) Utilization of high protein tissue powders as a binder/extender in meat emulsions. *J. Food Sci.*, **38**, 306–9.

Schachtel, A.P. (1981) Effects of preparative processes on the composition and functional properties of protein preparations from *Candida util*. *J. Food Sci.*, **46**, 377–82.

Sklar, L.A., Hudson, B.S. and Simoni, R.D. (1977) Conjugated polyene fatty acids as fluorescent probes: binding to bovine serum albumin. *Biochem.*, **16**, 5100–8.

Swift, C.E., Lockett, C. and Fryer, A.J. (1961) Comminuted meat emulsions: the capacity of meat for emulsifying fat. *Food Technol.*, **15**, 468–72.

Tadros, Th.F. and Vincent, B. (1983) Emulsion stability, in *Encyclopedia of Emulsion Technology*, Vol. 1 (ed. P. Becher), Marcel Dekker, New York.

Tejada, M., Borderías, A.J. and Jiménez-Colmenero, F. (1987) Myofibrillar and sarcoplasmic proteins: their role in alterations in certain functional properties of muscle during frozen storage, in *Proceedings of the IUFoST International Symposium Chemical Changes during Food Processing*, Instituto de Agroquímica y Tecnología de Alimentos (CSIS), Valencia, pp. 103–11.

Tornberg, E. (1980) Functional characterization of protein stabilized emulsions: emulsifying behaviour of protein in a sonifier. *J. Food Sci.*, **45**, 1662–8.

Tornberg, E. and Hermansson, A.-M. (1977) Functional characterization of protein stabilized emulsions: effect of processing. *J. Food Sci.*, **42**, 468–72.

Verwey, E.J.W. and Overbeek, J.Th.G. (1948) *Theory of Stability of Lyophobic Colloids*, Elsevier, Amsterdam.

Walstra, P. (1968) Estimating globule-size distribution of oil-in-water emulsions by spectroturbidity. *J. Colloid Interface*, **27**, 493–8.

Walstra, P. (1983) Formation of emulsions, in *Encyclopedia of Emulsion Technology*, vol. 1 (ed. P. Becher), Marcel Dekker, New York, pp. 57–127.

Walstra, P. (1987a) Overview of emulsion and foam stability, in *Food Emulsions and Foams* (ed. E. Dickinson), Royal Society of Chemistry, London, pp. 242–57.

Walstra, P. (1987b) Physical principles in emulsion science, in *Food Structure and Behaviour* (eds J.M.V. Blanshard and P. Lillford), Academic Press, London, pp. 87–148.

Washington, C. (1992) *Particle Size Analysis in Pharmaceutical and Other Industries*, Ellis Harwood, London.

Webb, N.B., Ivey, F.J., Craig, H.B., Jones, V.A. and Monroe, R.J. (1970) The measurement of emulsifying capacity by electrical resistance. *J. Food Sci.*, **35**, 501–3.

Williams, P.A. (1995) Depletion flocculation, in *Mixed Biopolymers*, (eds S.E. Harding, S.E. Hill and J.R. Mitchell) Nottingham University Press, Nottingham, pp. 161–72.

Xu, L. and Diosady, L.L. (1994) Functional properties of chinese rapeseed protein isolates. *J. Food Sci.*, **59**, 1127–30.

Yamauchi, K., Shimizu, M. and Kamiya, T. (1980) Emulsifying properties of whey protein. *J. Food Sci.*, **45**, 1237–42.

Yasumatsu, K., Sawada, K., Moritaka, S., Misaki, M., Toda, J., Wada, T. and Ishii, K. (1972) Studies on the functional properties of food grade soybean products. IV. Whipping and emulsifying properties of soybean products. *Agricultural and Biological Chemistry*, **36**, 719–27.

7 Determining water and fat holding

S. BARBUT

7.1 Introduction

Determining the water and fat holding capacity of protein-based food is of utmost importance to the food industry, both in marketing existing products and in developing new products. Water and fat holding determine not only the acceptability of a given food product (texture, juiciness, mouth feel) but also the profit margin. Since most conventional food products contain more than 50% water and sometimes up to 95% (gelatin desserts), good water holding capacity is essential. Consumers tend to avoid products that show 'free' water in the package such as in fresh meat trays, yogurt containers, etc. Furthermore, formulations that result in poor water and fat holding capacity (WHC/FHC) directly translate to liquid losses during processing (cooking, freezing, etc.).

Different methods have been suggested and used by various research groups over the years. It is interesting to note that several reviews have been published within the last decade on various methods for a specific food commodity. Kneifel *et al.* (1991) and Kneifel and Seiler (1993) reviewed WHC in dairy protein products, Trout (1988) and Hamm (1986) reviewed methods for fresh meat and meat products, while Offer and Trinick (1983) and Offer *et al.* (1989) reviewed the basis of WHC in meat. These reviews and the review by Hermansson (1986) on both WHC and FHC should provide background material for the reader.

It is not the intention of this chapter to cover all the material published about methods to determine water and fat holding. A recent (1995) electronic literature search, on the Agricola® System, identified almost a thousand articles in which key word combinations such as food, protein and water/fat holding capacity were used.

Overall the purpose of this chapter is to describe the major methods and to provide examples for different food commodities, such as fibrous protein structures (meat), fine strands (clear gels), aggregated protein structures (opaque gels) and powders, to allow the reader to compare different methods. In addition to reviewing methods published in the literature, one set of samples which shows a wide range of WHC was evaluated in the author's laboratory by some of the most common methods. The chapter also describes some of the new on-line methods which can be used to assess WHC and FHC in real time. Most of these methods have not yet been reviewed, and it is the intention here to provide information regarding the concepts behind developing such methods. Some of these new methods are very accurate while others just provide estimates. However, their major advantage is in providing fast on-line results, which can be used for feedback/feed-forward decision-making in real time.

It is hoped that this chapter will contribute to better standardization of test conditions in different laboratories. Although a major undertaking, this is essential to allow more uniformity and permit direct comparison of results from laboratories around the world, and sometimes even from the same institute. It is true that there is no single accepted method for all protein-based foods because of their diversity (e.g. powders, gels). However, agreeing on one or two concepts to be used for evaluating protein foods will assist all workers in this field. In the final section of the chapter a protocol is suggested for one method – the Net-Test – which has been recommended by various researchers over the years (Wierbicki *et al.*, 1956; Hermansson and Lucisano, 1982; Kocher and Foegeding, 1993; Barbut, 1995), but has not been adopted as a standard. The main advantages of this method are its minimal damage to structure, relative simplicity and good reproducibility.

7.2 Scientific basis

Food proteins differ in amino acid composition, molecular size, structure (globular, fibrous) and isoelectric point. During processing, the proteins are subjected to different environmental conditions required for extraction, purification or improved digestibility. Such chemical and physical treatments affect the functionality of the proteins. The term 'protein functionality' has been used to describe different characteristics of proteins. In this book alone there are references to solubility, viscosity, gelation, foam formation and emulsification; all are part of protein functionality.

WHC is one of the functionalities that has also been used to describe different properties, some of which have been used synonymously in the literature. Terms such as water binding, hydration capacity, water retention, water absorption, water imbedding, suction potential and swelling have been used to describe the phenomenon of WHC (Kneifel and Seiler, 1993).

Protein composition and conformation have significant effects on WHC and FHC, and have been reviewed by various researchers (Lillford *et al.*, 1980; Mohsenin, 1986; Kinsella *et al.*, 1989; Doi, 1993). In general terms, water held within a protein structure (gel, powder) can be divided into two categories. The first is the water bound to the protein molecule, which is no longer available as a solvent; this is referred to by some as the absorbed water (Kneifel *et al.*, 1991). The second is the water trapped within the protein matrix; this can be regarded as the retained water. While binding of the absorbed water is largely dependent on the physicochemical properties (type of amino acids, pH), the retained water is more affected by the structure of the matrix (pore size). Kinsella *et al.* (1989) distinguished between six basic 'forms' of water association within proteins: (1) structural water which is tightly bound to the protein molecules and unavailable for chemical reaction; (2) hydrophobic hydration water found around apolar residues of amino acids; (3) monolayer water which is the first layer of water absorbed to the protein groups and may be available for some reactions; (4) unfreezable water consisting of water that does not freeze at the first sharp transition temperature; (5) capillary water which is held by surface tension forces; and (6) hydrodynamic hydration water which loosely surrounds the proteins. The amount of water and degree of binding of each one of the six categories are affected by factors such as protein type, concentration, number of exposed polar groups, pH, salt(s) and temperature. The reader should also be aware of the fact that other models describing water associated with proteins have been suggested (Mohsenin, 1986; Harwalkar and Brown, 1989).

The major concepts to assess WHC and FHC in food systems include:

(a) exposing the sample to pressure (mild to severe);
(b) observing the sample behaviour during normal processing conditions;
(c) studying microstructure and relating it to WHC and FHC;
(d) using equipment such as NMR to study the state and position of water and/or fat in a sample.

As will be described in the following pages, the tests can be performed under different conditions and the results will depend on these conditions. This is a very important issue in selecting a method and will be discussed in detail in this chapter. In order to provide the reader with a true comparison of the major methods used to determine WHC, a comparison of one set of whey protein isolate (WPI) samples was evaluated by the following methods: press (high, low pressure), centrifugation (low, high), capillary suction, optical and microscopy. The samples were made with 10% WPI and different sodium concentrations (25–500 mM), and heated to 80°C for 30 min. The sodium range used here exhibits the transformation from a clear protein gel structure composed of a fine strand (Figure 7.1A) to an opaque aggregated structure (Figure 7.1D), or from excellent to very poor WHC, respectively. It is hoped that demonstrating the effect of a method on the results will assist readers in selecting the most appropriate method for their purposes.

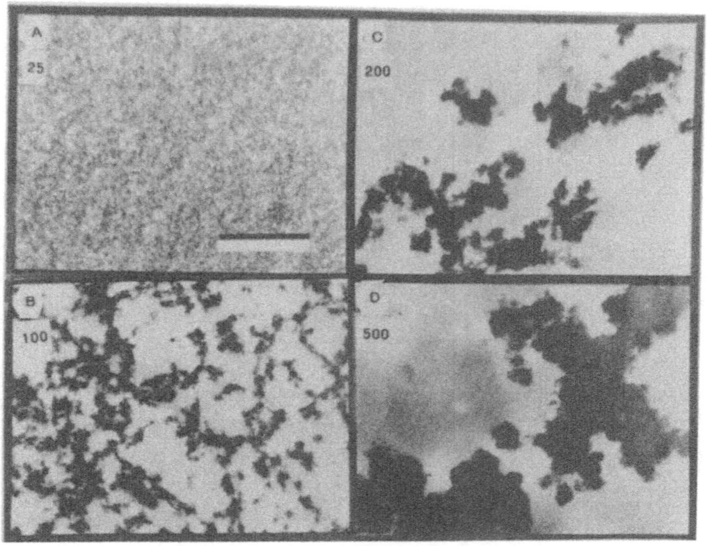

Figure 7.1 Transmission electron micrograph of WPI gels (10%) formed by heating (80°C, 30 min) in the presence of: (A) 25 mM Na; (B) 100 mM Na; (C) 200 mM Na; and (D) 500 mM Na. Bar = 1 μm. From Barbut (1995).

7.3 Water holding

7.3.1 Press method

This is one of the methods where an external force is applied to the sample. The pressure can be moderate to high, but usually a high force is used compared to the other methods. The sample is compressed between two parallel plates using equipment such as a hydraulic press or a universal testing machine (i.e. Instron, Nene). The released water is collected on a pre-weighed filter paper. The amount of fluid squeezed out from the sample is commonly expressed as a percentage of the total moisture in the samples (calculated after drying the sample) or as the absolute amount of water coming out from the sample.

The method has been used to evaluate milk protein concentrates (Brendl and Klein, 1972; Kabus, 1972; Ozimek and Poznanski, 1981), soy proteins (Furukawa and Ohta, 1982; Aljawad and Bowers, 1988) and various meat products (Brendl and Klein, 1970, 1972; Zhang *et al.*, 1993). However, a wide range of conditions has been reported in the literature. The data in Table 7.1 show that sample size varied from 0.3 to 1.5 g; force from 0.0098 to 44.4 kN; temperature from 4 to 23°C; and compression time from 1 to 20 min. In addition, different filter papers, with different drying capabilities, were used. The type of paper is not always mentioned; this can be

Table 7.1 Press method – test conditions previously used for WHC measurement in meat

Investigators	Meat type	Mass (g)	Force (kN)	Temperature (°C)	Compression time (min)	Salt (%)	Filter paper
Ackerman *et al.* (1981)	Frankfurter	1.5	6.2	23	1	*	Two Whatman #2 (D = 110 mm)
Gumpen and Martens (1977)	Sausage	0.3	0.39	4	5	0	Preweighed, predried**
Min and Ni (1989)	Ground muscle	0.5	0.34	0–4	1	##	Used**
Severini *et al.* (1984)	Ground pork	0.3	0.39	10	5	0	Used**
Suvakov *et al.* (1984)	Ground pork	0.3	0.39		5	4	Used**
Vaisey *et al.* (1975)	Ground beef	0.5	44.4	*	2	1.3	*
Zayas and Lin (1989)	Frankfurter with corn	0.3	0.0098	4	20	*	Whatman #1 stored over saturated KCl

##Meat was cooked at 90°C for 10 min before test. A 5% solution, which contained sodium pyrophosphate, sodium tripolyphosphate, sodium hexametaphosphate, NaCl and isolated soybean protein, was mixed with the meat.
*Not mentioned.
**Type of paper not mentioned.

Table 7.2 Press method – effect of force, press time, salt and sample size on WHC results for lean meat. Data from Zhang *et al.* (1993)

Test conditions	Mean value of WHC (decimal)			
	Overall	Sample mass (g)		
		0.5	1.0	1.5
Test time (min)				
1	0.41a	0.32a	0.39a	0.52a
2	0.33b	0.20b	0.33b	0.46b
3	0.28c	0.18c	0.27c	0.40c
Applied force (kN)				
10	0.43a	0.38a	0.41a	0.51a
20	0.33b	0.22b	0.32b	0.46b
30	0.26c	0.10c	0.27c	0.42c
Salt concentration (%)				
0	0.22c	0.09c	0.21c	0.35c
1	0.37b	0.28b	0.34b	0.49b
2	0.44a	0.32a	0.45a	0.54a
Sample mass (g)				
0.5	0.23c	–	–	–
1.0	0.33b	–	–	–
1.5	0.46a	–	–	–

Data followed by the same letter in a column are not significantly different at the 95% level.
Number of replications for each treatment value was four.
Overall mean values were the averages of 135 WHC measurements. All others were the averages of 45 WHC measurements.

a very important variable as will be seen in section 7.3.4. The fact that different test conditions have been used makes it difficult, if not impossible, to compare results among experiments and laboratories. The examples provided here deal with meat samples; however, the same is true for dairy products (Kneifel *et al.*, 1991, 1992) and other foods. In order systematically to compare the effects of these parameters, a common ground meat sample was used (Table 7.2). Three levels of applied force, press time, sample size and salt concentrations were studied since all affect the outcome. The filter paper used was Whatman#2, one per side. The results for WHC were expressed as described by Min and Ni (1989):

$$\text{bound water} = 1 - \text{free water}$$

The free water was calculated as:

$$\frac{\text{wet area of filter paper} - \text{compressed meat film area on paper}}{\text{initial water in meat sample} \times 2 \times \text{water filter paper absorbs unit area}}$$

It was determined that WHC decreased with additional applied force and compression time, and increased with larger sample size and salt concentration. Plastic, elastic and permanent deformations were used to explain the effects of applied force, viscoelastic properties and compression time. Overall, the recommended test conditions were a sample weighing 1 g and compression at 20 kN for 2 min at any of the salt concentrations tested (0, 1 or 2%). In many studies salt concentration is in fact part of the experimental treatment.

The effect of two different compression rates on the set of WPI samples is presented in Figure 7.2. The sample made with 25 mM Na formed a fine protein strand microstructure (Figure 7.1A) and showed the best WHC value when compressed under both conditions. By contrast, the highly aggregated sample formed with 500 mM Na (Figure 7.1D) showed the lowest WHC. The fact that a fine protein gel structure possesses a better WHC has been described by various researchers (Clark, 1992; Kalab, 1993; Barbut, 1994). For the comparison here it is important to examine the differences in the two curves and the distinction among the treatments [all samples, 19 mm in diameter, 5 mm high, were compressed for 30 sec between two filter papers (Whatman #1), one on each side]. The results reported, expressed as 1 − water loss from the sample (%), show that at the low pressure (1.43 kg/cm^2) the high WHC samples (25–75 mM) could not be distinguished as well as by using high pressure (10 kg/cm^2). The difference between the two pressure treatments was about four-fold (10% versus 40% expressed moisture at 25 mM Na). For the aggregated samples

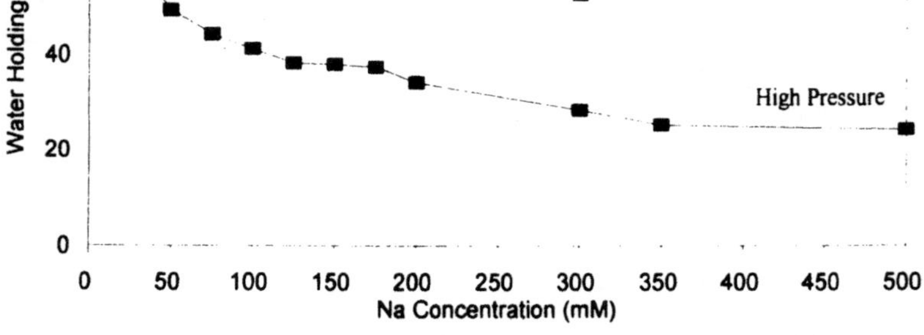

Figure 7.2 Water holding of WPI gels measured by the press method at high (10 kg/cm^2) and low (1.43 kg/cm^2) pressure.

(>150 mM Na), which are known to have poor WHC, the difference was narrowed down to two-fold. At the intermediate salt concentration the low-pressure conditions provided the best separation. An important point to remember is that the high pressure caused a permanent deformation (i.e. structural damage) to all the samples; the resulting samples were flattened thinly. This, as will be discussed later in section 7.3.3, is a very important consideration when selecting a test procedure.

7.3.2 *High-speed centrifugation*

In this test a centrifugal force is applied to a sample placed in a test tube. The method has been used to assess the WHC of dairy proteins such as whey (Sternberg *et al.*, 1976), casein (Tarodo and Alais, 1975), milk powders (Comer, 1979), soybean proteins (Thomas *et al.*, 1974; Luther *et al.*, 1983), egg white proteins (Quinn and Paton, 1979; O'Brien *et al.*, 1982) and wheat flour proteins (Sollars, 1973; Huber, 1982).

Again, various test conditions have been reported for various food products and often even for the same food product. Table 7.3 summarizes some of the test conditions used to evaluate fresh and processed meat samples. The same have been reported for other food protein systems (Kneifel *et al.*, 1991, 1993). Table 7.3 indicates that sample size varied from 1.5 to 20 g and force from 1500 to 190 000 *g*. Again, such large variations in test conditions (×13 in terms of mass and ×127 in *g* force) does not allow comparison of results.

In order to assess the effects of test conditions on the results, three levels of *g* force, time, temperature and salt were studied using a common ground meat sample (Table 7.4). In this test 8 ml of water or a salt solution was added to the samples (5 g) which were incubated (30 min) prior to centrifugation. This is a fairly common test procedure (Wardlaw *et al.*, 1973) when protein swelling is expected to occur. Salt and water are added in this test since they are commonly used by industry in the production of further processed meat products (Whiting, 1987; Offer *et al.*, 1989). In this case WHC was expressed as:

$$\frac{\text{solution added to sample} - \text{solution removed after centrifugation}}{\text{meat sample mass}} \times 100$$

At the high *g* force water was expelled from the sample (shown as a negative value), while at a lower force it was retained (Table 7.4). WHC increased with decreasing temperature and increasing salt content; both trends have previously been reported. The recommended test conditions (obtained by statistically evaluating each factor at every level) were 8630 *g* for 7.5 min at 20°C at all the salt concentrations tested (0–2.25%, corresponding to 0–0.6 M NaCl in the salt solutions added to the meat).

Table 7.3 Centrifugal method – range of test conditions previously used to measure WHC in meat

Investigators	Meat type	Mass (g)	Force (kN)	Temperature (°C)	Time (min)	Salt (%)
Bouton and Harris (1972)	Ovine muscle	3–4	100 000	*	60	Ascorbic acid
Bouton *et al.* (1972)	Ground muscle	3–4	120 000–190 000	*	30–60	*
Dagbjartsson and Solberg (1972)	Ground lobster	1.0	12 000	*	10	*
Thomas *et al.* (1973)	Sausage (48.2% mutton, 21.7% pork fat)	10	3000 rpm	0	20	2
Hermansson and Akesson (1975)	Beef and pork added with protein (raw)	20	18 000	*	30	*
Jauregui *et al.* (1981)	Ground fish, chicken and beef	1.5	7710	2	10	0
		1.5	17 300	2	15	0
		1.5	30 900	2	30	0
Eide *et al.* (1982)	Minced fish	2	1500	10	5	*

*Not mentioned.

Table 7.4 Centrifugal method – effect of *g* force, time, temperature and salt level on WHC results of lean beef. Data from Zhang *et al.* (1995)

Test conditions	Mean value of WHC (decimal)			
	Overall	Centrifugal force (*g*)		
		959	8630	34 500
Test time (min)				
7.5	4.0a	23.7a	4.1a	−15.7a
15	1.1b	25.0a,b	−3.7b	−17.9b
22.5	1.2b	27.0b	−4.9b	−18.5b
Test temperature (°C)				
2	6.5a	28.3a	6.0a	−14.9a
10	1.8b	26.6a	−3.3b	−18.0b
20	−1.9c	20.7b	−7.1c	−19.2c
Salt concentration (M)				
0	−2.2c	19.3c	−4.7c	−21.3c
0.3	1.9b	25.3b	−1.9b	−17.6b
0.6	6.6a	31.0a	2.1a	−13.2a
Centrifugal force (*g*)				
959	25.2a	–	–	–
8630	−1.5b	–	–	–
34 500	−17.4c	–	–	–

Data followed by the same letter in a column are not significantly different at the 95% level. Number of replications for each treatment value was five.
Overall mean value were the averages of 135 WHC measurements. All others were the averages of 45 WHC data. Negative values indicate loss of original water.

The effect of high- and medium-speed centrifugation on the WHC of 5% plasma protein was reported by Hermansson and Lucisano (1982). They showed that increasing the force increased the amount of water released (Table 7.5). The two important factors demonstrated by the authors were: (1) that the increase was not linear; (2) that the relationship between the size of the test tube and the sample is critical in obtaining meaningful results (i.e. if the sample touches the walls of the tube, not all the water will be drained).

The use of three different *g* forces for the WPI samples is shown in Figure 7.3. Again, as centrifugal force increased (200, 790 and 7800 *g*) the amount of liquid separated increased. In this case the samples (19 mm diameter, 5 mm high) were placed in 25 mm diameter test tubes. The effect of trapping the expressed water is shown as the difference between samples spun at 790 *g* without a net (placed at the bottom) and with the net (samples raised from the bottom). In the case of the 7800 *g*, samples were spun with and without filter paper (placed at the bottom) because the net was not strong enough to hold the samples at this *g* force. In any case, all

Table 7.5 Effect of experimental conditions on moisture loss values of 5% plasma protein at pH 9.0 heated at 82°C. Data from Hermansson and Lucisano (1982)

Forge (*g*)	Sample weight (g)		
	1	2	5
High speed	(small tube – 10 mm diameter)		
5090	6.9	3.8	1.1
9750	13.1	6.6	2.0
30 000	41.1	38.3	36.9
High speed	(large tube – 27 mm diameter)		
5090	–	–	20.0
9750	–	–	28.1
30 000	–	–	51.1
Net test			
465	41	20	15
790	41	18	19
1045	42	20	15
1290	41	22	20

– Not determined

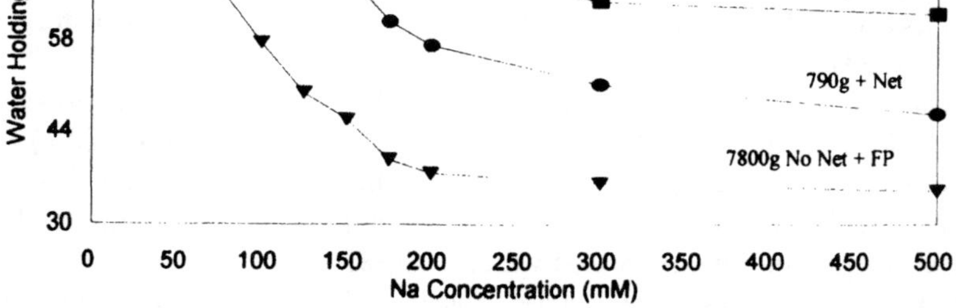

Figure 7.3 Water holding of WPI gels (25–500 mM Na) measured by a centrifugal method (200, 790 and 7800 *g*) with (+ filter) and without (no filter; NF) trapping the released water. Net = filter unit with wire. FP, filter paper.

these settings have been used in the past. The results show a very big difference depending on the presence of filter paper or a net. This illustrates the importance of removing the exudate from the sample in order to obtain meaningful WHC estimates. In some cases, the presence or absence of a filter paper resulted in a two-fold difference between identical treatments. Using 7800 *g* resulted in good distinction among the samples when the exudated water was trapped by a filter paper. However, it should be pointed out that 7800 *g* caused permanent sample deformation, indicating structural breakdown. As can be seen, both 200 *g* and 790 *g* provided good differentiation of the treatments. This is an important point since under these relatively low *g* force conditions, gel structure is usually not damaged.

7.3.3 Low-speed centrifugation

For this test a low centrifugal force (100–800 *g*) is applied to protein samples (usually gels) and the amount of water released is measured. The test has been used for meat research (Wierbicki *et al.*, 1957; Hermansson and Akesson, 1975; Ockerman and Crespo, 1982), for soybean (Ochiai-Yanagi *et al.*, 1978) and to study blood plasma gels (Hermansson and Lucisano, 1982). The main advantage is that sample deformation is minimal and that no or minimal structural damage occurs. This contrasts with the massive structural breakdown observed in the press or the high centrifugal force and tests has been used as the main argument to support the low-speed test, referred to by some as the net test because the sample is placed in a net or bucket within a larger test tube. Wierbicki *et al.* (1957) were among the first to recommend the test. They used it to evaluate WHC changes in post-mortem meat obtained from different beef animals. Later, Hermansson and Lucisano (1982) used it for plasma proteins (Table 7.5) where low *g* forces (465–1290) did not result in major WHC differences when the sample was small (1 g). However, when sample size was increased the results were affected by the *g* force (illustrating the importance of physical constraints).

Kocher and Foegeding (1993) also used the test to evaluate WPI under different *g* forces (153–3700), different sample size and different centrifugation times (2.5–25 min). They looked at the effect of centrifugation force (Figure 7.4) and geometry of the gel in a micro-centrifugation test. The rate of water loss could be modelled as the summation of two exponential decay functions, and the degree of compression was correlated with the held water in the gel. After the gels had been subjected to the various *g* forces they were put in water and their rehydration ability, as affected by the deformation/damage occurring during centrifugation, was studied. The goal was to find conditions that did not damage the gel while still allow distinction between experimental treatments.

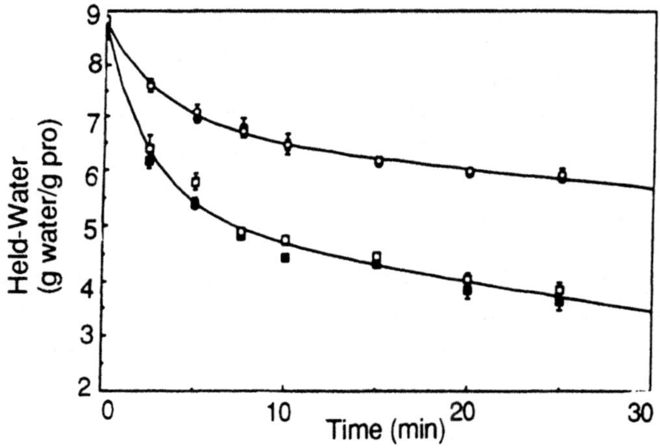

Figure 7.4 Held water kinetic curves derived from holding the force constant and increasing the time. Circles, 153 *g*; squares, 2451 *g*. Open and closed symbols differentiate between each replicate. The data from each replicate were pooled and fitted to an equation. From Kocher and Foegeding (1993).

7.3.4 Capillary suction methods

An original piece of apparatus called a capillary volumeter was developed by Hofman (1975) for use with meat samples. The equipment is based on the ability of a porous gypsum corpus to absorb moisture from a compressed meat sample. The gypsum is pressed to a small meat sample and the released fluid displaces air in an overflow pipe connected to the device. The result can be expressed as water released per cm^2 or water per g/sample. Hofman (1975) reported a maximum deviation of ±6%. Another device was described by Labuza and Lewicki (1978) in which a gel sample is placed at the bottom of a polypropylene cup, covered with a filter paper (with a known moisture content) and sealed with a rubber stopper to which a capillary glass pipe is inserted (to avoid pressure during closing). After equilibrating the cup (6°C for 72 h) the weight of the filter paper is determined. The authors tested gelatin, carrageenans, and starch samples. They reported that gel concentration, moisture of the filter paper and temperature were the main factors affecting the results; the coefficient of variation was about 2.5%.

To evaluate the effect of the suction test, WPI samples were placed on two different filter papers for various times (Figures 7.5 and 7.6). During the test, samples (19 mm diameter, 5 mm high, weighing 1.4 ± 0.05 g) were placed on a filter paper and the diameter of the water ring produced around them was recorded. The test was performed at room temperature and the samples were covered with a glass lid. The fine paper (Whatman #42) had a 2.5 μm particle retention rating and the coarse paper

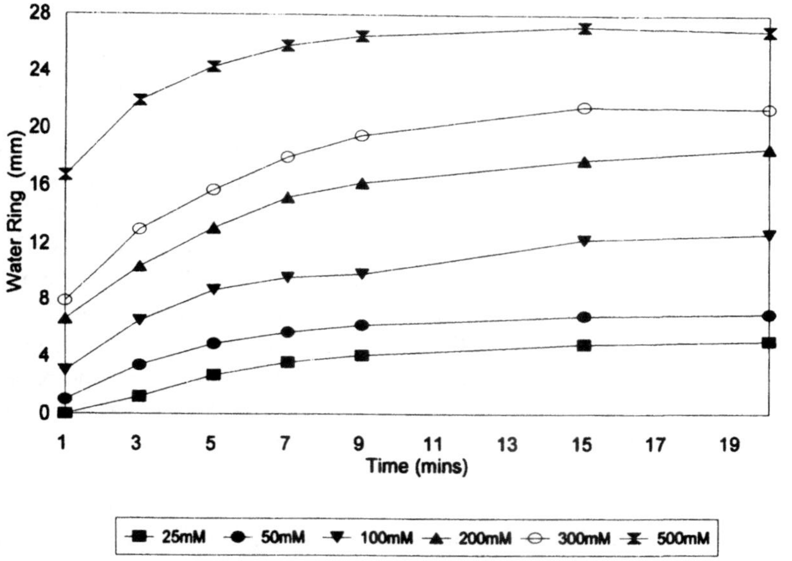

Figure 7.5 Water holding capacity of WPI gels measured by capillary action using a coarse filter paper.

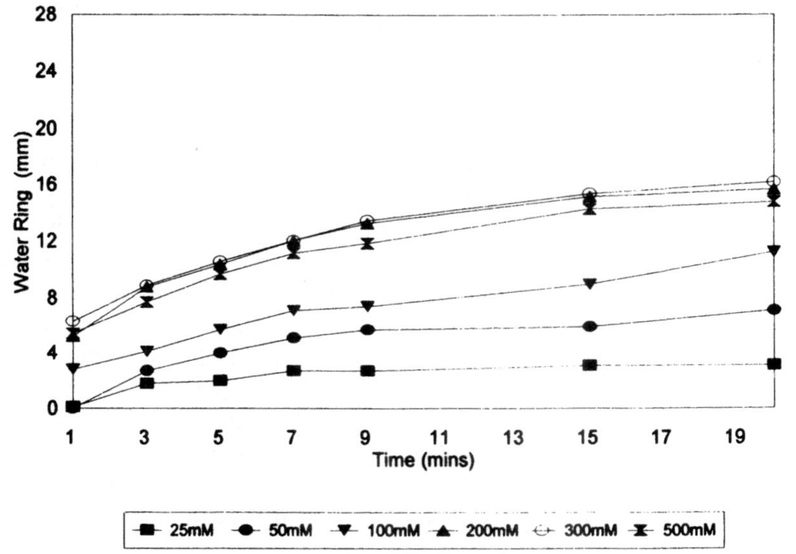

Figure 7.6 Water holding capacity of WPI gels measured by capillary action using a fine filter paper.

(Whatman #4) a 20–25 μm rating. Differences in the hydration force of the papers were observed (Figures 7.5 and 7.6); however, the curves were parallel indicating the same trend. Overall, one should select conditions (in this case a filter paper) that will provide greater distinction among the samples. In this case the coarse paper provided a much better distinction among the treatments, especially at the high salt concentrations (200–500 mM Na). The results obtained using the fine paper did not allow differentiation between the 300 and 500 mM treatments (Figure 7.6). The coarse paper also provided higher WHC values indicating its higher absorption potential. Another important consideration is the time taken to reach the equilibrium point. In this particular case equilibrium was reached after 10 min with the coarse paper and about 20 min with the fine paper. Overall, the test is very simple, relatively fast and does not require any special equipment. Therefore, it can be used to assess WHC of a variety of samples very quickly and inexpensively. However, one should remember that some gels (such as 5% gelatin prepared at different pH values; 3–10) cannot be distinguished using this test. In this case, differences in WHC would be only seen with lower protein concentration or using a higher driving force (the centrifugal test) to remove the water out of the samples (Barbut, unpublished data). In any case, the test can be used to compare samples based on equal protein concentrations.

7.3.5 *Absorption/rehydration methods*

This group of methods is used to measure water uptake by protein powders such as whey protein, casein, soybean and wheat flours. The maximum water uptake can be determined by different instruments, changes in light absorption or rheological parameters, and the results are usually expressed as g water absorbed by 1 g protein.

One popular way to follow rehydration is by using a spectrophotometer. According to this procedure, continuous measurements of the changes in light transmission are plotted as a function of time (De Wit and Klarenbeek, 1986). A sample is mixed with water and pumped into a flow cell connected to a spectrophotometer. Changes in transmission are read at 600 nm and the optical index can be used to define the kinetic relationship between the re-constitution properties and absorption characteristics of the powder. A further modification of this test was reported by De Wit (1989), where milk protein samples were removed at different times and placed in cuvettes for the measurements. Overall, two steps can be distinguished during the test; the first is wetting the sample and the second is dissolving it. Acceptable reproducibility of the method has been reported by the authors.

Another way to monitor water absorption is to follow the change in viscosity of a protein–water solution. A rotation viscometer has been used to calculate hydration of casein solutions (Korolczuk, 1982), casein micelles (Dewan *et al.*, 1973) and whey protein concentrates (McDonough *et al.*, 1974). The method can show the increase in viscosity as a function of hydration time. Korolczuk (1982) used an equation to calculate WHC and showed that cross-linked casein, of high viscosity, exhibited a higher hydration level than untreated casein. He evaluated over 60 casein samples and demonstrated the validity of the equation over a broad range of protein concentrations. Changes in viscosity can be also measured by the Brabender farinograph where water absorption, by flours and doughs, is studied as a change in Brabender units (usually development time to reach a level of 500 Brabender units). The procedure can be used to measure stability time, tolerance index and percentage water absorbed as reported by Knightbridge and Goldman (1975), who used the equipment to classify the suitability of different milk proteins as dough ingredients. Later, Short (1980) used the method to evaluate heat-precipitated whey proteins and showed correlations between WHC and protein content of the samples.

Another popular method, commonly used for dairy proteins, is the Baumann apparatus (Baumann, 1967). According to this procedure, a small sample (20–500 mg) is sprinkled on a wetted filter paper which is placed on a glass filter and positioned on top of a funnel (temperature controlled by a water jacket) filled with water. The uptake of water is read from a graduated capillary tube attached to the funnel. The results are expressed as ml of water absorbed by 1 g of dry matter after equilibrium is reached. The time to reach equilibrium depends on the sample and can take from 20 min for whey protein concentrate, to 150 min for soybean proteins. The basic principle of this test was detailed by Enslin (1933) who developed the equipment to characterize biological and non-biological substances. Over the years, the Baumann apparatus has been used to evaluate absorption capabilities of WPC (De Wit and De Boer, 1975), water absorption of different food proteins (Hermansson, 1972), and to predict water holding of hydrocolloids to be used in low fat meat batters (Wallingford and Labuza, 1983). Hermansson (1972) reported that soybean protein was much superior to sodium caseinate and whey in terms of WHC. Wallingford and Labuza (1983) showed that the Baumann apparatus provided results correlating with viscosity, cell dough plasticity behaviour and consistency coefficients for the hydrocolloids. It should be remembered that the values obtained by this test are relative values. In addition, one should be very careful in obtaining representative results because of the very small samples being tested.

The equilibrium of water absorption at a certain relative humidity can also be used to assess water holding of different dry protein samples. Protein samples can be placed over saturated salt solutions such as NaCl, KCl and $CaCl_2$ to determine sorption isotherms. The results are usually expressed as g of water per 100 g of protein sample (Grufferty and

Mulvihill, 1990). Moisture equilibrium is attained within 24 h; however in some cases (i.e. *p*-casein) a few days might be required (Geurts *et al.*, 1974). One should also be aware of the fact that samples such as whey powders which contain lactose will show different absorption curves compared to lactose-free samples (Kinsella, 1984). Results for the test have been reported for casein (Hagenmaier, 1972; Mellon *et al.*, 1974; Rüegg *et al.*, 1974) and skim milk powders (Ozimek *et al.*, 1992).

Another method that can be used for powders is the filtration method, by which a known amount of water is mixed with the sample and the volume of released water after equilibration is measured. Such a rehydration index, reported by Rustad and Ness (1983), was obtained by rehydrating 1 g samples in 50 ml water for 1 min. The excess water was removed by a filter funnel, under moderate suction, and the weight of the filter cake was determined. A slightly modified procedure was reported by Kneifel *et al.* (1990) who determined the WHC of caseinate by sampling only an aliquot of the sample (pipetted onto the filter after equilibration). The process was timed for 5 min and showed good reproducibility.

7.3.6 Optical methods

Although some optical methods can provide very rapid estimates of WHC, they are not so widely used by the food industry. An exception is in the meat industry where colour measurements are used to sort out the pale, soft, exudative (PSE) pork meat. The relationship between colour and WHC in meat is very complex and not fully understood. However, for practical application it was found that the lightness value can be used to predict the PSE condition (Bendall and Swatland, 1989). Development of this particular optical method has not been straightforward and will be used here to illustrate the considerations for developing a new method. Overall, the evaluation of meat reflectance, used to predict WHC, correlates well with the violet and red light region (Swatland, 1995). However, in the past, various researchers have used the conventional International Commission on Illumination (CIE) system and obtained poor results. The CIE system puts much more emphasis on green light since this is the region where the human eye is most sensitive. Equipment that does not put emphasis on the green region usually shows much better correlations. Overall, if one examines the changes in meat spectral data in the visible range (400–700 nm) one can see that the middle point does not change much as WHC changes; however, the high and low ends of the curve (red and violet) change much more. Therefore, any mono-chromatic light meter set in the red region would be preferred to assess WHC. Another point that one should be aware of is the determination of the 'type' of water measured in the meat. Bendall and Swatland (1989) discussed in great detail the effect of measuring water coming from the myofibrial spaces compared with water coming from the inter-fibre and inter-fisiculae spaces, as well as the relationship between WHC and pH. Depending on the 'type' of water, the curve can either be linear (increasing from pH 5 to 7), or step-wise with a change at around pH 6. In order to compare results they summarized data from different experiments and showed that the physical location of the water determines the apparent behaviour of the curve. Therefore, when optical measurement is used, one should be careful in understanding the origin of the water. Overall, water within the myofibrial spaces is highly pH-dependent whereas inter-fibrillar water is much more dependent on capillary forces (Offer *et al.*, 1989; Hamm, 1994).

Recently, Swatland and Barbut (1995) reported that near-infra red (NIR) birefringence was correlated with WHC of raw turkey breast meat samples ($r = 0.85$, $P < 0.0005$), and with fluid losses during cooking ($r = -0.82$, $P < 0.005$). The newly developed NIR birefringence probe was almost as useful as pH, and better than a visible colorimeter paleness measurement, for predicting WHC and cooking losses of the turkey meat samples.

In the case of semi-transparent gels light reflection can also be used. Figure 7.7 shows preliminary results from the development of a method to distinguish between good and poor WHC in WPI samples (the same set of samples previously used to demonstrate the effect of the other methods). In this case, the transformation from a clear to an opaque gel structure was measured by light reflectance (L-value). As shown in the micrograph (Figure 7.1), increasing sodium resulted in increasing protein strand size up to the point where aggregation occurs. Light reflection from gels produced with 25–150 mM Na showed good separation of the treatments. These treatments showed large differences in WHC (Figures 7.2, 7.3, 7.5 and 7.6) and the colour results were significantly correlated with WHC values obtained in the other tests. Above the massive aggregation point (>150 mM Na), the gels were opaque and the response reflectance line plateaued. Overall, this example illustrates that a measurement (at the appropriate zone) which takes a few seconds can be used to assess WHC. Such an evaluation can be very useful in a production environment where rapid feedback is required to control ion concentration, protein concentration and/or process conditions.

Optical methods can also be used to monitor a specific component important in WHC. For example, collagen in meat batters can be monitored with a quartz probe which can evaluate the fluorescence intensity. Collagen (part of the connective tissue of the muscle) can be detrimental to WHC if present at high levels because its conversion to gelatin during cooking can result in excessive cooking losses. Therefore, a fast collagen content determination (based on a very characteristic fluorescence pattern) has successfully been used to predict the cooking

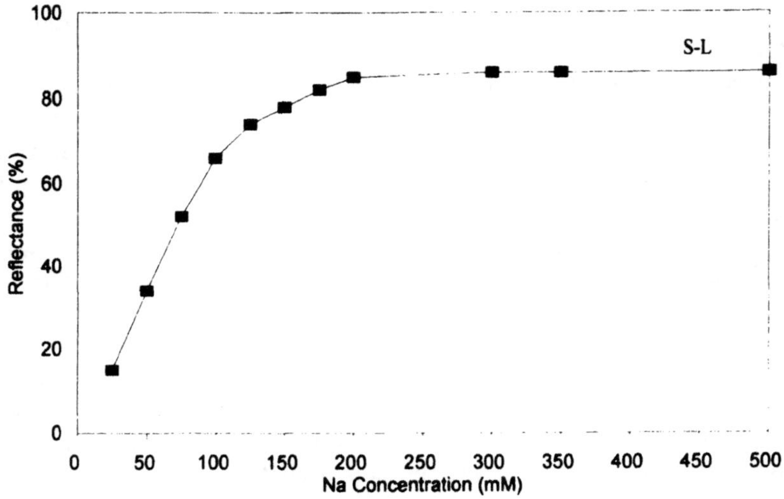

Figure 7.7 Reflectance of WPI gels showing the transformation from a clear to an opaque gel which can be used to estimate water holding.

losses ($r = 0.99$, $P < 0.005$) and WHC (Swatland and Barbut, 1991). When the latter was compared to a common centrifugation method (i.e. raw meat mixed with salt and centrifuged at 7000 *g*; Wardlaw *et al.*, 1973), a good correlation ($r = -0.92$, $P < 0.005$) was obtained.

Overall, there are many opportunities in the food industry for employing fast responding sensors and computers to reduce production costs and optimize product quality (Krol *et al.*, 1988; Swatland and Barbut, 1991; Swatland, 1995). The development of new optical sensors for monitoring WHC could save money by allowing fast feedback control of product composition, which is currently not so commonly available.

7.3.7 Microscopy

As indicated in the previous section, the structure of a gel can indirectly be assessed by light reflection/absorption. Optical methods can provide rapid assessment of strand/aggregate size and distribution. However, more precise microscopical examination (scanning or transmission) can reveal much more detail (Figure 7.1). The study of the relationship between structure and functionality is a growing area of research (Kalab, 1993; Russ, 1994) which helps provide food scientists with a better understanding of the factors affecting gelation and water holding. Proteins can form different gels depending on concentration, pH, ionic strength, temperature, etc. In general, protein gels show two distinct structures. The first category is the fine-stranded gels, produced by association of small-diameter molecules to form an ordered network (such as in WPI formed with

25 mM Na; Figure 7.1A). These gels are usually clear and have good WHC (Figure 7.3; Hermansson, 1986; Clark, 1992; Barbut, 1996). The second category is aggregated gels, typically formed by egg white proteins or casein. Aggregated gels comprise relatively large particles bound to one another to form a gel network (Figure 7.1D). These gels are usually opaque (because of the relatively large aggregates), and have low WHC. There are also mixed gels and/or gels showing different degrees of transition between the first and second categories.

The examples provided in the previous sections deal with WPI; however, the same has been shown in blood plasma, soy proteins and other preparations of whey, where increasing the salt resulted in increased random aggregation (Hermansson, 1986). Such information can be used to show how microstructure affects WHC. Increasing the pore sizes above 0.5 μm seems to show the greatest effect on WHC. Capillary forces can exhibit high force at a small pore diameter. Calculating the height of a water column that can be drawn by capillary forces (due to surface tension) shows that a 0.1 μm radius capillary will draw water up to 150 m, 1.0 μm to 15 m, 10 μm to 1.5 m, and 100 μm to 0.15 m. The corresponding water activities will be 0.90, 0.99, 0.999 and 0.9999, respectively (Hermansson, 1986).

Changing the pH or salt level in a meat batter can result in the formation of a finer protein structure. Wang and Smith (1992) showed that salt-soluble proteins (0.6 M NaCl) formed a finer structure at pH 6.5 and 7.5 than at pH 4.5; the latter showed large aggregates. Hermansson *et al.* (1986) showed that myosin heated with 0.25 M KCl (at pH 5.5) formed a fine-strand structure. Increasing the salt to 0.6 M resulted in an aggregated gel structure. Such a change will also affect the WHC. In finely comminuted meat batters, Gordon and Barbut (1992a) showed that adding urea caused a reduction in pore size, as compared to a control (2.5% salt only) and hence resulted in a much higher WHC (0.4 versus 4.8%, respectively). Overall, the relationship between a protein gel structure (i.e. pore size) and WHC still requires more work to better understand the forces governing WHC.

7.3.8 Special instruments (NMR, DSC, X-ray)

Special instruments such as nuclear magnetic resonanace (NMR) and differential scanning calorimetry (DSC) can be used indirectly to measure the degree of WHC in various foods. These instruments can provide precise information on water molecule relaxation time (by NMR), and the degree of hydration and/or amount of freezable water in a protein system (by DSC). Overall, the equipment is much more expensive than that used for the methods described earlier, and requires more extensive technical knowledge both to operate and to interpret the results. However, fundamental information on the relationship between proteins and WHC can be provided.

NMR spectroscopy measures water binding indirectly by scanning a small homogeneous sample (usually 1 to 5 g placed in an NMR tube) several hundred times, to produce a representative value for the relaxation times of the water molecules within the sample. The values indicate the times required for the magnetic nuclei of the water molecules to return to their original energy level after being excited to a higher energy level (by high-frequency radio waves in the presence of a magnetic field). The three nuclei with magnetic moments that can be used in an NMR study of water are the proton, the deutron and oxygen-17; the last two are most commonly used. After excitation, the nuclei show two relaxation times (T_1 and T_2). These represent the times required for the longitudinal (rotational) and transfer (gyrational) motions of the nuclei to return to normal. Usually the T_2 values are used to determine WHC in food systems such as muscle because they show greater changes. In pure water, T_2 values are in the order of 1–2 sec, and when compounds such as proteins are present the time is reduced by a factor of 10–150. This is because proteins can absorb some of the energy from the water, resulting in a faster return to the original low-energy state. In a muscle system, a large portion of the water is held in pores; therefore, it has a shorter distance to diffuse to the water–protein interface than in a pure water system. As a result, the water in the pores shows a much lower T_2 value. For more detailed information on the fundamentals of NMR, the reader is referred to McCarthy (1994).

NMR has been used to investigate protein–water interaction during the formation of a cheese curd (Lelievre and Creamer, 1978); the state of water in regular cheese (Callaghan *et al.*, 1983); and the kinetic properties of water in different milk proteins (Kuntz, 1971; Farrell *et al.*, 1989). Lillford *et al.* (1980) proposed a model which relates relaxation times of water molecules, in a meat system, to the size of the pores within raw, cooked, and cooked and drained samples. The data were subsequently related to WHC of the muscle and will here be used to demonstrate the value of NMR. In a raw muscle sample, water showed four distinct T_2 values (0.23, 2.5, 47.2 and 160). The first three T_2 values (<50 ms), accounting for 94% of the total water in the muscle, were attributed to intracellular water; the fourth to extracellular water. In the cooked sample, about 36% of the water showed a relaxation time greater than 1120 ms. This water was actually expressed from the muscle during cooking, since it was easy to remove by decanting. When a cooked and drained muscle sample was tested, the distribution of T_2 values was similar to that of the raw muscle. Other examples related to food systems can be found elsewhere (McCarthy, 1994).

DSC has traditionally been used to study protein denaturation, fat melting/crystallization and starch gelatinization, and can be used to detect first-order (denaturation/melting) and second-order (glass transition) transitions in foods. The theory behind the DSC equipment was reviewed by Harwalkar and Ma (1990). DSC was found to be useful when studying hydration in milk proteins (Rüegg and Blanc, 1976), and in estimating freezable water in whey protein concentrates (Berlin *et al.*, 1973). The method can also be used to evaluate the effect of salt or sugar on protein–water interactions in products containing a single protein, or in protein mixtures such as baked goods (Barbut and Findlay, 1990).

X-ray diffraction techniques have been used to evaluate structures and later relate them to WHC. Irving *et al.* (1989) used long-angle X-ray diffraction to determine the lateral separation of myofilaments in pork meat samples ranging from pale, soft, exudative (PSE) meat to dark, firm, dry (DFD) meat. They compared the results to reflectance spectra obtained by fibre optics (used to separate PSE meat, as described in section 7.3.6) and sarcomere length measurements obtained by diffraction of laser light. The separation of myofilaments ranged from 39.3 nm at a sarcomere length of 2.0 μm (mild PSE), to 45.6 nm at a sarcomere length of 1.40 μm (mild DFD). The rank order of PSE–DFD was correlated with the rank order of filament spacing ($r = 0.78$, $P < 0.05$), the rank order of 400/700 nm reflectance ($r = 0.72$, $P < 0.01$) and the rank order of sarcomere length ($r = -0.90$, $P < 0.005$). As the authors indicated, "using X-ray allowed to directly confirm, perhaps for the first time, one of the classical tenets in meat science – that myofilaments of PSE pork are more closely packed than in DFD pork".

Another technique is the use of neutron scattering and X-ray to identify water molecules within a crystalline protein structure (Schnepf, 1989). For this test, the protein should be in its crystalline form (dry). The technique does not provide practical information regarding the majority of the water in a food protein system, and is expensive to perform, although it does provide very useful fundamental information for research purposes.

7.4 Fat holding

7.4.1 Review of methods

Fat holding capacity (FHC) in a protein-based food is governed by the protein's matrix structure (strand size, pore size) as well as the type of fat and its distribution (fat droplet size, inclusion/absence of an emulsifying agent). Interactions between the proteins and fat are also very important since they can affect the gel structure, as well as the distribution and/or emulsification of the fat. This topic is covered in Chapter 4 of this book, and in other reviews (Lee, 1985; Hermansson, 1986; Gordon and Barbut, 1992b), therefore only selected references regarding these interactions will be provided here. Some of the techniques previously described for WHC are also used to measure fat holding; however, not all are appropriate. In food systems, the fat is usually dispersed within an aqueous protein phase (at least prior to cooking) and its flocculation/creaming is undesirable. Therefore, fat distribution patterns, fat globule size and the presence of an interfacial protein film (coating the fat) or cellular structure (fat cells in meat) are of major importance in FHC, as will be seen in the description of the various methods.

7.4.2 Cooking/processing

A common method of evaluating FHC is to observe fat losses during the cooking/processing steps by using a test, sometimes called the fat stability test, to monitor fat retention in a given product. The procedure usually duplicates the steps used on a production line but on a smaller scale, with a test sample of 100–1000 g. Cooking yields can also be calculated by collecting all the fluids (water and fat).

The test has been used to evaluate the effect of fat content on fat retention in hamburgers (Figure 7.8), different protein binders in meat

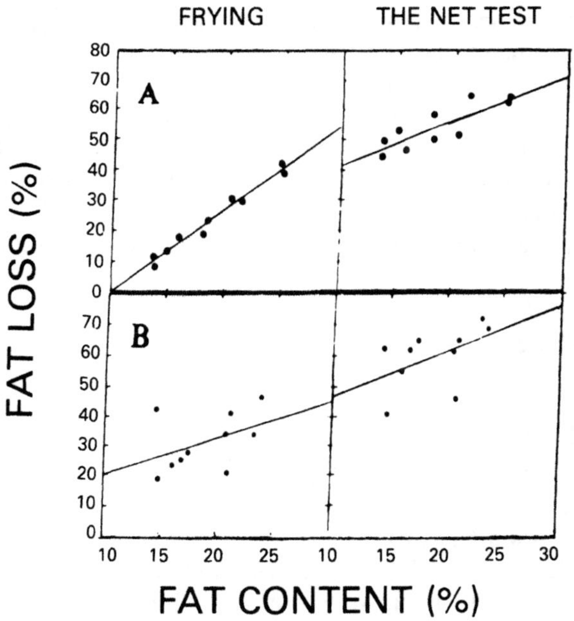

Figure 7.8 Fat loss (%) on frying (left) and as determined using the net test (right) as a function of fat content (%) for hamburger patties with: (A) adipose tissue; and (B) rendered fat. From Olsson and Tornberg (1991).

systems (Comer and Allan Wojtas, 1988), the effect of lipid composition on fat retention (Whiting, 1987), emulsifying schedules (Jones and Mandigo, 1982), and salt level on batter stability (Barbut, 1989). Whiting (1987) used the test to study the effect of a broad spectrum of chemical compounds (nonionic detergents, alcohols, monoglycerides, chelating agents, etc.) on fat holding in a food protein system. He showed that cations from groups IA and IIA of the periodic table equalled or surpassed the stability obtained in meat batters prepared with sodium chloride (the standard ingredient used to extract meat proteins). The introduction of cations such as zinc greatly reduced fat holding. Nonionic detergents, alcohols and monoglycerides were detrimental for both FHC and WHC. Other compounds such as urea, which stabilized hydrophobic and peptide groups, helped to improve FHC. The cooked fat holding test was beneficial, in Whiting's study, to provide more information about the mechanisms associated with fat binding. Olsson and Tornberg (1991) used the test for a more practical application concerning the relationships between fat level and fat retention in fried hamburgers. This relationship is of importance to meat processors, restaurant operators and the consumer in terms of yield and product acceptability. Figure 7.8 shows that fat losses were minimal at low fat content and increased linearly with increasing fat content ($r = 0.98$). When rendered fat (fat boiled at 80°C for 30 min and then filtered to remove cell wall material) was used fat losses, even at the low fat level, were much higher than when regular adipose tissue was used. The reason for this was the absence of the cell wall structure (connective tissue) to hold the fat in place during frying. The authors also used the net test and light microscopy to study this phenomenon; these will be discussed in sections 7.4.4 and 7.4.8, respectively.

One major advantage of this test is that a relatively large representative sample is used. In addition, the results can be used to predict the yields from a large-scale production line. Some of the disadvantages are that the method is not as rapid as other techniques and can only be used to evaluate cooked products.

7.4.3 High-speed centrifugation

High-speed centrifugation can be used to accelerate fat separation in various foods and provide an estimate of the expected rate of fat separation or stability.

With dairy proteins the test has been used to evaluate the stability of fat in various products. Aguilera and Kessler (1989) used centrifugation to obtain a syneresis index in protein gels (WP and skim milk) containing different levels of cream (0, 7, 20%). In the meat industry the test is not very popular; however, it has been used to assess the effects of different chloride salts on meat batter stability. A g force of 18 000 was used to separate the fat in raw meat batters prepared with NaCl, $MgCl_2$ and $CaCl_2$ (Gordon and Barbut, 1989). The test showed that Mg caused significantly higher fat release than the other divalent cation, Ca (7.9 versus 6.0%) and that fat destabilization in meat batters, by divalent salts, is therefore not only salt-dependent as previously reported but also depends on the type of ion used.

7.4.4 Low-speed centrifugation

Low-speed centrifugation, or the net test, is more commonly used in studying fat stabilization. As described for WHC, the test employs low centrifugation of the product in order to mildly 'stress' the system without causing any textural damage to the structure.

A comparison of results obtained by the net test and the cooking (frying) test for hamburger patties is provided in Figure 7.8. Fat losses during frying were minor in low-fat hamburgers and increased linearly with fat content ($r = 0.98$). The corresponding net test results also showed a linear increase ($r = 0.88$), but fat losses were significantly higher (about 30%; $P < 0.01$) than for the frying test. The authors suggested that the higher values observed in the net test were due to the centrifugation step which followed cooking. In their net test, 10 g samples were cooked in test tubes (77°C for 35 min) and later centrifuged for 20 min at 500 *g*. On the other hand, fat losses during frying were determined by frying (175°C for 3.5 min on each side; centre temperature 70°C). The authors also suggested that the longer cooking time used for the net test allowed fat to coalesce more and therefore separate more easily from the product. The two lower graphs in Figure 7.8 show the effect of using rendered fat. In this treatment, fat losses were higher for both the frying and the net test as compared to the non-treated fat. The rendered fat could easily migrate out of the product, and this was the most important factor governing FHC. Overall, comparing the two methods emphasizes that in order to develop a test procedure, the methods should first be evaluated independently. Only later can they be correlated so that the right conversion factor will be determined. Another example of using the net test with meat samples was provided by Ofstad *et al.* (1993) who studied the effect of heating temperature on both FHC and WHC in cod and salmon. They reported using 210 *g* for 15 min at 10°C.

Centrifugation force (900 *g*) was used to assess the creaming rate in milk (Walstra and Oortwijn, 1975). The so-called 'creaming pipettes' were filled with milk and centrifuged at 40°C. Creaming was expressed as the proportion of fat that had reached the cream layer (upper 5 ml). Walstra and Oortwijn (1975) showed that with increasing fat content, creaming rate first increased and then decreased. For gravity creaming (samples kept in the same pipettes but at 20°C for 24 h), creaming rate diminished with increased fat content and was further decreased for very small fat globules.

The same creaming tests were used by Lee *et al.* (1992) for caseinate and whey proteins. Haque and Kinsella (1988) used centrifugation (769 *g* at 25°C for 2, 4, 8 and 16 min) to measure turbidity (at 600 nm) of a sample mixed with sodium dodecyl sulfate (SDS)-containing buffer, and calculated the interfacial surface area of protein (BSA) used to prepare the emulsion.

7.4.5 *Emulsification test*

In this test a homogenizer (or sometimes a food processor) is used to emulsify oil added slowly (at a constant rate, through a burette) to a known protein solution. The fat is emulsified within the protein solution up to a point where a 'breakdown' occurs when the oil and protein phases separate. This separation can be seen by the abrupt change in viscosity and/or by a change in the sound of the mixer (Maurer *et al.*, 1969). Changes in electrical conductivity can also be monitored since a continuous fat phase has much higher resistance to electrical conductivity. Crenwelge *et al.* (1974) improved the test precision by using an electrical circuit to determine the emulsion breakdown point. They introduced a multivolt meter which monitored the steady rise in amp of the blender motor as the viscosity increased because of oil addition. When no more oil could be emulsified, a sudden drop in viscosity could be measured. This point correlated with a drop in amp and was used to determine precisely the emulsion breakdown point. Bovine haemoglobin produced the best emulsions and was followed by cottonseed flour, soy concentrate and nonfat dry milk.

The test is usually carried out at 20–25°C (temperature can have a significant effect) and a vegetable oil (such as cottonseed oil) is slowly added at a rate of about 1 ml/sec. The test has been used to evaluate emulsifying capacity and stability of soy, caseinate and nonfat dry milk (Pearson *et al.*, 1965). Maurer *et al.* (1969) used the test to characterize proteins from chicken breast muscle. They showed that salt-soluble proteins had a higher emulsifying capacity compared to salt-soluble proteins from which salt was removed by dialysis prior to the test, or the same proteins to which salt was added again after dialysis. Acton and Saffle (1972) compared results obtained from different studies using this test and showed that the same general type of meat (cow, pork, beef, etc.) differed among studies because of variations in equipment design, blending speed, rate of oil addition, etc. They indicated that the relative rank in emulsifying capacity, within a given group of meats, will vary depending on the basis by which emulsification capacity is expressed. These general concerns have been already discussed in this chapter and highlight the need to develop standardized methods.

7.4.6 *Solvent extraction*

Hexane extraction has been reviewed by Tinbergen and Olsman (1979). The basic assumption used in this test is that fat which is not stable, by being surrounded by an interfacial protein film or cellular membrane, will be extracted by the solvent. The method has been used for dried milk products (Graf and Bauer, 1976), as well as for meat (Tinbergen and Olsman, 1979). The latter authors observed relatively good correlations between the amount of fat separating during cooking and fat extracted by hexane. Further investigation of the samples by microscopy (after extraction) showed that the remaining fat was actually entrapped in fat cells. Tornberg *et al.* (1989) reported that 60–70% of the fat in hamburger (formulated with 20% fat) was unstable and could be extracted by hexane. This value was about the same as that measured by the net test (Figure 7.8), suggesting that most of the unstable fat was released from the product during centrifugation in the net test.

Tornberg and Ediriweera (1988) also used the test to assess the emulsifying capacity of caseinates, soy, whey and blood plasma proteins. They used hexane to estimate the degree by which oil could be extracted. The results correlated with oil separation by centrifugation (an 'accelerating' technique, as they also referred to it) and the increase in fat droplet size. They indicated that even though the extraction method is empirical, it is capable of measuring coalescence stability over a wide range, whereas the centrifugation method only provided information about the final stage of emulsion instability. Overall, they reported that caseinates provided the highest stability followed by whey and blood plasma (medium stability) and soy (minimal stability).

7.4.7 *Optical methods*

Spectrophotometry is not commonly used to assess fat binding in food products but is relatively widely used to assess the absolute fat content in foods by using infrared (Norris, 1984). However, the drive to develop rapid methods to assess conditions of food processing lines has sparked interest in such methods. One example is the development of a fibre optic probe to predict lipid content and processing losses of comminuted meat batters. A bifurcated light guide was developed to measure reflectance of different lean beef and fat mixtures (Swatland and Barbut, 1990). In the development work a range of 400–1000 nm was investigated. Reflectance at 1000 nm correlated best with lipid content of the meat/fat mixtures ($r = 0.99$; $P < 0.005$). At 930 nm, centrifugation fluid loss was significantly correlated ($r = 0.77$; $P < 0.005$) with fat content of the mixtures. Such a probe can be calibrated against a reference and provides rapid response to control composition of meat slurries in a continuous processing line. As described in section 7.3.6, the advantages of light measuring devices/probes are their very fast responses, their convenient use for on-line measurements, and their relatively simple operation.

7.4.8 Microscopy

Examining the microstructure of a food product can provide an indication of the fat stability within the product (Lee, 1985; Heertje, 1993; Kalab, 1993). However, it should be mentioned that there might be a difference between the microstructure viewed at the raw state and cooked state. This comment is made since data show that examining the microstructure of a raw product might indicate an acceptable fat dispersion, which might not hold during the cooking stage.

In the case of fat stability, one looks for both the protein and the fat phase stability. In various fat/protein mixtures an intermediate layer separates the fat and an aqueous phase (an emulsifier) is seen. Examining the microstructure can reveal information on the interactions between the fat and the aqueous phase. These interactions are very important in understanding some of the methods used to determine FHC and thus will be highlighted here.

In the study previously mentioned, comparing fat holding capacity in hamburgers, Olsson and Tornberg (1991) used light microscopy to examine the raw and cooked structures. They showed that when regular fat trimmings were used, intact fat cells were dispersed within the protein matrix. However, rendered fat resulted in large fat pools and only a few small fat droplets. After frying, connective tissue around the fat cells (i.e. cell walls, stained with aniline blue) still existed to a certain degree. This, as evaluated by the frying and net tests, was an important factor in determining FHC. In the rendered fat treatment hardly any connective tissue around the fat cells was seen (since cell walls were removed during rendering). This resulted in higher fat losses compared to the fat trimming treatment (Figure 7.8). Besides being able to assess fat losses from the sample, microscopical studies can be used to learn more about the reason(s) for fat separation and to help in suggesting ways to improve the product. Some of the new image analysis systems can be employed to measure automatically and calculate fat globules size, area and distribution (Russ, 1994).

Jost *et al.* (1989) showed that microscopy can be beneficial in evaluating fat stability of oil-in-water emulsions (Figure 7.9A). They prepared sunflower oil emulsions (15% v/v) with whey protein concentrate (WPC) (7.5%) by homogenizing the mixture and then heating it (90°C for 30 min). In the uncooked emulsions perfectly spherical fat globules could be seen. Following cooking the structure changed and rough coating on the fat globules (described as 'cauliflower like') could be seen. This was the result of the WPC coagulating around the fat globules and forming a stabilizing interfacial protein film. When film formation was disrupted by adding lecithin (during the homogenization step) the lipid droplets showed smooth surfaces. This was the result of devoiding protein adsorption to the fat and resulted in considerably weaker gels compared to the ones produced with WPC. Jost *et al.* (1989) used both scanning and transmission electron microscopy for their evaluations. This stresses the need for using comparative methods to evaluate the microstructure of food products. This point cannot be over stressed in food microstructure studies, mainly because of the potential artefacts associated with each of the microscopy

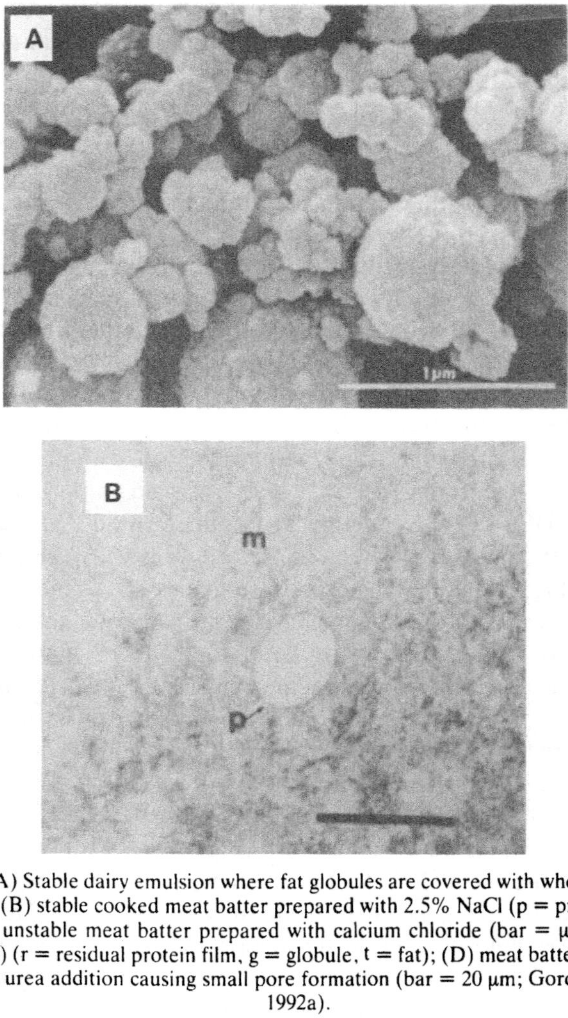

Figure 7.9 (A) Stable dairy emulsion where fat globules are covered with whey proteins (Jost *et al.*, 1989); (B) stable cooked meat batter prepared with 2.5% NaCl (p = protein film, m = matrix); (C) unstable meat batter prepared with calcium chloride (bar = µm; Gordon and Barbut, 1989) (r = residual protein film, g = globule, t = fat); (D) meat batter with very high WHC due to urea addition causing small pore formation (bar = 20 µm; Gordon and Barbut, 1992a).

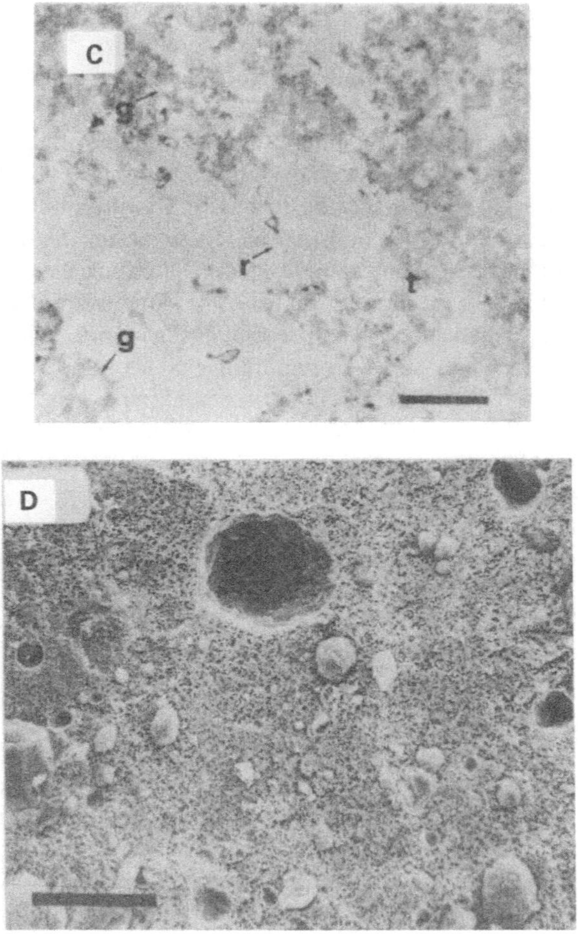

Figure 7.9 *Continued*

methods mentioned so far. Jost *et al.* (1989) also used microscopy to show that repeated homogenization (no lecithin present) increased the surface area available for protein adsorption (seen as smaller fat droplets), which actually resulted in increased gel strength. Under these conditions the small fat globules assumed the role of filler particles.

Aguilera and Kessler (1989) demonstrated the importance of fat in creating filler particles. They prepared single-component, mixed (two proteins) and filled protein (protein and fat) gels from WPI, skim milk and cream. Mixing the two proteins resulted in higher gel strength; adding homogenized fat further strengthened the gels. The important factors in achieving maximum strengthening from the fat component (as viewed by SEM and TEM) were the reduction of fat globule size, the coating of the fat with protein and, later on, the compatibility of the fat globules with the protein matrix.

In products such as finely comminuted meat batters, there has been a long debate regarding the mechanism(s) affecting FHC (Lee, 1985; Gordon and Barbut, 1992b,c). The first theory is the interfacial protein film which, according to the emulsion theory, separates the fat and water. The second is the physical entrapment theory which suggests that the fat is physically held within the protein matrix. The debate is not yet over and microstructural evidence has been presented to support each theory. In a stable meat batter a distinct protein film can be seen around the fat globules (Figure 7.9B) whereas in unstable batters, a coherent film is lacking (Figure 7.9C). In the latter case, the use of a divalent cation resulted in significantly lower protein extraction (3.0 versus 12 mg/g) and hence a lower FHC (0.15 versus 0.55%).

The characteristics of the protein matrix itself are very important in FHC. The addition of urea to a meat batter, prepared with 2.5% NaCl, resulted in a marked reduction in the protein matrix pore size (Figure 7.9D) and hence better FHC and WHC. The values obtained for FHC were 0.03% for the urea and 0.18% for the control treatment. As mentioned earlier (section 7.3.2), Whiting (1987) reported similar results for urea addition; however, in his study no microscopy was used.

Jones and Mandigo (1982) showed that the presence of 'rupture holes' in the protein envelope covering the fat globules in a meat batter can be used as a sign of instability. They showed that by varying the end-point chopping temperatures, the thickness of the protein layer around fat globules varied. When thick and elastic coats were produced (at high end-point chopping temperature), more rupture holes were observed. The authors proposed that the function of these holes was as a 'pressure release mechanism' which can assist in maintaining the expending fat (due to heating) inside the envelope.

7.4.9 *Special instruments*

NMR and DSC can be used to map fat distribution and to study the effect of processing condition on fat melting point, respectively. Duce *et al.* (1994) used NMR to map the spatial distribution of lipid protons in sunflower oil and in a meat fat emulsion. They mapped the fat directly from the NMR images in the 16 emulsions they prepared (5–70% fat) by using a phase correction algorithm to compensate for regional inhomogeneities of the magnetic field. For other examples and a detailed explanation of the procedure the reader is referred to McCarthy (1994).

DSC has been used to examine the melting points of different fat sources and to evaluate the effect of heating conditions (rate and final processing temperature) on fat stability in food products (Barbut and Findlay, 1990).

As mentioned in section 7.3.8, using such methods is very expensive and requires highly trained people to measure and interpret the results. However, the information obtained is useful for more fundamental research.

7.5 Development of a test protocol

In this section detailed descriptions of the net test for determining WHC and FHC of protein gels, and the rehydration test for protein powders will be provided. The nature of each test and references to previous works have been mentioned earlier in the chapter.

Over the years, the net test has been recommended by various research groups (Wierbicki *et al.*, 1957; Hermansson and Lucisano, 1982; Kocher and Foegeding, 1993). Since different conditions were evaluated in each of these studies, they will be used as a base for recommending test conditions. Overall, the net test employs mild centrifugation, in order to estimate WHC of a gel structure. The main advantage of the test is that it allows the removal of water (under mild pressure) without damaging the gel structure. This is a very important consideration when assessing WHC. The net or basket in which the sample is placed prevents reabsorption of the water after centrifugation. This is very important since some gels can easily reabsorb the moisture after centrifugation (Figure 7.3) and provide incorrect results. One such example is a heat-induced WPI gel made with 10% protein and 500 mM sodium. In this soft gel, moisture can easily be released when the gel is gently compressed on a flat surface. When pressure is released, all the moisture will be reabsorbed into the structure, just as if it were a household sponge. The effect of trapping, or not trapping, the released water (with a filter paper or placing the gel on a screen) at different *g* forces is presented in Figure 7.3, along with the results for the other gels (25–500 mM Na).

The test is carried out by placing a piece of the gel in an inner tube with a nylon mesh net (200 μm) at the bottom which allows drainage of the water through the bottom section (Figure 7.10). The size and weight of the gel will depend on the size and diameter of the inner test tube. In the three references given above the recommended sizes were 30, 16 and 4.8 mm, respectively, and the diameters were 30, 16 and 5.5 mm, respectively (the first two were centrifuged in the same tube they were made). For the test conditions suggested here (Figure 7.10) a 1.5 g sample (19 mm diameter × 5 mm high) is placed in a 25 mm diameter inner tube. These dimensions allow good water drainage without any blocking. The use of microcentrifuge

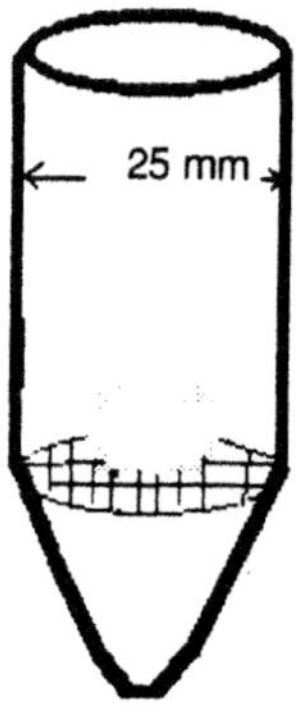

Figure 7.10 A procedure for the net test.

tubes is not discouraged but one should remember that cutting a cylindrical-shaped specimen 5 mm in diameter can be very difficult in some cases (e.g. WPI gel made with 500 mM Na) but not in others (e.g. WPI made with 50 mM Na). It is recommended that a piece of a filter paper will be placed on top of the net to prevent clogging, and to make cleaning much easier. The inner tube is positioned in a larger centrifuge tube by placing an O-ring between the middle and bottom sections of the larger tube. Different companies are now selling this kind of test tube arrangement (originally intended for protein separation). Wierbicki *et al.* (1957) and Hermansson and Lucisano (1982) built their own test tubes, but Kocher and Foegeding (1993) purchased microcentrifuge tubes from Lida Corp. This author has built some test tubes and has also purchased some from Millipore. With the wide use of microcentrifuge tubes today, prices have been substantially reduced and various companies sell disposable test tubes at reasonable prices.

The centrifugation forces recommended in the three papers referred to above were 170, 790 and 153–750 *g*, respectively. Kocher and Foegeding (1993) mentioned that the WHC curve (as a function of *g* force) could be divided into two zones: above and below 750 *g*. A *g* force of 750 and not 790 is therefore recommended The recommended centrifugation time is 10 min, suggested by Wierbicki *et al.* (1957) and shown by Kocher and Foegeding (1993) to be an optimal time (Figure 7.4). The results presented here for the WPI set of samples (Figure 7.3) were obtained by placing samples weighing 1.5 ± 0.05 g (19 mm diameter, 5 mm high) on a screen (200 μm mesh) and centrifuging at 200 and 790 *g* for 10 min at room temperature. The test can be used to determine both WHC and FHC in protein gels, as demonstrated by Ofstad *et al.* (1993) who used it for fish meat gels.

For powders the most popular test is the solubility test (Morr and Ha,

1993) which determines the percent soluble proteins and is important in assessing processing parameters such as protein extraction and drying (Morr *et al.*, 1985). However, the test is not directly related to WHC. It is interesting to note that in a small survey done by the author at seven major North American universities, most research groups indicated that when comparing WHC of protein powders, they would first prepare a gel (commonly 10% protein) and then run the net or an equivalent test. In any case, the method developed by De Wit and Klarenbeek (1986) to determine both rehydration and solubility of powders is very useful. This method was developed to estimate the re-constitution properties of protein-rich powders. It is based on a continuous spectrophotometric measurement of the changes in transmission of dispersed protein solutions as a function of time. The method is combined with the AOCS (1985) method. The device consists of a cylindrical tube (2 cm diameter) with a fritted glass bottom (150–200 μm pore size) on which 100–200 mg dry protein sample (quantity depending on the hydration properties) is placed. The glass tube is connected to a flow-through cell (5 cm) in the spectrophotometer (set at 600 nm). The time–density relationships are plotted and an optical index is calculated. To perform the test, 20 ml of distilled water at 20°C is added and the spectrophotometer reading is zeroed. A peristaltic pump then circulates the water at a rate of 6 ml/min. The powder is wetted by the circulating water via a 1.25 m plastic tube (3 mm inner diameter) which maintains an air-tight connection between the flow-through cell and the side of the cylindrical tube. During the test, two steps are observed: (i) wetting of the powder; (ii) dispersion and dissolving of the powder. The internal transmission density of the dispersion, denoted as $D(t)$, is recorded over a period of 20 min. Values of $D(t)$ are related to the final transmission density, indicated as $D(\infty)$, which is derived from the protein solubility test. The reconstitutability is then defined as an optical index according to the ratio:

$$\mathrm{OI}(t) = D(t)/D(\infty) \times 100$$

in which OI = optical index, indicating the rate of dispersion of a powder in an aqueous solution. In a later modification, De Wit (1989) used interval sampling of the protein–water mixture, by withdrawing with a syringe, to a cuvette, for reading at 600 nm.

7.6 Conclusions

Establishing a common method or methods for WHC and FHC will not happen overnight and will obviously require the contribution of various research groups. Some international scientific organizations are trying to undertake this challenge; however it seems that regional research groups are better suited for the cooperative work required for such tasks. For example, the USDA North Eastern working group on protein functionality has coordinated a joint effort to establish a standard test for protein foaming (Phillips *et al.*, 1990) and protein solubility (Morr *et al.*, 1985). It is currently working on standardizing textural analysis of protein-based foods (coordinated by Dr C. Lee) and the WHC test (coordinated by Dr J. Regenstein). The group (of which the present author is a member) first discusses the various methods available for a particular test, after which the participants evaluate a common protein sample (supplied by one of the members). Data and comments are collected and refinements to the procedure are suggested. The discussion, evaluation and re-evaluation process usually takes a few years but seems to be worth the effort, as has been demonstrated in the case of the foam test. European groups, e.g. that coordinated by Dr Honikel, are currently establishing a WHC test for meat products. The fact that it is currently difficult to compare results among laboratories is a problem and we should all strive to standardize the methods. Before adopting a specific method, many considerations (such as simplicity, maintaining sample integrity, and reproducibility) should be taken into account. One recommendation for a standard test is the net test which has been described in this chapter. This test has been used by different research groups since the early 1950s, and appears to address many of the concerns raised in regard to the other methods.

References

Ackerman, S.A., Cohen, C.E. and Benedict, R.C. (1981) Evaluation of the press method for frankfurter juiciness estimation. *Fleischwirtschaft*, **61**, 1169–70.

Acton, J.C. and Saffle, R.L. (1972) Emulsifying capacity of muscle protein: phase volumes at emulsion collapse. *J. Food Sci.*, **37**, 904–6.

Aguilera, J.M. and Kessler, H.G. (1989) Properties of mixed and filled type dairy gels. *J. Food Sci.*, **54**, 1213–17.

Aljawad, L.S. and Bowers, J.A. (1988) Water-binding capacity of ground lamb–soy mixtures with different levels of water and salt and internal end-point temperatures. *J. Food Sci.*, **53**, 376–8.

AOCS (1985) *Official and Tentative Methods of the American Oil Chemists' Society, Official Methods Ba 11–65*, revised edn., American Oil Chemists Society, Champaign, IL.

Barbut, S. (1996) Effect of calcium level on the structure of pre-heated whey protein isolate gels. *Lebensn-Wiss-Tech.* In press.

Barbut, S. (1989) Microstructure of reduced salt meat batters as affected by polyphosphates and chopping time. *J. Food Sci.*, **53**, 1300–4.

Barbut, S. (1994) Protein gel ultrastructure and functionality, in *Protein Functionality in Food Systems* (eds N. Hettiarachchy and G.R. Ziegler), Marcel Dekker, New York, pp. 383–434.

Barbut, S. (1995) Effect of sodium levels on the microstructure and texture of whey protein isolate gels. *Food Res. Int.*, **28**, 437–43.

Barbut, S. and Findlay, C.J. (1990) Thermal analysis of egg proteins, in *Thermal Analysis of Foods* (eds V.R. Harwalker and C.Y. Ma), Elsevier, London, pp. 126–48.

Baumann, H. (1967) Apparatur nach Baumann zur Testimmung der Flussigkeitsaufnahme von pulvrigen Substanzen. Glas-und Instrumenten-Technik. *Fachzeitschrift für das Laboratorium*, **11**, 540.

Bendall, J.R. and Swatland, H.J. (1989) A review of the relationship of pH with physical aspects of pork quality. Meat Sci., **24**, 85–126.

Berlin, E., Kliman, P.G., Anderson, B.A. and Pallansch, M.J. (1973) Water binding in whey protein concentrates. *J. Dairy Sci.*, **56**, 984–7.

Bouton, P.E. and Harris, P.V. (1972) A comparison of some objective methods used to assess meat tenderness. *J. Food Sci.*, **37**, 218–21.

Bouton, P.E., Harris, P.V. and Shorthose, W.R. (1972) The effects of ultimate pH on ovine muscle: water-holding capacity. *J. Food Sci.*, **37**, 351–5.

Brendl, J. and Klein, S. (1970) Das rehydrationsvermogen gefriergetrockneten Rindfleisches in Natriumbhlorid- und Natriumhexametaphosphat-Losungen. *Fleischwirtschaft*, **50**, 331.

Brendl, J. and Klein, S. (1972) Der Einfluß von Milcheiweißzusatzen auf das Wasserbindungsvermogen des Fleisches. *Fleischwirtschaft*, **52**, 339.

Callaghan, P.T., Jolley, K.W. and Humphrey, R.S. (1983) Diffusion of fat and water in cheese as studied by pulsed field gradient nuclear magnetic resonance. *J. Coll. Interface Sci.*, **93**, 521–9.

Clark, A.H. (1992) Gels and gelling, in *Physical Chemistry of Foods*, (eds. H.G. Schwartzberg and R.W. Hartel), Marcel Dekker, New York, pp. 263–305.

Comer, F.W. (1979) Functionality of fillers in comminuted meat products. *Can. Inst. Food Sci. Technol. J.*, **12**, 157.

Comer, F.W. and Allan-Wojtas, P. (1988) Functionality and microstructure of fillers in comminuted meat products. *Food Microstruct.*, **7**, 25–46.

Crenwelge, D.D., Dill, C.W., Tylor, P.T. and Landmann, W.A. (1974) A comparison of the emulsification capacity of some protein concentrates. *J. Food Sci.*, **39**, 175–7.

Dagbjartsson, B. and Solberg, M. (1972) A simple method to determine the water-holding capacity of muscle foods. *J. Food Sci.*, **37**, 499–500.

Dewan, R.K., Bloomfield, V.A., Chudgar, A. and Morr, C.V. (1973) Viscosity and voluminosity of bovine milk casein micelles. *J. Dairy Sci.*, **56**, 699–705.

De Wit, J.N. (1989) Modified method for the determination of rehydration properties of milk protein products. Internal method, De Melkindustrie Veghel BV, Product Research and Analysis Division, Veghel, The Netherlands.

De Wit, J.N. and De Boer, R. (1975) Ultrafiltration of cheese whey and some functional properties of the resulting whey protein concentrate. *Neth. Milk Dairy J.*, **29**, 198.

De Wit, J.N. and Klarenbeek, G. (1986) Method for the determination of rehydration and solubility of powdered protein-rich milk products. *Milchwissenschaft*, **41**, 463–7.

Doi, E. (1993) Gels and gelling of globular proteins. *Trends Food Sci.*, **4**, 1–5.

Duce, S.L., Ablett, S., Guiheneuf, T.M., Horsfield, M.A. and Hall, L.D. (1994) Quantitative determination of water and lipid in sunflower oil and water/meat/fat emulsions by nuclear magnetic resonance imaging. *J. Food Sci.*, **59**, 808.

Eide, O., Borresen T. and Strom, T. (1982) Minced fish production from capelin (*Mallotus villosus*). A new method for gutting, skinning and removal of fat from small fatty fish species. *J. Food Sci.*, **47**, 347–54.

Enslin, O. (1933) Uber einen Apparat zur Messung der Flussigkeitsaufnahme von quellbaren und porosen Stoffen und zur Charakterisierung der Benetzbarkeit. *Die Chem. Fabrik*, **6**, 147.

Farrell, H.M., Pessen, H. and Kumosinski, T.F. (1989) Water interactions with bovine caseins by hydrogen nuclear magnetic resonance relaxation studies. *J. Dairy Sci.*, **72**, 562–74.

Furukawa, T. and Ohta, S. (1982) Mechanical and water-holding properties of heat-induced soy protein gels as related to their structural aspects. *J. Text. Study*, **13**, 59.

Geurts, T.J., Walstra, P. and Mulder, H. (1974) Water binding to milk protein with particular reference to cheese. *Neth. Milk Dairy J.*, **28**, 46–72.

Gordon, A. and Barbut, S. (1989) The effect of chloride salts on the texture, microstructure and stability of heat batters. *Food Microstruct.*, **8**, 271–83.

Gordon, A. and Barbut, S. (1992a) Effect of chemical modification on the microstructure of cooked meat batters microstructure. *Food Struct.*, **11**, 133–46.

Gordon, A. and Barbut, S. (1992b) Mechanisms of meat batter stabilization: a review. *Crit. Rev. Food Sci. Nutr.*, **32**, 299–332.

Gordon, A. and Barbut, S. (1992c) The effect of chloride salt on protein, extraction and interfacial protein film formation on meat batters. *J. Sci. Food Agric.*, **58**, 227–38.

Graf, E. and Bauer, H. (1976) Milk and milk products, in *Food Emulsions* (ed. S. Friberg), Marcel Dekker, New York, p. 122.

Grufferty, M.B. and Mulvihill, D.M. (1990) Hydration related properties of protein isolates prepared from heated milks. *J. Soc. Dairy Technol.*, **43**, 99–103.

Gumpen, S. and Martens, H. (1977) On water-binding capacity (WBC) and texture properties. *23rd European Meeting of Meat Research Workers (Moscow)*, pp. N6, 8–13.

Hagenmaier, R. (1972) Water binding of some purified oilseed proteins. *J. Food Sci.*, **37**, 965.

Hamm, R. (1986) Functional properties of the myofibrillar system and their measurements, in *Muscle as Food* (ed. P. Bechtel), Academic Press, Inc., Orlando, FL, pp. 135–99.

Hamm, R. (1994) The influence of pH on the protein net charge in the myofibrillar system, American Meat Institute, Recipient Meat Conference, University Park, PA, pp. 5–9.

Haque, Z. and Kinsella, J. (1988) Emulsifying properties of food proteins: bovine serum albumin. *J. Food Sci.*, **53**, 416–20.

Harwalkar, V.R. and Brown, R.J. (1989) Symposium: water–protein interactions. Introduction. *J. Dairy Sci.*, **72**, 561.

Harwalkar, V.R. and Ma, C.Y. (1990) *Thermal Analysis of Foods*. Elsevier, London.

Heertje, I. (1993) Microstructural studies in fat research. *Food Struct.*, **12**, 77–94.

Hermansson, A.-M. (1972) Functional properties of proteins for food-swelling. *Lebensm.-Wiss. u. Technol.*, **5**, 24–9.

Hermansson, A.-M. (1986) Water and fatholding, in *Functional Properties of Food Macro-Molecules* (eds J.R. Mitchell and D.A. Ledward), Elsevier Applied Science, London, p. 273–314.

Hermansson, A.-M. and Akesson, C. (1975) Functional properties of added proteins correlated with properties of meat systems. *J. Food Sci.*, **40**, 595.

Hermansson, A.-M. and Lucisano, M. (1982) Gel characteristics – waterbinding properties of blood plasma gels and methodological aspects on the waterbinding of gel systems. *J. Food Sci.*, **47**, 1955–9.

Hofman, K. (1975) Ein neues Gerät zur Bestimmung der Wasserbindung des Fleisches: des 'Kapillar-Volumeter.' *Fleischerwirtschaft*, **55**, 25.

Huber, H. (1982) Water binding of rye flour during its processing, in *Proceedings of the 7th World Cereal Bread Congress*, The Cereal Technologists Organization, Prague, Czechoslovakia, p. 765.

Irving, T.C., Swatland, H.J. and Millman, B.M. (1989) X-ray diffraction measurements of myofilament lattice spacing and optical measurements of reflectance and sarcomere length in commercial pork loins. *J. Anim. Sci.*, **67**, 152–6.

Jauregui, J.M., Regensten, J.M. and Baker, R.C.A (1981) A simple centrifugal method for measuring expressible moisture. *J. Food Sci.*, **46**, 1271–3.

Jones, K.W. and Mandigo, R.W. (1982) Effect of chopping temperature on the microstructure of meat emulsions. *J. Food Sci.*, **47**, 1930.

Jost, R., Danneberg, F. and Rosset, J. (1989) Heat set gels based on oil-in-water emulsions: an application of whey protein. *Food Microstruct.*, **8**, 23–8.

Kabus, M. (1972) Wasserbindung bei Milcheiweißpraparaten. *Milchforsch. Milchprax.*, **14**, 44.

Kalab, M. (1993) Practical aspects of electron microscopy in dairy research. *Food Struct.*, **12**, 95–104.

Kinsella, J.E., Whitehead, D.M., Brady, J. and Bringe, N.A. (1989) Milk proteins: possible relationships of structure and function, in *Developments in Dairy Chemistry – 4. Functional Milk Proteins* (ed. P.F. Fox), Elsevier Applied Science, London, pp. 55–95.

Kneifel, W. and Seiler, A. (1993) Water holding properties of milk protein products – a review. *Food Struct.*, **12**, 297–308.

Kneifel, W., Abert, T. and Luf, W. (1990) Influence of preheating skim milk on water-holding capacity of sodium salts of caseinates and coprecipiates. *J. Food Sci.*, **55**, 879–80.

Kneifel, W., Paquin, P., Abert, T. and Richard J.-P. (1991) Water-holding capacity of proteins with special regard to milk proteins and methodological aspects – a review. *J. Dairy Sci.*, **74**, 2027–41.

Kneifel, W., Abert, T., Sendai, M. and Seiler, A. (1992) Water-holding properties of milk protein products as characterized by three different methods. *Milchwissenschaft*, **47**, 139–42.

Knightbridge, J.P. and Goldman, A. (1975) Water absorptive capacity of dried milk products. *N.Z. J. Dairy Sci. Technol.*, **10**, 152–7.
Kocher, P.M. and Foegeding, E.A. (1993) Microcentrifuge-base method for measuring water holding of protein gels. *J. Food Sci.*, **58**, 1040–6.
Korolczuk, J. (1982) Hydration and viscosity of casein solutions. *Milchwissenschaft*, **37**, 274–6.
Krol, B., van Roon, P.S. and Houben, J.H. (1988) *Trends in Modern Meat Technology 2*, Wageningen Press.
Labuza, T.P. and Lewicki, P.P. (1978) Measurement of gel water-binding capacity by capillary suction potential. *J. Food Sci.*, **43**, 1264–73.
Lee, C.M. (1985) Microstructure of meat emulsions in relation to fat stabilization. *Food Microstruct.*, **4**, 63–72.
Lee, S.-Y., Morr, C. and Ha, E.Y.W. (1992) Structural and functional properties of caseinate and whey protein isolate as affected by temperature and pH. *J. Food Sci.*, **57**, 1210–13.
Lelievre, J. and Creamer, L.K. (1978) An NMR study of the formation and syneresis of renneted milk gels. *Milchwissenschaft*, **33**, 73–6.
Lillford, P.J., Clark, A.H. and Jones, D.V. (1980) in *Water in Polymers*, ACS Symposium Series #127, American Chemical Society, Washington, D.C., p. 177.
Luther, H., Weber, E. and Schuster, E. (1983) Ein Beitrag zur Charkterisierung der Wasserbindungskapazitat von Proteinen. *Nahrung*, **27**, 265.
McCarthy, M.J. (1994) *Magnetic Resonance Imaging in Foods*, Chapman & Hall, New York.
McDonough, F.E., Hargrove, R.E., Mattingly, W.A., Posati, L.P. and Alford, J.A. (1974) Composition and properties of whey protein concentrates from ultrafiltration. *J. Dairy Sci.*, **57**, 1983.
Maurer, A.J., Baker, R.C. and Vadehra, D.V. (1969) Kind and concentration of soluble protein extract and their effect on the emulsifying capacity of poultry meat. *Food Technol.*, **23**, 575–7.
Mellon, E., Korn, A.H. and Hoover, S.R. (1947) Water absorption of proteins. I. The effect of free amino groups in casein. *J. Am. Oil Chem. Soc.*, **69**, 827.
Min, L. and Ni, C. (1989) Research in improving WHC of meat in sausage products. In *Proceedings of the 35th International Congress of Meat Science (Denmark)*, Vol. 3, International Meat Science Association, pp. 781–6.
Mohsenin, N.N. (ed) (1986) Retention of water in food and agricultural materials, in *Physical Properties of Plant and Animal Materials*, Gordon and Breach, New York, p. 55.
Morr, C.V. and Ha, E.Y. (1993) Whey protein concentrates and isolates: processing and functional properties. *CRC Crit. Rev. Food Sci.*, **33**, 431–76.
Morr, C.V., German, B., Kinsella, J., Regenstein, J., Van Buren, J., Kilara, A., Lewis, B. and Mangino, M. (1985) A collaborative study to develop a standardized food protein solubility procedure. *J. Food Sci.*, **50**, 1715–18.
Norris, K.H. (1984) Reflectance spectroscopy, in *Modern Methods of Food Analysis* (eds K.K. Stewark and J.R. Whitakar), AVI, Westport, CT.
O'Brien, S.W., Baker, R.C., Hood, L.F. and Liboff, M. (1982) Water-holding capacity and textural acceptability of precooked, frozen, whole-egg omelets. *J. Food Sci.*, **47**, 412.
Ochiai-Yanagi, S. and Miyauchi, H. (1985) *AOCS: Official and Tentative Methods of the American Oil Chemists' Society, Official Method Ba 11–65*, revised edn. Modified soybean protein with high water holding capacity. *Cereal Chem.*, **55**, 157.
Ochiai-Yanagi, S., Miyauchi, Saio, K. and Watanabe, T. (1978) Modified soybean protein with high water holding capacity. *Cereal Chem.*, **55**, 157.
Ockerman, H.W. and Crespo, F.L. (1982) Physicochemical changes occurring during storage of precured beef blends at different temperatures and two levels of salt. *J. Food Sci.*, **47**, 849.
Offer, G. and Trinick, J. (1983) On the mechanism of water holding in meat: the swelling and shrinking of myofibrils. *Meat Sci.*, **8**, 245–81.
Offer, G., Knight, P., Jeacocke, R., Almond, R., Cousins, T., Elsey, J., Parsons, N., Sharp, A., Starr, R. and Purslow, P. (1989) The structural basis of the water-holding, appearance and toughness of meat and meat products. *Food Microstruct.*, **8**, 151–70.
Ofstad, R., Kidman, S., Myklebust, R. and Hermansson, A.-M. (1993) Liquid holding

capacity and structural changes during heating of fish muscle: cod (*Gadus morhua* L.) and salmon (*Salmo salar*). *Food Struct.*, **12**, 163–74.

Olsson, A. and Tornberg, E. (1991) Fat-holding in hamburgers as influenced by the different constituents of beef adipose tissue. *Food Struct.*, **10**, 333–44.

Ozimek, G. and Poznanski, S. (1981) Influence of an addition of textured milk proteins upon physicochemical properties of meat mixtures. *J. Food. Sci.*, **47**, 234.

Ozimek, L., Switka, J. and Wolfe, F. (1992) Water sorption properties of ultrafiltration retentate skim milk powders. *Milchwissenschaft*, **47**, 751–4.

Pearson, A.M., Spooner, M.E., Hegarty, G.R. and Bratzler, L.J. (1965) The emulsification capacity and stability of soy sodium proteinate, potassium caseinate, and non-fat dry milk. *Food Technol.*, **19**, 1841.

Philips, L.G., German, J.B., O'Neill, T.E., Foegeding, E.A., Harwalker, V.R., Kilane, A., Lewis, B.A., Mangino, M.E., Morr, C.V., Regenstein, J.M., Smith, D.M. and Kinsella, J.E. (1990) Standardizing procedure for measuring foaming properties of three proteins. A collaborative study. *J. Food Sci.*, **55**, 1441–5.

Quinn, J.R. and Paton, D. (1979) A practical measurement of water hydration capacity of protein materials. *Cereal Chem.*, **56**, 38.

Rüegg, M. and Blanc, H. (1976) Effect of pH on water vapor sorption by caseins. *J. Dairy Sci.*, **59**, 1019–24.

Rüegg, M., Luscher, M. and Blanc, B. (1974) Hydration of native and rennin coagulated caseins as determined by differential scanning calorimetry and gravimetric sorption measurements. *J. Dairy Sci.*, **57**, 387.

Russ, J.C. (1994) *Image Processing Handbook*. CRC Press, London.

Rustad, T. and Nesse, N. (1983) Heat treatment and drying of capelin mince. Effect on water binding and soluble protein. *J. Food Sci.*, **48**, 1320–2; 1347.

Schnept, M. (1989) Protein–water interactions, in *Water and Food Quality* (ed. T. Hardham), Elsevier Applied Science, London, pp. 135–68.

Serverini, M., Vizzani, A. and Cenci, C. (1984) The relationship between water holding capacity, pH and glycolysis rate in the swine muscle post mortem. *30th European Meeting of Meat Research Works (Bristol)*, pp. 158–9.

Short, J.L. (1980) The water absorption capacity of heat precipitated whey proteins. *N.Z. Jl. Dairy Sci. Technol.*, **15**, 167–76.

Sollars, W.F. (1973) Fractionation and reconstitution techniques for studying water retention properties of wheat flours. *Cereal Chem.*, **50**, 708.

Sternberg, M., Chiang, J.P. and Eberts, N.J. (1976) Cheese whey proteins isolated with polyacrylic acid. *J. Dairy Sci.*, **59**, 1042.

Swatland, H.J. (1995) *On-Line Evaluation of Meat*. Tecnomic, Lancaster, PA.

Swatland, H.J. and Barbut, S. (1990) Fibre-optic spectrophotometry for predicting lipid content, pH and processing loss of comminuted meat slurry. *Int. J. Food Sci. Technol.*, **25**, 519–26.

Swatland, H.J. and Barbut, S. (1991) Fluorimetry via a quartz-glass rod for predicting the skin content and processing characteristics of poultry meat slurry. *Int. J. Food Sci. Technol.*, **26**, 373–80.

Swatland, H.J. and Barbut, S. (1995) Optical prediction of processing characteristics of turkey meat using UV fluorescence and NIR birefringence. *Food Research International*, **28**, 227–32.

Suvakov, M.D., Radetic, P.M. and Panin, J.T. (1984) Water-holding capacity (WHC) of meat: a possible parameter in heat process calculations. *30th European Meeting of Meat Research Workers (Bristol)*, pp. 302–4.

Tarodo de la Fuente, B. and Alais, C. (1975) Solvation of casein in bovine milk. *J. Dairy Sci.*, **58**, 293.

Tinbergen, B.J. and Olsman, W.J. (1979) Fat cell rupture in a comminuted meat batter as a determinative factor of heat stability. *J. Food Sci.*, **44**, 693–5.

Thomas, M.A., Baumgartren, P.A., Board, P.W. and Gipps, P.G. (1973) Evaluation of some non-meat proteins for use in sausage. *J. Food Technol.*, **8**, 175–84.

Thomas, M.A., Baumgartner, P.A. and Hyde, K.A. (1974) Study on some of the functional properties of calcium co-precipitates in a model system. *Aust. J. Dairy Technol.*, **29**, 59.

Tornberg, E. and Ediriweera, N. (1988) Factors that influence the coalescence stability of protein stabilised emulsions, estimated from the proportion of oil extracted by hexane. *J. Sci. Food Agric.*, **46**, 93–113.

Tornberg, E., Olsson, A. and Persson, K. (1989) A comparison in fat holding between hamburgers and emulsion sausages, in *Proceedings of the 35th International Congress of Meat Science and Technology (Denmark)*, pp. 752–9.

Trout, G.R. (1988) Techniques for measuring water-binding capacity in muscle foods – a review of methodology. *Meat Sci.*, **23**, 235–52.

Vaisey, M., Tassos, L., McDonald, B.E. and Youngs, C.G. (1975) Performance of fababean and field pea protein concentrates as ground beef extenders. *Can. Inst. Food Sci. Technol. J.*, **8**, 74–8.

Wallingford, L. and Labuza, T.P. (1983) Evaluation of the water binding properties of food hydrocolloids by physical/chemical methods and in low fat meat emulsion. *J. Food Sci.*, **48**, 1–5.

Walstra, P. and Oortwijn, H. (1975) Effect of globule size and concentration on creaming in pasteurized milk. *Neth. Milk Dairy J.*, **29**, 263–78.

Wang, S.F. and Smith, D.M. (1992) Functional properties and microstructure of chicken breast salt-soluble protein gels as influenced by pH and temperature. *Food Struct.*, **11**, 273.

Wardlaw, F.B., McCaskill, L.H. and Acton, J.C. (1973) Effect of postmortem muscle changes on poultry meat loaf properties. *J. Food Sci.*, **38**, 21.

Whiting, R.C. (1987) Influence of lipid composition on the water and fat exudation and gel strength of meat batters. *J. Food Sci.*, **52**, 1130–2.

Wierbicki, E., Kunkle, L.E., Cahill, V.R. and Deatherage, F.E. (1956) Post mortem changes in meat and their possible relation to tenderness together with some comparisons of meat from heifers, bulls, steers, and diethylstilbestrol treated bulls and steers. *Food Technol.*, **11**, 69.

Wierbicki, E., Kunkle, L.E. and Deatherage, F.E. (1957) Changes in the water-holding capacity and cationic shifts during the heating and freezing and thawing of meat as revealed by a simple centrifugal method for measuring shrinkage. *Food Technol.*, **11**, 69–73.

Zayas, J.F. and Lin, C.S. (1989) Effect of the pretreatment of corn germ protein on the quality of frankfurters. *J. Food Sci.*, **54**, 1452–6.

Zhang, M., Mittal, G.S. and Barbut, S. (1993) Optimum conditions to measure water holding capacity of beef products by press method. *Journal Muscle Food*, **4**, 255–68.

Zhang, M., Mittal, G.S. and Barbut, S. (1995) Effect of test conditions on the water holding capacity of meat by a centrifugal method. *Food Sci. Technol.*, **28**, 50–5.

8 Measurement of surface hydrophobicity

S. NAKAI, E. LI-CHAN and G.E. ARTEAGA

8.1 Introduction

It is generally accepted that hydrophobicity, especially surface or effective hydrophobicity, plays an important role in the function of food proteins. Therefore, information on surface hydrophobicity is essential for understanding protein functionality.

To discuss the measurement of surface hydrophobicity, it is important to define protein hydrophobicity, especially the effects of macromolecular structure of proteins on the surface hydrophobicity. The significance of protein hydrophobicity in the mechanism of functional properties of food proteins will be explained using the concept of QSAR (quantitative structure–activity relationships).

8.1.1 Definition of hydrophobicity

The term 'lyophobic' is used to describe a solute that has little or no affinity for the solvent medium in which it is placed. When the medium is composed of water or an aqueous solution, the more specific term 'hydrophobic' is used as the descriptor. Similarly, 'lyophilic' or 'hydrophilic' are used to indicate the affinity of a solute for solvent or water.

According to Edsall and McKenzie (1983), in 1959 Kauzmann introduced a new concept of entropy-driven forces to explain protein hydrophobicity, thus delegating a minor role to the involvement of van der Waals forces in hydrophobic interaction, which had been suggested by Linderstrom-Lang in 1951. Even today, this controversy has not been completely settled.

The theory of defining the hydrophobic interaction as entropy-driven is based on the hypothesis that the hydrophobic effect is purely the result of the phobia of water from contact with hydrocarbons. This theory does not acknowledge the effect of mutual affinity between the non-polar chains through van der Waals forces, therefore the presence of water is essential for hydrophobic effect to be active.

A different approach was proposed by van Oss and his group to explain hydrophobic affinity (Nakai and Li-Chan, 1988). According to van Oss, van der Waals interaction between two different solutes depends on the Hamaker coefficient (A_{ij}). Hydrophobic effect is an attraction in aqueous medium between macromolecules, by which more hydrophobic sites, upon approaching one another, preferentially interact with each other. This attraction frequently originates from van der Waals interactions, which tend to be attractive between low surface energy materials immersed in high energy solvents such as water, i.e. when the relationships $A_{33} > A_{11}$ and $A_{33} > A_{22}$ exist, where 1 and 2 represent solutes and 3 is the water in this case.

The recent theory of van Oss (1992) states that hydrophobic attraction occurs when the interfacial free energy change defined by the following equation, is negative:

$$G^{IF}{}_{iwi} = -2\gamma^{LW}{}_{iw} - 2\gamma^{AB}{}_{iw} \tag{8.1}$$

This equation demonstrates that when two identical solutes (i) are placed in water (w), the interfacial free energy change G^{IF} consists of free energies contributed by Lifshitz–van der Waals (LW) forces and Lewis acid–base (AB) forces (electron–acceptor/electron–donor interaction) represented by the corresponding surface tensions (γ). Hydrogen-donor/hydrogen-acceptor interactions (Bronsted system) are a subset of the Lewis system. The surface tension of the solute in water is computed from contact angle measurement using the relationship:

$$\gamma_{iw} = (\sqrt{\gamma_i} - \sqrt{\gamma_w})^2 \tag{8.2}$$

Surface tension components and parameters for non-polar and polar forces are also computed from contact angles measured with non-polar liquids and at least two polar liquids, respectively. However, the contribution of the Lifshitz–van der Waals forces is appreciably smaller than the Lewis acid–base forces. In the above balance of apolar forces and polar forces, polar interactions are major forces in the so-called hydrophobic interaction. In other words, hydrophobic interaction is the case when polar interaction is weak (or conversely when polar repulsion is strong).

The most popular definition for the hydrophobicity of a protein is based on the sum or average hydrophobicity of its constituent amino acids expressed as free energy of transfer from water to ethanol, which is computed from solubility difference of the amino acids in the two solvents. The hydrophobicity of chemical compounds has also been computed using hydrophobic fragmental constants of the component fragments, derived from the distribution between two immiscible solvents, e.g. water and octanol (Rekker, 1977). Meanwhile, polarity of solvents is usually computed from distribution coefficients of selected solutes in a solvent and its immiscible counterpart (Snyder, 1978). Therefore, polarity as well as hydrophobicity is in principle based on solvation effects between solutes and solvent. This linear relationship is generally accepted due to the well-recognized theory for the linearity of free energy. Good agreement of the elution behavior of peptides on reverse phase chromatography with the predicted hydrophobicity values computed from constituent amino acid residues supports this linear relationship.

However, chromatographic behaviour of proteins is not as straightforward as it is for small peptides. The difficulty in quantifying protein hydrophobicity using chromatography is due to the fact that under any eluting conditions, i.e. changes in pH, ionic strength or solvent polarity, it is extremely difficult to maintain the native structure of protein molecules.

8.1.2 Protein structure and surface hydrophobicity

Interest in the hydrophobicity of proteins has long been focused on its contribution to the stabilization of molecular structure of native proteins and on the mechanisms of molecular folding. To achieve the minimum free energy in folding of macromolecules, non-polar or hydrophobic groups should be restricted to the interior of folded molecules, thus not being exposed to the solvent water molecules. In fact, crystallographic study of the three-dimensional structures of proteins has revealed that many hydrophobic groups are at least partly exposed on the surface of the protein molecules. The resulting hydrophobic patches play a key role in intermolecular interactions, such as the binding of small ligands or the association with other macromolecules, including protein–protein or protein–lipid interactions. However, because of the objective of this chapter, this aspect of protein hydrophobicity will not herein be discussed in detail.

There is ample evidence that the spatial folding of polypeptide chains is responsible for the differences between the NMR spectrum observed for a native globular protein and the spectrum computed for the random coil form of the polypeptide chain, which corresponds closely to the measured spectrum for denatured protein. Lecomte *et al.* (1982) stated that resonance linewidth can afford a rough estimate of the deviation from the chemical shift for random coil form. The line sharpening of NMR spectra of casein was related to an increase in flexibility of side chains (Griffin and Roberts, 1985). By applying computer-aided simplex optimization to curve-fit model NMR patterns, reasonable estimation of the exposure of the side chains was achieved (Nakai *et al.*, 1991). Surface hydrophobicity close to the theoretical value could be computed from this information. However, changes in chemical shift frequently occurring upon folding may interfere with the accuracy of assessing the exposure of side chains based on line sharpening. For proteins showing significant changes in chemical shifts upon folding, the irregularity of NMR peaks caused by changes in environment, e.g. appearance of new peaks, may be used as an index of the departure from theoretical patterns of the completely unfolded molecule. At any rate, surface hydrophobicity of protein molecules is a collective term rather than representing the exposure of hydrophobic side chains of individual amino acid residues in the protein molecules. Therefore, a crude calculation of surface hydrophobicity from NMR patterns may be justified in evaluating the reliability of different empirical measures of protein surface hydrophobicity currently in use (Arteaga, 1994).

Measurements using vibrational spectroscopy have indicated that protein molecules can show varying degrees of unfolding of tertiary and secondary structures, with different exposure of hydrophobic groups to the aqueous environment, depending on the conditions used to induce denaturation. For example, Casal *et al.* (1988) have demonstrated using infrared spectroscopy that alkaline and thermal denaturation of β-lactoglobulin proceed via a common intermediate structure, but the final structures differ. On increasing the pH from 8 to 11, the α-helical content decreases dramatically while β-sheets responsible for exposed β-strands unfold, leading to an increase in random coil. On further increase to pH 13, the remaining β-sheets unfold, although not completely. Thermal denaturation at 60°C yields an infrared spectrum for β-lactoglobulin which is superimposable on the spectrum of β-lactoglobulin at pH 11, while heating at 90°C results in aggregation and polymerization without any further significant increase in random coil, in contrast to alkaline denaturation. Using Raman spectroscopy to monitor changes during thiol- or heat-induced gelation of lysozyme, Li-Chan and Nakai (1991) similarly showed that the extent of hydrophobic group exposure and secondary structural changes in lysozyme depended on the conditions of denaturation.

8.1.3 Quantitative structure–activity relationships

Another important aspect of protein hydrophobicity is an explanation of the function of proteins, especially quantitative structure activity relationships (QSAR). Functional properties of proteins depend on the interaction of the protein molecules with other components in the food system, including water, macromolecules (e.g. proteins, complex carbohydrates and lipids) and small molecules (e.g. salts, simple sugars and flavor compounds). These interactions may locate in the bulk phase, or at the surface or interface.

QSAR or simply SAR may be the most common concept in chemistry which explains the function of chemical compounds based on their structure. The uniqueness of QSAR among general concepts of the structure–function relationships is in mathematical expressions of the relationships, in most cases in the form of equations to model and predict the specific activity of compounds as functions of independent variables representing the chemical/physical properties of the compounds.

According to Hansch and Klein (1991), Hammett (1970) first introduced the use of σ-constants, representing the electronic effect of substituents in the structure of a group of compounds in a quantitative approach to the structure–activity relationships. Taft (Hansch and Leo, 1979) was then able to separate the electronic and steric effects of structural changes (E_s constant) on reaction rates as well as factor field/inductive and resonance effects of substituents. Although biochemists began to use σ-constants to rationalize the effects of substituents on the interaction of ligands with enzymes, some biochemists found that good correlations were the exception. Octanol–water partition coefficients (P) were then used by Hansch and his group (Fujita *et al.*, 1964) to define the hydrophobic character of compounds (log P), and π-constants were devised to define the relative hydrophobicity of a substituent. However, until the progress of computers facilitated multivariate data analysis in the 1960s, the work on QSAR was not greatly promoted. A new concept of computer-assisted QSAR to explain biological functions of chemical compounds was published by Stuper *et al.* (1979) by applying, for the first time, pattern recognition techniques using multivariate analysis. In their monograph, 'Hansch analysis' was defined as linear free energy relations using a first or second order linear model of parameters in relation to the hydrophobic, steric and electronic changes caused by making modifications to the molecular structure.

As a consequence of the complexity of protein molecules as macromolecules, strong criticism and resistance have frequently appeared in the literature against the application of the QSAR concept in protein chemistry, especially the application of the Hansch analysis to protein structure–function relationships. It is more or less reasonable to admit the difficulty in QSAR interpretation of protein functional properties; although being practical, in QSAR of small molecular compounds, modification of specific sites in the protein by mutation in the nucleotide sequence followed by expression and characterization of the modified proteins is time-consuming. Furthermore, published data of proteins altered by recombinant DNA techniques for mutagenesis which can probably be used in QSAR study are still presently limited. Direct modifications of proteins by chemical or enzymatic techniques are much easier; however, this approach suffers from lack of specificity of the modifications. In addition, it is quite difficult to collect sufficient numbers of samples of pure proteins from which reliable prediction equations can be derived using the QSAR approach.

Probably because of the same reasons mentioned above, most examples used in the articles of Hansch and Klein (1991) were on small molecular ligands and not on proteins *per se*. They stated that the structure–activity relationship could be investigated via two independent methods: molecular graphics and QSAR. The crystallographic studies can be used to confirm or refute the deduction about structure–activity relationships obtained for a purified enzyme in buffer, or an enzyme in a living cell or possibly in a whole organism. The second point of view provided by computer graphic models of enzymes is extremely important for confirming the QSAR results.

In an earlier review, Hansch (1978) discussed parameters to be used in QSAR. The recommended steric parameters were mostly for small molecular compounds and not apparently applicable to proteins, except for molar refractivity (MR) as an expression of polarizability of substituents of ligands in the interaction with macromolecules. After listing 92 references on QSAR of enzymes, Hansch commented that often, too few data points have been used and not enough attention paid to the problem of collinearity, and few investigators have considered the possible importance of MR. It is uncertain if MR could be a significant factor in QSAR of food proteins. Klein *et al.* (1986) reported compositional or physicochemical properties of an amino acid sequence (attribute) in 1603 protein sequences to find correct classification into 26 protein functions using nonlinear discriminant analysis. The results showed that three or four attributes were generally sufficient to distinguish each of 26 functional categories from the remainder of the database. The attributes used in this work were related to hydrophobicity, charge and distribution, e.g. frequency of occurrence, a periodicity of appearance, as well as secondary structure parameters (average propensity to form α-helix, β-sheet or β-turns).

Activity of single-site mutants of tyrosyl-tRNA synthesis and T4 lysozyme were correlated with intermolecular forces (IMF) equations as functions of polarizability, electrical effects, hydrogen bonding parameter, lone pairs, ionic side chains, charge transfer side chains and steric parameters derived from van der Waals radii. Polarizability, ionic side chains, and steric effects were the major factors in transition state binding of ATP and tyrosine by tyrosyl-tRNA synthetase (Charton, 1986). The IMF equation was derived from data for mutant enzymes, with single side chain substitutions at different sites in the sequence. The contribution from hydrogen bonding was small. In the case of T4 lysozyme (Charton, 1990), Pro86 was replaced with each of the other 19 amino acids. Bacterial cell lysis was affected by hydrogen bonding, dispersion forces and steric effects in this order. In both cases, the coefficient of determination R^2 were mostly greater than 0.9. Usually, the molecular structure is not appreciably altered by replacing one amino acid residue in the sequence, therefore simple steric effect is adequate for obtaining a high R^2 value; however, among different proteins with considerable difference in the dimensional structures, high R^2 may not always be expected. Also, when molecular structure is not appreciably altered, the extent of exposure of hydrophobic sites should not be significantly changed, thus the involvement of hydrophobicity in functions could be insignificant.

In the case of food proteins, important functionality, e.g. emulsification, foaming, gelation and breadmaking, are all affected by molecular structure levels higher than tertiary structure, which are difficult quantitatively to describe, except for rheological parameters, e.g. viscosity. Furthermore, in the case of most food proteins either their sequences are not known or they are mixtures of several fractions. Therefore, detailed IMF information on every side chain such as the case of Charton QSAR is difficult to obtain. Charge density, hydrophobicity, sulphydryl/disulphide (SH–SS) relationships and hydrogen bonding may be the parameters which can be used as the intermolecular forces for QSAR study of food proteins. It is however possible that the effect of hydrogen bonding can be included in hydrophobicity as the two are closely related as explained by van Oss (1992).

Nakai and Li-Chan (1993) reviewed the recent advances in QSAR of food proteins. Emulsifying ability was closely related with hydrophobicity; incorporation of hydrophobicity to solubility as the independent variables or factors improved R^2 of regression analysis. Foaming activity required hydrophobicity and factors pertaining to the adsorption of proteins at the interface to obtain adequate foam lamella strength. Hydrophobicity as well as other factors relating to the intermolecular interactions, e.g. Ca and SH, was involved in thermally induced gelation. For breadmaking, although no extensive QSAR work has been carried out, the important function of high molecular glutenin subunits was confirmed; notably, the critical function of hydrophobicity in breadmaking was also demonstrated.

8.2 Measurement of surface hydrophobicity

The important roles of hydrophobicity, especially surface hydrophobicity, in the mechanisms of functional properties of proteins have been confirmed in many recent investigations. However, measurement of surface hydrophobicity is still quite controversial so that no standard method has ever been established. The following is a discussion on the current status of this important problem.

8.2.1 Hydrophobicity scales for proteins

Proteins found in nature are composed of twenty amino acids. Numerous scales have been proposed to assess the relative hydrophobicities of these amino acids. Broadly speaking, these scales can be classified into: (1) those that are based on the solubility behavior in solvents of different polarity; and (2) those that are calculated using crystallographic or other data of location of amino acid residues in the molecular structure, assuming that hydrophobic residues will locate in the interior of the molecule.

The former scales include those based on the free energy of transfer of the amino acid residues or their derivatives from water to an organic solvent or vapor, or the partition coefficients measured as a solubility ratio between water and a non-polar immiscible organic solvent (Wolfenden *et al.*, 1981), as well as those based on relative retention times of amino acids and peptides during reverse-phase liquid chromatography (Wilson *et al.*, 1981).

The latter scales include those based on accessible or buried areas of the residues (Chothia, 1984; Lesser and Rose, 1990), or on the location of amino acid residues in proteins assessed either in terms of the distance from the protein center of mass and average orientation of the side chain (Meirovitch *et al.*, 1980) or in terms of the average surroundings of residue types (Manavalan and Ponnuswamy, 1978). A hydropathy scale has been reported based on both the average extent of buriedness and the energy of transfer from water to vapor (Kyte and Doolittle, 1982). Calculation of hydrophobicity values based on the hydrophobic fragmental constant of constituent fragments has also been proposed (Rekker, 1977).

Using the hydrophobicity scales of the constituent amino acids, various approaches have been proposed for calculating the values for proteins. An equation has been formulated to calculate average hydrophobicity values of proteins based on Tanford's scale for free energies of transfer of amino acid side chains from an organic to an aqueous environment (Bigelow, 1967). Bigelow's average hydrophobicity values may be calculated using information on the amino acid composition of the protein. In contrast, most other values require knowledge of the amino acid sequence or primary structure of the proteins as well. Computer programs are now commonly used to calculate histograms or plots showing 'hydrophobicity profiles' (Rose *et al.*, 1985), 'hydropathy profiles' (Krystek and Anderson, 1985) or 'amphipathic helix patterns' (Segrest and Feldmann, 1977).

Despite the potential validity of these approaches, adequate consideration should be given to the effects of location of amino acid side chains in the three-dimensional structure of protein molecules. There must be a topographic restriction requiring a certain distance between lipophilic and hydrophilic terminals in molecules to maintain affinity of oil with water. Furthermore, it is difficult to accept the concept of simple summation of hydrophobicity scales of amino acid residues to calculate corresponding hydrophobicity values or profiles for proteins. It is essential to consider the extent of exposure to the environment or availability of hydrophobic side chains to the function of the proteins.

A crucial drawback to the use of hydrophobicity scales of amino acid residues to calculate corresponding hydrophobicity values for proteins is the lack of consideration for the effect of three-dimensional structure of each individual protein on the extent of exposure of the residues. The inadequacy of this approach is especially true for calculations using scales based on the behaviour of amino acids or small peptides. Scales that were formulated on the basis of location or buriedness of residues measured in different proteins were attempted to address this problem. However, this approach would still be limited until the distribution of hydrophilic and hydrophobic residues in the molecules is taken into consideration.

The use of hydrophobicity scales is of restrictive use in universality of application, due to the need to extrapolate the behaviour of residues to proteins other than those for which data are available. Cid *et al.* (1992) proposed that different hydrophobicity coefficients should be computed from different structural classes of proteins. They compared Ponnuswamy's original hydrophobicity values, based on 21 proteins, to values based on a database containing 60 proteins. They also computed four sets of hydrophobicity values for each of four structural classes of the 60 proteins (α α, β β, $\alpha + \beta$ and α/β). The differentiation by structural changes introduced changes in the hydropathic character of a few amino acid residues and also several changes in the absolute values of the hydrophobic coefficients.

The problem is usually even more complex, due to the simultaneous occurrence of many different types of proteins. The calculation of a net or average value of hydrophobicity for all of the proteins present requires calculation of a hydrophobicity value for each protein as well as knowledge of how each value might change through possible interactions between the proteins. Furthermore, the calculated values fail to take into account the effects of processing on buriedness or surface exposure of residues. For these reasons, various methods of measuring empirical parameters that may relate to the hydrophobicity of complex protein systems such as food proteins are usually favoured over calculation of average or overall value or profiles based on the constituent amino acids.

8.2.2 Empirical scales

Since sequence and structure are not known for all proteins, empirical methods for measuring hydrophobicity are commonly used. In this section, the most popular methods for empirical assessment of hydrophobicity are discussed.

8.2.2.1 Partition in aqueous two-phase systems. The most popular method for assessing the polarity of a solute involves determining the distribution constant between water and a water-immiscible solvent, e.g. octanol. However, since difficulty in dissolving protein samples in the solvent does not allow the conventional approach, a two-phase aqueous system containing dextran and poly (ethylene glycol), or PEG, and a salt such as potassium sulfate was developed (Shanbhag and Axelsson, 1975). Esterification of PEG with a fatty acid such as palmitate is used to alter non-polarity of the system. The difference in partition of proteins in phase systems with and without the fatty acyl hydrocarbon group bound to PEG is taken as a measure of hydrophobic interaction. Length of the hydrocarbon chain may be varied to investigate the effects on the partition behavior of different proteins. For example, the effects of increasing chain length from C_{12} to C_{16} on the partitioning behaviour in an aqueous two-phase system were used in monitoring the hydrophobic surface properties of rabbit muscle myosin (Pinaev *et al.*, 1982). The method has yielded useful data (Keshavarz and Nakai, 1979), but suffers from the tediousness of the procedure and difficulties in solubilizing certain proteins, frequently floating at the interface of the two phases. Details of the construction of various aqueous phase systems for partitioning of proteins as well as other macromolecules and cells are given by Albertsson (1986).

Other aqueous two-phase systems (ATPS) which have been suggested to partition proteins according to their hydrophobicity include mixtures of detergents such as Triton X-100, Triton X-114 and Triton X-45 (e.g. Ganong and Delmore, 1991). The role of three-dimensional structure and exposure of hydrophobic residues on the partitioning behaviour of proteins has been recognized. For example, the effects of amino acid sequence on peptide and protein partitioning in PEG 3400/potassium phosphate/water systems were reported by Diamond *et al.* (1990). An equation was proposed to predict values of partition coefficients for peptides or proteins, using the partition behaviour not of free amino acids but of dipeptides. The investigators were successful in predicting the partition coefficients for di- and tri-peptides and the ratio of partition coefficients for homologous proteins. However, the effect of secondary and tertiary structure of proteins resulted in failure accurately to predict partitioning behaviour on the basis of amino acid sequence alone.

8.2.2.2 High performance liquid chromatography (HPLC). The retention time of solutes during chromatography has been used as a measure of hydrophobicity. According to the solvophobic effect theory (Melander and Horvath, 1977), the retention time of peptides depends mainly on their non-polar and polar surfaces and thus may be a good measure of hydrophobicity. In fact, this hypothesis has proved true for the behaviour of small peptides (Wilson *et al.*, 1981). However, the chromatographic behaviour of proteins is not always predictable. Possible denaturation of proteins under the harsh solvent conditions often required in reverse-phase chromatography has led to recommendations to use the milder conditions of hydrophobic-interaction chromatography. At any rate, the relevance of these retention data to hydrophobicity of native protein molecules has still been questioned. Nevertheless, differences in chromatographic retention between proteins have been demonstrated. It has been suggested that differences observed during the binding to aliphatic versus aromatic types of adsorbent (Hofstee and Otillio, 1978) may confirm the need to differentiate between these types of interaction in the hydrophobic effect

(Mozhaev and Martinek, 1984; Burley and Petsko, 1985; 1988; Hayakawa and Nakai, 1985).

Despite good explanation of the elution behavior of peptides on a reverse-phase (RP) column, difficulty in dissolving some proteins in eluent with a low polarity precludes the use of RP-HPLC for defining protein hydrophobicity. Although hydrophobic-interaction chromatography (HIC), which uses salt solutions in a negative gradient, is milder in selection of eluent than is RP-HPLC, the elution behaviour of proteins has not been fully understood. Melander *et al.* (1989) proposed a three-parameter equation:

$$\log k' = A - B \log m_s + C\, m_s \tag{8.3}$$

to explain the effect of salts on the retention behaviour of proteins in electrostatic and HIC, where k' is the retention factor and m_s is the molarity of the salt in the eluent; parameters B and C are the electrostatic interaction parameter and the hydrophobic-interaction parameter, respectively A is a constant encompassing all characteristic parameters. From $\log k'$ versus $\log m_s$ plots, values for B and C can be computed.

However, Barford *et al.* (1988) reported that the binding of salts to proteins altered the number and distribution of protein surface groups, including charged groups, thereby causing anomalous elution behaviors of proteins in HIC. Damodaran and Kinsella (1982) also discussed changes in electrostatic as well as hydrophobic forces in proteins by ions, thus inducing conformational changes in the proteins. Elution of proteins with a gradient of decreasing ionic strength and polarity may cause the adsorbent to bind the more hydrophobic proteins of the mixture so tightly that desorption may not be possible without denaturation, and the adsorbent may not bind less hydrophobic proteins. Therefore, Hofstee and Otillio (1978) suggested the use of a hydrophobicity gradient which consisted of a series of interconnected columns, e.g. from C_4 to C_8. In this case, each protein was held by the adsorbent which provided the minimum degree of hydrophobicity required for binding. Following washing, each column was disconnected and separately eluted with a mild eluent such as water–ethylene glycol mixture.

In addition to the effect of salts on protein conformation, the hydrophobicity of the stationary phase can also affect the protein conformation (Wu *et al.*, 1986; Szepesy and Rippel, 1992). Szepesy and Rippel (1992) compared HIC columns with phenyl, propyl and polyether bonded phases and eluting salts of ammonium sulfate and sodium citrate. HIC is usually conducted with weakly hydrophobic surfaces compared to RP-HPLC (RP packings commonly have ten to one hundred times the density of hydrophobic groups compared to HIC packings according to Goheen and Engelhorn, 1984) and in general with high concentrations of stabilizing salts, e.g. ammonium sulfate. However, the conformational effects can be exhibited depending on the mobile phase, the column temperature, the hydrophobicity of the stationary phase and especially the lability of the protein. It was shown that the hydrophobicity of the stationary phase is an important parameter in maintaining native state upon elution from the chromatographic column. In addition, column operation at subambient temperature was suggested as a useful procedure for HIC.

According to Szepesy and Rippel (1992), in order to exploit the potential of HIC to modulate selectively for the separation of proteins, the combined effect of the stationary phase and the type of salt should be taken into account because of the effects of both on protein conformation. This is critical for assessing the surface hydrophobicity of protein samples because the mildest conditions without affecting the native state of protein structure are difficult to find.

8.2.2.3 Binding methods. Various methods have been proposed for quantitating the binding of a non-polar or hydrophobic ligand to proteins as a measure of the protein hydrophobicity. The ligands used include aliphatic and aromatic hydrocarbons (Mohammadzadeh *et al.*, 1969a,b), sodium dodecylsulfate (Kato *et al.*, 1984), simple triglycerides (Smith *et al.*, 1983), and corn oil (Tsutsui *et al.*, 1986). Commonly, a mixture of protein solution and ligand is incubated for a specified time interval to allow interaction, followed by removal of any unbound or free ligand by dialysis, extraction, microfiltration or other similar techniques. The protein-bound ligand may then be quantitated by gas chromatographic or radioactive count analysis, spectrophotometry or by a fluorescence probe method. Although the procedures are time-consuming, these binding methods may be useful in determining empirical parameters that are relevant in assessing interaction of proteins with various non-polar compounds in foods, such as flavor compounds, vitamins, fatty acids or triglycerides.

As a detergent binding method like the SDS binding method of Kato *et al.* (1984), polyoxyethylene sodium monooleate (Tween 80) was used as a ligand (Lieske and Konrad, 1994). Instead of measuring SDS bound to protein, the interference of dye binding in a protein assay due to covering hydrophobic sites of the proteins with Tween 80 was measured.

8.2.2.4 Contact angle measurement. The quantitative determination of the Lifshitz–van der Waals (LW) and electron donor–acceptor or Lewis acid–base (AB) interactions contributing to surface tension has been extended to the case of proteins. The method is based on Young's equation:

$$\gamma_{SV} - \gamma_{SL} = \gamma_{LV} \cos \theta \tag{8.4}$$

where γ is the interfacial tension between solid (S), liquid (L) and vapour

(V). This equation describes the relationship of the advancing contact angle (θ) of drops of a liquid on a flat solid surface (van Oss *et al.*, 1981). The solid surface in this case consists of protein sample prepared as flat layers. The advancing contact angles of droplets of three different well-characterized liquids (e.g. water, glycerol and α-bromonaphthalene) on the protein surface are measured. Based on Young's equation and the known surface tensions of the three liquids, the contact angles are then used to calculate the contributions of LW and AB interactions to the surface tensions of the liquids on the layered protein.

This approach has been used to obtain values for native hydrated proteins using one liquid only, namely drops of saline water, which correlated with the hydrophobicity of proteins indicated by relative retention on hydrophobic chromatography (van Oss *et al.*, 1981). However, a much higher surface tension value is obtained by contact angle measurement of a drop of protein solution on a solid polymer surface, compared to the value obtained by measuring the angle for the air-dried protein layer, or by other measurements for surface tension at an air–liquid interface, such as platinum ring tensiometry, Wilhelmy plate and pendant drop shape. This suggests that when protein is exposed at the interface, it may reorient to a much more hydrophobic configuration than its native hydrated state (Absolom *et al.*, 1981).

It is known that the most common food emulsifiers have a tendency to crystallize during use, forming particles at the interface. These solid particles must be located at the interface in order to stabilize the emulsion by preventing coalescence of the droplets. The wetting energy needed to move an adsorbed solid particle into the most wetting phase, in order to protect the emulsion from coalescence, strongly depends on the contact angle between the liquid phase and solid particle. A contact angle of or close to 90° gives the most stable emulsion.

8.2.2.5 Intrinsic fluorescence. The intrinsic fluorescence spectrum of proteins is primarily attributed to the aromatic amino acid residues of tryptophan, tyrosine and phenylalanine. In practice, tryptophan fluorescence is the commonly studied aspect of the spectrum, because phenylalanine has a low quantum yield and tyrosine fluorescence is frequently weakened due to quenching by ionization or by interactions with amino, carboxyl or tryptophan residues (Freifelder, 1976). The fluorescence of both tryptophan and tyrosine depends significantly on their environment, both the magnitude of fluorescence intensity as well as the wavelength of maximum fluorescence emission intensity being sensitive to the polarity of the environment.

Three classes of tryptophan residues in proteins have been reported. These are the residues that are completely buried in non-polar regions of the molecule, those that are completely exposed to the surrounding water, and those that have limited contact with water and are probably immobilized at the protein surface. The wavelengths of maximum emission for these three groups of tryptophan residues are 330–332 nm, 350–353 nm and 340–342 nm, respectively (Burstein *et al.*, 1973). Measurements of intrinsic fluorescence can thus give information on buriedness of aromatic residues and on the effect of interactions with other molecules through the effect on the spectrum, including quenching (Eftink *et al.*, 1977). However, it is often difficult directly to relate this information to hydrophobic interactions, as the fluorescence characteristics may also be altered by general changes in conformation.

A new attempt was made to obtain more information on protein conformation by recording intrinsic emission spectra at different excitation wavelengths (Myers, 1988). Assuming no significant phenylalanine contribution, only tyrosine excites at 270 nm and, if its emission is efficiently relayed to tryptophan, then only the tryptophan emission is apparent. The width of the peak is a measure of the efficiency of the energy transfer; the smaller the width the greater the energy transfer. At 297 nm excitation, only tryptophan residues are excited with its emission wavelength as low as 325 nm if the residues are in a non-polar environment. However, when the tryptophan is exposed, its emission will be at about 350 nm. If a conformational change has occurred, thereby moving tyrosine residues away from tryptophan, the energy relay may be broken and both emission bands become evident using excitation at 280 or 290 nm. Thus, the fluorescence intensity can be used for assessing conformational changes due to the exposure of a fluorophore to the solvent causing a decrease in fluorescence intensity by solvent-quenching.

Figure 8.1 shows the two-dimensional fluorescence spectra of native and urea-denatured bovine α_{s1}-casein. As expected, upon unfolding, the emission spectrum shifts to longer wavelength; at 297 nm excitation the maximum emissions occur at 341.3 and 354.5 nm for the native and urea-denatured proteins, respectively. The width of emission band increases upon denaturation; at 290 nm the full-widths at half-height are 62.3 and 69.5 nm for the native and urea-denatured proteins, respectively. The decrease in the maximum fluorescence intensity also suggests the exposure of aromatic residues to solvent upon urea denaturation. The effect of urea on emission peak position of α_{s1}-casein and α-lactalbumin at two different excitation wavelengths is shown in Figure 8.2. The position of emission band in the native proteins would indicate the exposure of Trp + Tyr residues to the solvent. For α_{s1}-, β- and κ-caseins, the wavelengths for maximum emission at 297 nm excitation and pH 7.5, were 341, 340 and 337 nm, respectively, compared to 327 and 335 nm for α-lactalbumin and β-lactoglobulin, respectively. These results suggest that Trp and/or Tyr are more exposed in the caseins than in globular whey proteins.

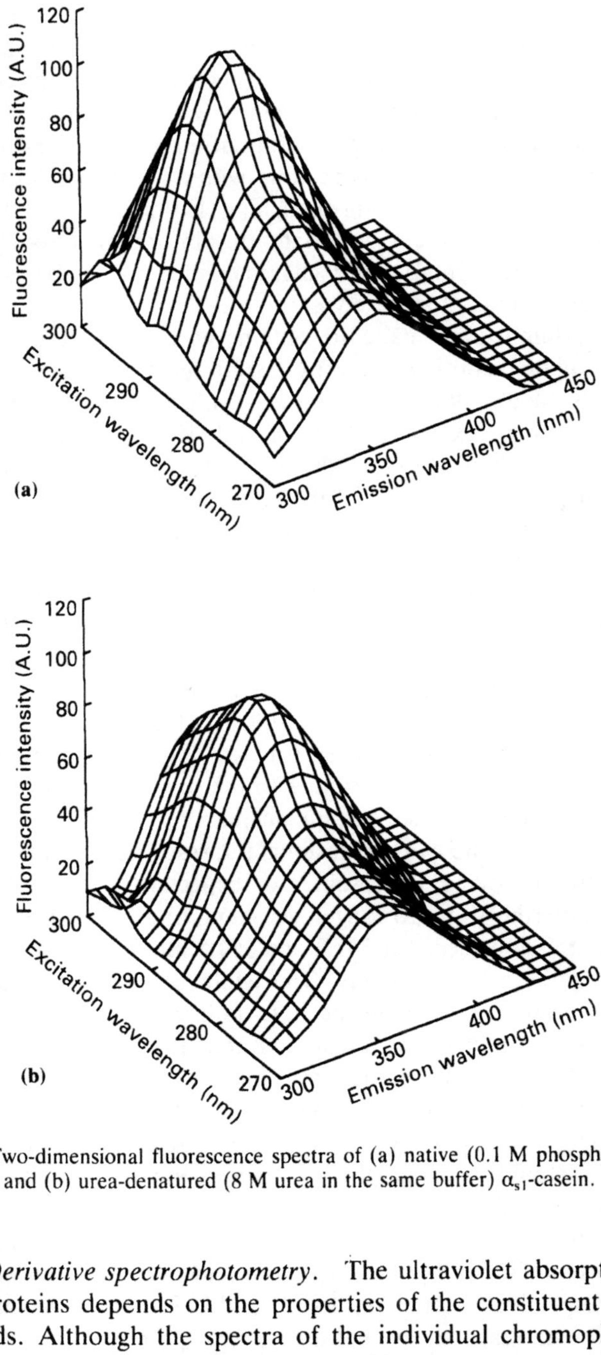

Figure 8.1 Two-dimensional fluorescence spectra of (a) native (0.1 M phosphate, pH 7.5) and (b) urea-denatured (8 M urea in the same buffer) α_{s1}-casein.

8.2.2.6 Derivative spectrophotometry. The ultraviolet absorption spectrum of proteins depends on the properties of the constituent aromatic amino acids. Although the spectra of the individual chromophores are different, it is difficult to resolve their contributions in the resulting broad absorption spectrum. Derivative spectroscopy, particularly of the second

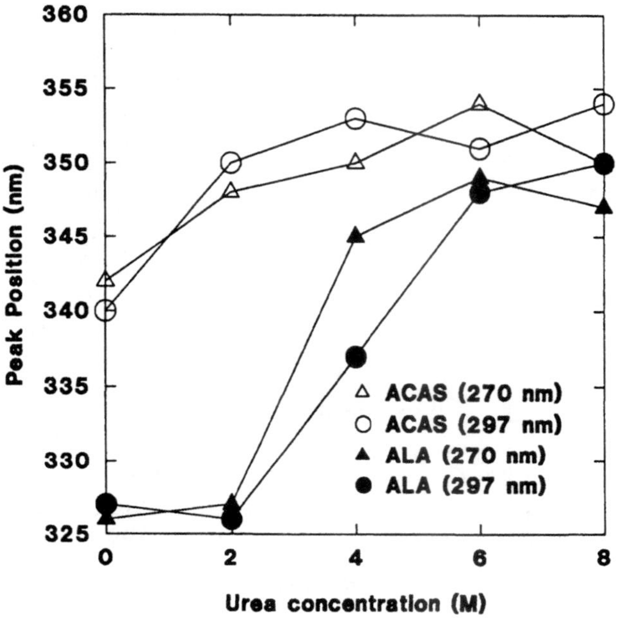

Figure 8.2 Effect of urea on the fluorescence peak position at two excitation wavelengths (270 and 297 nm) of α_{s1}-casein (ACAS) and α-lactalbumin (ALA).

order ($d^2A/d\lambda^2$) and fourth order ($d^4A/d\lambda^4$), has the ability to resolve overlapping bands in the original spectrum. Over the last decade the commercial availability of spectrophotometers with built-in derivative mode has spurred on application of derivative spectroscopy for analysis of protein absorption spectra. It has been used quantitatively to determine the contents of phenylalanine, tyrosine and tryptophan residues and to probe for changes in polarity of the microenvironment around the chromophores and thus evaluate the extent of exposure or buriedness of these aromatic amino acid residues (Ichikawa and Terada, 1979, 1981). Changes in the state of aromatic amino acid residues in soybean globulin proteins as a function of heating or sulphhydryl modification were monitored by using difference second-derivative absorption as well as spectrofluorimetric spectra of native and heated or modified proteins (Yamagishi *et al.*, 1981, 1982).

Fourth-derivative spectroscopy was used to investigate the characteristics of aromatic amino acid residues in protein (Padros *et al.*, 1982, 1984; Mozo-Villaria *et al.*, 1991). Fourth-derivative spectroscopy provides better resolution than the first or second derivatives with the same maximum wavelengths as for the original absorption spectrum (Padros *et al.*, 1984). The fourth-derivative spectra of native, SDS/heat-treated and urea-denatured *β*-lactoglobulin (Figure 8.3) demonstrate blue shifts due to a

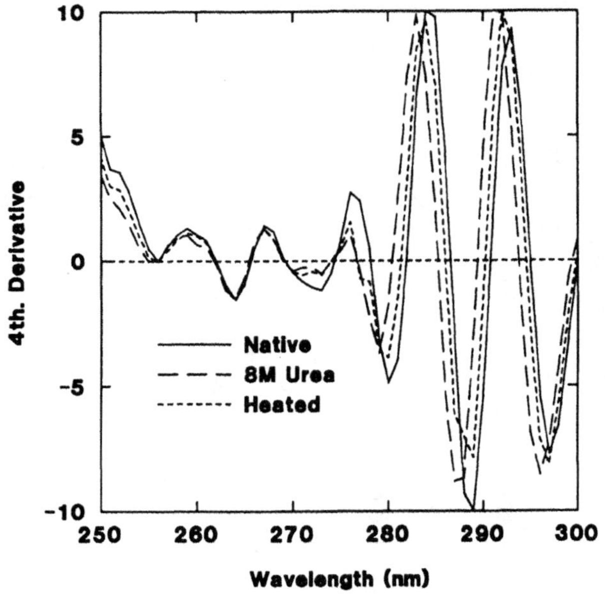

Figure 8.3 Fourth-derivative spectra of native (0.1 M phosphate buffer, pH 7.5), urea-denatured (8 M urea in the same buffer) and heated* β-lactoglobulin. *The protein in 5% SDS-phosphate buffer was heated for 10 min at 90°C and cooled for 5 min in an ice bath. In all cases protein concentration was 0.5 mg/ml.

transfer of buried aromatic residues from a non-polar to a polar environment. It should be noted that not only polarity change but also changes in the strength and type of hydrogen bonds in tyrosine residues affect the band shift.

8.2.2.7 Probe spectrofluorometry. Compounds for which quantum yields of fluorescence and wavelength of maximum emission depend on the polarity of their environment have been used as probes to assess the hydrophobic or non-polar nature of proteins. The most popular types include the anionic probes of the aromatic sulfonic acid class, such as the amphiphilic 1-anilinonaphthalene-8-sulfonate (ANS), or its dimeric form bis-ANS. These probes have high quantum yields of fluorescence in organic solvents but not in water; they thus fluoresce when bound to membranes or relatively hydrophobic cavities in many proteins (Weber and Young, 1964; Stryer, 1965). However, the direct interpretation of ANS fluorescence with hydrophobicity should be made with great caution, as enhancement of fluorescence and blue shift of the emission maximum have also been observed even in strong aqueous $MgCl_2$ solution (Penzer, 1972). It has been suggested that solvents or environments that are not necessarily non-polar but that favour the rigid, planar configuration of the ANS molecule may influence fluorescence (Ainsworth and Flanagan, 1969).

Another group of anionic fluorescence probes is of the fatty acid analog type, including *cis*-parinaric acid (CPA) and *trans*-parinaric acid, which has been used as a probe for proteins and biological membranes in particular (Sklar *et al.*, 1975, 1976, 1977). The parinaric acids are among the few non-aromatic fluorophores known; their similarity to native fatty acids, non-fluorescence in water and good Stokes' shift characteristics are among the advantages of their use to probe for hydrophobic regions that may be important in protein–lipid interactions in food systems. Good correlations were obtained between the relative hydrophobicity values of proteins determined by CPA fluorescence, and properties related to protein–lipid interactions such as interfacial tension and emulsifying activity (Kato and Nakai, 1980).

Limitations in using anionic probes such as ANS and CPA to determine protein hydrophobicity include the possibility that electrostatic as well as hydrophobic interactions may contribute to the interaction with probes. The use of neutral or uncharged probes may circumvent this problem. Diphenyl-hexatriene (DPH), which labels the interior of membranes, is the most commonly used neutral lipophilic fluorescent probe. The primary application of DPH has been to estimate membrane fluidity and the effects of various drugs or treatments on the membrane structure. Fluorescence of DPH outside membranes or non-polar solvents is low. Unfortunately, the non-polar nature of DPH restricts its solubility in aqueous systems and thus limits its uses as a probe for protein solutions. An approach has been proposed to use DPH by first dissolving it in a corn oil–heptane mixture; after removal of the heptane by evaporation, the resultant DPH in corn oil is then allowed to interact with aqueous protein solutions. The emitted fluorescence may be measured as an indication of binding of DPH-in-oil by the protein molecules and used as an index of relative hydrophobicity (Tsutsui *et al.*, 1986). Recently, the authors have also used DPH in tetrahydrofuran directly for probing protein hydrophobicity by adding an aliquot of this stock DPH solution to the aqueous protein solution.

Cardamone and Puri (1992) stated that ANS binding and the resultant K_a measured by Scatchard- or Klotz-plots and, to a lesser extent, ANS quantum yield, may be used as a measure of the relative surface hydrophobicity of proteins. The ANS results differ from those based on RP-HPLC or using the method of average hydrophobicity of Bigelow (1967), where no allowances are made for the tertiary structure of a given protein. Titration of protein solutions with increasing concentrations of the fluorescence probe can provide information on both the number and the affinity of binding sites. This may be useful in knowing whether high values of fluorescence result from the presence of many binding sites of only moderate hydrophobic character, or from the existence of a high affinity site with considerable hydrophobic character. Examples of this type of application of fluorescence probes include the determination of F_{max} and K_D of ANS–milk protein complex (Bonomi *et al.*, 1988), quantitation of binding sites and K_D of ANS–myosin (Cheung and Morales, 1969) and CPA–myosin (Borejdo, 1983) complexes.

The application of fluorescence probes to study the proximity of the probe binding site(s) and aromatic groups has also been investigated by measuring the efficiency of energy transfer. Transfer of excitation energy from aromatic chromophores to bound probe at an adjacent hydrophobic site causes quenching of the aromatic side chain's intrinsic fluorescence. This transfer efficiency has been measured as an index of surface hydrophobicity with CPA (Kato and Nakai, 1980) and ANS (Cheung and Morales, 1969).

A few limitations of using probe fluorometry should be noted. Changes in relative fluorescence intensity of these probes have frequently been used to monitor surface hydrophobicity of proteins as a function of varying conditions such as pH, temperature or in the presence of denaturing agents. The effects of these conditions on the probes should be considered. The quantum yield of ANS fluorescence has been reported to be unaffected by pH in the range of pH 2 to 8. In the case of *cis*-parinaric acid, however, interpretation of results obtained at pH values lower than 5 should be made with caution due to poor solubility of the probe in the undissociated acid form.

Some researchers have reported changes in surface hydrophobicity resulting from proteolytic modification. Although changes in fluorescence intensity at the initial stages of proteolysis may be interpreted as changes in the exposure of the hydrophobic binding sites, a sharp decrease in fluorescence intensity resulting from extensive proteolysis cannot be interpreted as 'low surface hydrophobicity' since these probes may have low quantum yields even when bound to small, hydrophobic peptides.

8.2.2.8 Other methods. Most of the spectroscopic methods outlined above provide information on the contribution to hydrophobic interactions from aromatic amino acids; there is a great dearth of methods for aliphatic amino acids. A number of other spectroscopic methods provide useful information on protein structure and environment of constituent amino acid residues in general. These include circular dichroism (CD), optical rotatory dispersion (ORD), nuclear magnetic resonance (NMR), infrared spectroscopy (IR), and laser Raman spectroscopy (LRS). Each of these techniques has its particular advantages and limitations; more specialized volumes on the theory and applications of the individual techniques are available. Unfortunately, with the present state of knowledge, there are no general empirical rules to explain observed spectra in terms of the specific secondary or tertiary structure, or electrostatic or hydrophobic interactions. The interpretation of spectra must be made individually for each protein, based on data obtained from several methods coupled with known sequence and, whenever possible, the crystallographic data of the protein.

8.2.3 Comparison of relative hydrophobicity values of proteins

Compilation of data measured by different methods is listed in Table 8.1. Among these methods some pairs show relatively good correlations while others do not. Despite the limitation that the sample proteins used for the analysis were not always from the same sources, the hydrophobicity values measured by different methods have been compared (Nakai and Li-Chan, 1988). While there is good correlation between the triglyceride binding capacity (TBC) and CPA hydrophobicity ($r = 0.394$, $P < 0.01$, $n = 56$), the correlation between TBC and ANS is not significant ($r = 0.268$, $P > 0.05$, $n = 35$). There is a significant correlation between heptane-binding capacity and TBC hydrophobicity ($r = 0.587$, $P < 0.01$, $n = 22$). Also, a significant correlation is found between CPA and HIC hydrophobicities ($r = 0.452$, $P < 0.01$, $n = 55$). HIC data include the elution volume when eluted with increasing concentrations of propylene glycol (Keshavarz and Nakai, 1979) and not with decreasing concentrations of salts. The TBC method appears to be reasonable for selection as a standard method; CPA or ANS hydrophobicity methods can be alternatives to the TBC method because of convenience and simplicity in their analytical procedures. The use of CPA below pH 5.0 is not, however, recommended due to drastic change in the quantum yield of CPA when protonated.

An NMR cross-saturation experiment was carried out (Arteaga, 1994). When a weak radiofrequency field is applied to the sample, resonance from the residues would be less saturated than signals from buried or rigid residues. The built-in decoupler of the spectrometer usually provides the resaturation pulse of desired length and frequency. Thus, the ratio of the integrated NMR areas of a normal NMR spectrum and the integrated areas obtained after a cross-saturation pulse may be used to estimate exposure. Correlation found between the changes in NMR integrated areas and CPA hydrophobicity of eight proteins was more highly significant than that with ANS hydrophobicity.

Unfortunately at present, there is no consensus on a standard, absolute method for hydrophobicity measurement. In comparing these values, differences between average, total and surface-exposed hydrophobicity should be kept in mind. In addition, variability in published data may arise from different sources and purity of protein, and different conditions during measurement (e.g. pH, ionic strength, buffer salts). Until a more theoretically acceptable, universal method is established, simple methods which give a measure of the hydrophobicity that is actually available for

Table 8.1 Relative hydrophobicity values of some proteins measured by different methods[a]. From Li-Chan (1991) with permission

Protein	$H\phi_{av}$	Fluorescence probes			DPH probe TG binding	Chromatography		Binding			Partition $\Delta \log K$
		S_0ANS	S_0CPA	S_eCPA		RPC	HIC	heptane	SDS	TG	
Albumin, bovine	1120	1000	100	100	100	100	100	100	100	100	100
Albumin, chicken	1110	7	2	96	15	0	9–86	2	14	3	1–25
Casein, α-	1200	57[b]	6–30	/	150	400	/	/	/	160	/
Casein, β-	1320	60[b]	50	107	/	315	/	/	/	/	/
Casein, κ-	1210	83	13–93	89	/	340	/	/	30	/	/
Chymotrypsin, α-	1030	0	3	/	41	/	57–139	9	/	3	9–33
Globulin, soy 7S	1090	47	9–77	/	50	165	/	/	21	/	/
Globulin, soy 11S	950	27	2–17	/	25	150	/	/	/	/	/
Lactalbumin, α-	1150	33	9–54	/	98	225	10	/	/	/	/
Lactoglobulin, β-	1230	13	54–146	80	49	0	67–102	100	62	130	41–78
Lysozyme	970	0.7	7	/	/	/	42–113	0	19	1	0
Ovomucoid, chicken	920	/	0	22	/	/	/	24	18	108	/
Pepsin	1063	0.7	2	57	32	/	92	/	/	95	/
Ribonuclease A	870	/	1	24	/	/	4–71	17	18	2	/
Transferrin, chicken	1080	/	4	108	/	/	16–31	29	20	118	9–20
Trypsin	940	8	3	34	/	/	/	/	13	/	9–27
Trypsin inhibitor	1040	0	0	/	1	0	/	/	/	/	/

[a]The majority of the data are expressed relative to '100' for bovine albumin to facilitate comparison of values. Where varying values were reported by investigators using essentially the same method, a range of values is presented.

[b]530 and 800 for α- and β-casein, respectively, at the isoelectric pH.

Abbreviations: $H\phi_{av}$ = average hydrophobicity calculated by the method of Bigelow (1967); S_0 = initial slope of relative fluorescence intensity versus protein concentration plot, using native proteins; $S_e = S_0$ measured for protein solutions after treatment with 1.5% SDS at 100°C for 10 min; ANS = 1-anilinonaphthalene-8-sulfonic acid; CPA = *cis*-parinaric acid; DPH = 1,6-diphenyl-1,3,5-hexatriene; TG = triglyceride: RPC = reverse-phase chromatography; HIC = hydrophobic-interaction chromatography; SDS = sodium dodecyl sulfate; / = no data available.

Table 8.2 Hydrophobicity values of some food proteins measured by the fluorescence probe method. From Li-Chan (1991) with permission

Protein[a]	ANS S_0	CPA S_0
Albumin, bovine serum	1600	2750
Canola isolate	110	370
Casein, bovine	150	400
Egg albumen	40	135
Gelatin	0	0
Lactalbumin	145	1700
Lactalbumin, α-	90	280
Lactoglobulin, β-	40	7000
Lysozyme, hen	1	15
Muscle, chicken breast	75	350
Myosin, chicken breast	100	/
Ovalbumin, hen	10	40
Pea protein isolate	280	825
Soy protein isolate	250	1000
Sunflower protein isolate	155	910
Whey protein concentrate	70	3600
Zein	410	390

[a]Measurements for S_0 determination were carried out for protein solutions in 0.01 M sodium phosphate buffer, pH 7.0–7.4; 0.6 M NaCl and 0.3 M NaCl were included in the buffer for salt extracts of chicken breast muscle and for isolated chicken breast myosin, respectively. Abbreviations are as noted in the footnote to Table 8.1.

participation in functionality may be the methods of choice. An example is given in Table 8.2, which lists typical values obtained by measurement with aliphatic and aromatic probes for some food proteins, including complex systems composed of several molecular species. As discussed in section 8.2.2.2, it has been suggested that differences may exist between hydrophobic interactions caused by aliphatic and aromatic side chains (Hayakawa and Nakai, 1985; Burley and Petsko, 1988).

8.2.4 Prospect of measurement of protein surface hydrophobicity: exposure of hydrophobic side chains

In the folded form of proteins, non-polar residues tend to be buried in the interior with polar residues at or near the surface. However, many non-polar side chains are in fact in contact with water at the surface. Richards (1979) defined accessible surface area geometrically as the extent to which the surface of a protein is accessible to the surrounding solvent. Holbrook *et al.* (1990) also reported the prediction of surface exposure of amino acids from protein sequence using an algorithm proposed by Kabash and Sander (1983). However, it is not likely that the geometrically defined surface area is the surface area actively involved in the function of protein molecules, especially when unfolding of the molecules may take place at the surface or interface as in the case of foaming or emulsification.

When determining the surface hydrophobicity based on hydrophobic side chains exposed on the surface of three-dimensional proteins of known structure, the distribution and balance of hydrophilic and hydrophobic side chains on the surface of protein molecules should also be taken into consideration. No methods capable of measuring such a balance have been established, with the possible exception of NMR as mentioned in section 8.2.3. Such a method is urgently needed in order accurately to define protein surface hydrophobicity, otherwise full understanding of protein functions will never be achieved. The cross-saturation NMR method is advantageous, as there is no interference by contaminating lipids such as in the case of hydrophobic probe methods.

Recently, Mant *et al.* (1989) were successful in applying the retention parameters derived from small peptides to the correlation of the overall hydrophobicity of proteins, with their retention time behaviour during RP-HPLC when HPLC was run under the conditions of complete denaturation of the proteins. It was recognized that polypeptide chain length had to be taken into account when attempting to equate retention time behaviour of proteins and peptides larger than 15–20 residues with their overall hydrophobicity. Complete denaturation of a protein is required for full expression of its overall hydrophobicity. The hydrophobicity of a protein in its native conformation is considerably different from that of its unfolded state, since the hydrophobic side chains tend to be buried during the folding process.

Lau *et al.* (1984) demonstrated that the primary cause of protein denaturation during RP-HPLC is the hydrophobicity of the stationary phase, which disrupts the hydrophobic interactions stabilizing the native conformation. These workers showed that even an ultra-short chain length (C_3) sorbent with low carbon loading was able to denature very stable synthetic model proteins. These model proteins consisted of two-stranded α-helical coiled coils in which the quaternary structure was stabilized by hydrophobic interactions between the two α-helices. These proteins are probably the most stable proteins yet reported, as indicated by temperature and denaturation studies in 0.1% aqueous trifluoroacetic acid, which is a starting solvent for the chromatography. Similar correlation of protein retention time with polypeptide chain length and hydrophobicity was obtained for each of three different columns used with hydrocarbon chain lengths of C_4, C_8 and C_{18}.

It is uncertain if the surface hydrophobicity of native proteins should closely correlate with functional properties of the proteins. This is because denaturation of the protein at the interface is well documented, such as the cases of emulsification and foaming. Furthermore, in most of the functional properties of food proteins, surface hydrophobicity alone cannot explain their mechanism. As discussed in section 8.1.3, the distribution and periodicity of appearance of hydrophobic sites, or the

Table 8.3 Classification of analytical methods for protein surface hydrophobicity

Method	Serum albumin		Ovalbumin	Lysozyme	RNAse	Chymo-trypsin	Chymotryp-sinogen	Cyto-chrome C	Reference
	Bovine	Human							
Group 1									
ANS, S_0	1000		7	1		0			Hayakawa and Nakai, 1985; Tsutsui *et al.*, 1986
CPA, S_0	100		2	7	1	3			Kato and Nakai, 1980; Hayakawa and Nakai, 1985; Tsutsui *et al.*, 1986
Binding									
heptane	100		2	0	17	8	7		Mohammadzadeh *et al.*, 1969
TG	100	115	3	1	2	3	2	8	Smith *et al.*, 1983
Partition	100	100	1–25	0		9–33		1	Keshavarz and Nakai, 1979; Kato and Nakai, 1980
Group 2									
HIC	100		86	110	59	139		0	Gohcen and Engelhorn, 1984
	100		24–80	40–110	4–70	80–120	30–120	3–40	Summary data
CPA, S_e	100		96		24				Townsend and Nakai, 1983
RP-HPLC	80–100		76–160	28–120	8–20				Cardamone and Puri, 1992
Contact angle		145		150					van Oss, 1992
Intermediate									
Binding, SDS	100		14	19	18				Kato *et al.*, 1984
ANS, K_a	82		76	8	1				Cardamone and Puri, 1992
DPH	92		14			38			Tsutsui *et al.*, 1986
Interfacial tension									
	6.0		6.6	9.3		22.1			Keshavarz and Nakai, 1979

location and distance between hydrophobic and hydrophilic sites may be important parameters which are not represented by the simple overall value of surface hydrophobicity. It is worth noting that the relationships of surface hydrophobicity with individual functions of proteins may not be linear, as in the case of HLB values of low molecular food emulsifiers, with an optimum HLB value existing for each combination of protein and lipid. The QSAR approach may therefore be recommended as discussed in the review article by the authors (Nakai and Li-Chan, 1993).

When different methods for assessing the surface hydrophobicity of proteins are compared, they can be classified into two groups (Table 8.3). Group 1 methods yield a low value for lysozyme, less than 10% of the hydrophobicity value of BSA. This group also demonstrates a trend to yield low values (less than 10%) for ovalbumin, chymotrypsin (also chymotrypsinogen A) and RNAase. Group 2 methods yield high values for lysozyme (over 80%) and ovalbumin (over 30%). S_0 (using ANS and CPA; as defined in Table 8.1), partition, and binding with heptane and triglyceride (TG) belong to group 1, while HIC and contact angle belong to group 2. HIC yielded values of 60–120% and 40–70% of that of BSA for chymotrypsin and RNAase. Binding with SDS and TG using DPH, and K_a using ANS are intermediate. It is interesting to note that the interfacial pressure between corn oil and protein solution (Keshavarz and Nakai, 1979) correlates well with the values of group 2, with the exception of chymotrypsin, but not with those of group 1.

8.3 Experimental protocols

Of the approaches for measuring the surface hydrophobicity of proteins already discussed, intrinsic fluorescence and derivative spectrophotometry are not included here, since they measure only aromatic side chains and not the overall hydrophobicity including that contributed by aliphatic hydrophobic side chains. However, it should be remembered that the extent of exposure of aromatic side chains could be a good index of the surface hydrophobicity. With regard to the binding methods, hydrocarbon binding capacity and triglyceride binding capacity quantitated using GLC, which were described in a previous monograph (Nakai and Li-Chan, 1988), have also not been included in this section. The main reason for this omission is the complexity of procedures and the high costs of gas chromatography. Although HPLC is also costly, it is included here as it is more likely to be found in protein laboratories than is GLC.

8.3.1 Probe spectrofluorometry

Hydrophobicity is determined using hydrophobic fluorescence probes, such as 1-anilinonaphthalene-8-sulfonate (ANS) or *cis*-parinarate (CPA).

The magnesium salt of ANS is prepared from the technical grade sodium salt according to the method of Weber and Young (1964). CPA can be purchased from Molecular Probes (Junction City, Oregon). The purity of CPA is critical in obtaining reproducible results; the crystals should be completely soluble in ethanol. Measurements are performed essentially according to the method of Kato and Nakai (1980). Each protein sample (2 ml) is serially diluted with 0.01 M phosphate buffer (pH 5.5 to 7.4, containing 0 to 0.6 M NaCl depending on the purpose of analysis; for most routine tests, pH 7.0 without NaCl has been used) to obtain a protein concentration typically from 0.005 to 0.030%. Following this, 10 μl ANS (8.0 mM in 0.1 M phosphate buffer, pH 7) or CPA (3.6 mM in absolute ethanol containing equimolar butylated hydroxytoluene) solution are added. Relative fluorescence intensity (RFI) is measured with a spectrofluorometer at excitation and emission wavelengths (λ_{ex}, λ_{em}) of 390 and 470 nm, and 325 and 420 nm for ANS and CPA, respectively. The RFI reading is standardized by adjusting the reading of the fluorometer for ANS in methanol, and for CPA in decane. The net RFI for each protein concentration is computed by subtracting the RFI measured for each solution without probe from that with probe. The initial slope (S_0) of the net RFI versus protein concentration (%) plot is calculated by linear regression analysis (or the slope at the origin of quadratic curve fitting for more precise measurement) and used as an index of the protein hydrophobicity.

Plotting the net RFI versus protein concentration prior to computation of S_0 is essential for detecting non-linearity or outliers. The present authors usually work with five or six protein concentrations, each one in triplicate. The coefficient of variation within triplicates is normally below 5%. If an outlier is found and cannot be measured again, this point should not be included in regression analysis.

For most proteins, the RFI increases linearly as the protein concentration is increased (Figure 8.4). Unfolding, for example by urea denaturation, usually significantly increases the slope of the RFI versus protein concentration plot (Figure 8.5). For some proteins, the titration curve reaches a plateau at protein concentrations higher than 0.02% (Figure 8.6). In this case, two methods for calculating S_0 are recommended. If more than four points are available in the low concentration range with pseudo-linearity, linear regression can be applied. Otherwise, quadratic regression is recommended.

Cardamone and Puri (1992) used the association constant of ANS binding with proteins as an index of the surface hydrophobicity. A 50 μl portion of a 0.3% protein in 0.1 M Tris/HCl buffer, pH 8.0 containing 0.2 M NaCl, was added to one set of 5 ml tubes containing ANS ranging from 0 to 500 μM; 50 μl of buffer were added to another identical set of tubes. After vortexing the tubes and equilibrating at 25°C for 2 h,

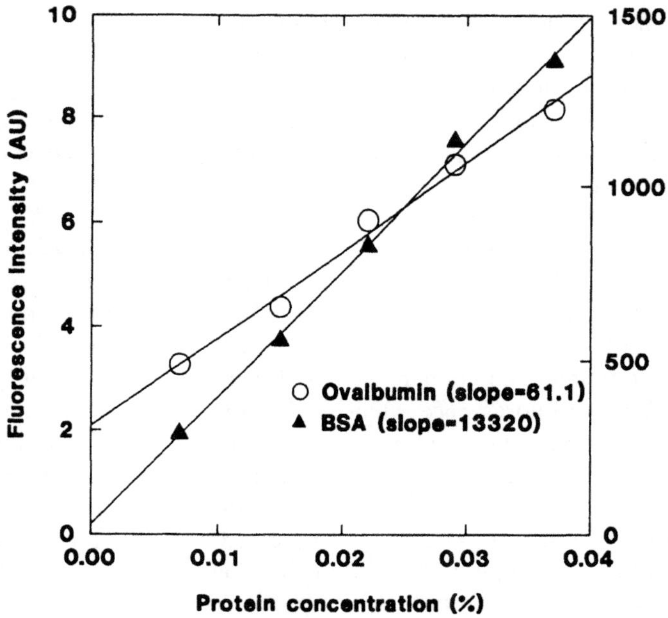

Figure 8.4 ANS hydrophobicity of ovalbumin and bovine serum albumin (BSA). Left and right y-axes are for ovalbumin and BSA, respectively. Solid lines are the least squares regression lines.

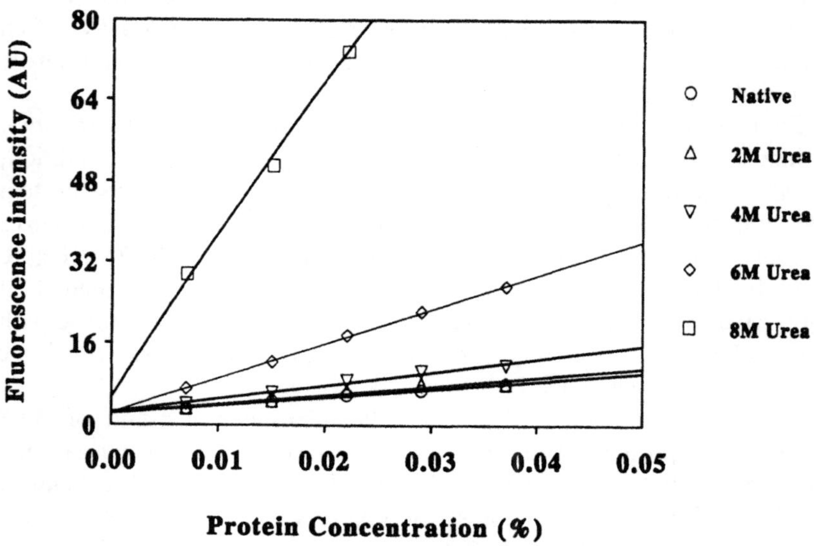

Figure 8.5 Effect of different urea concentrations on the ANS hydrophobicity of ovalbumin.

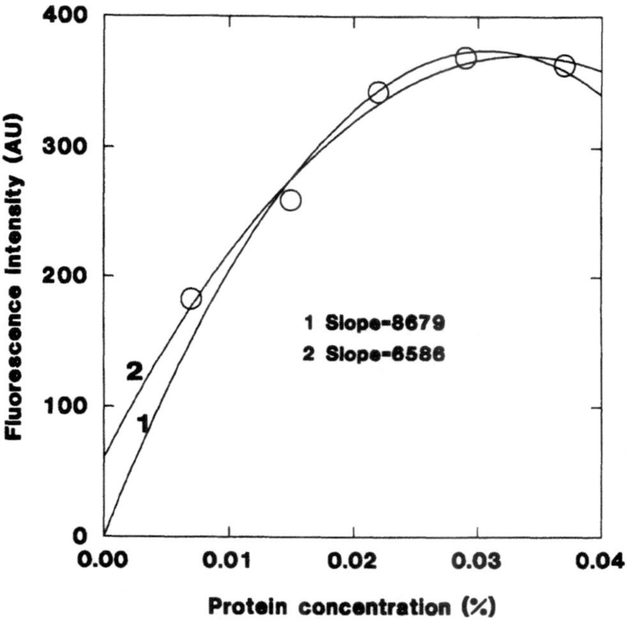

Figure 8.6 CPA hydrophobicity of bovine serum albumin. A quadratic model was fitted to the RFI versus protein concentration plot. Curve 1 was obtained by including the data point for blank (buffer + probe) in the regression analysis, while in curve 2 this point was not included. Curve 2 gives a larger y-intercept and less steep slope.

fluorescence intensity was measured using a λ_{ex} of 350 nm. Fluorescence emission was monitored at the wavelength of maximum intensity, which in each case was evaluated from a variable wavelength scan of the protein/ANS mixture. The fluorescence measurements were corrected for primary absorbance effects usually caused by the ligand using the procedure of Birdsall *et al.* (1983). The association constant (K_a) was computed using the graphical representation of Klotz and Hunston (1971), $1/v$ versus 1/added ANS, where v is mol ANS bound per mol protein, which is calculated from fluorescence intensity.

The standard protocol is usually applicable to most proteins; however, in some cases it may be necessary to change the range of protein concentration as well as the amount of probe to be added. For proteins which bind greater quantities of probe and have steeper titration curves, preliminary experiments are recommended to ensure that the amount of probe is adequate for the protein. For such proteins, protein concentrations lower than 0.01% may have to be used to avoid the self-quenching phenomenon which may arise at high fluorescence intensity.

8.3.2 *Hydrophobic-interaction chromatography*

Proteins are dissolved in solvent A containing 1.7 M ammonium sulphate in 0.1 M sodium phosphate buffer (38 mM NaH_2PO_4 and 62 mM Na_2HPO_4) at pH 7.0 and applied to a TSK-Phenyl 5-PW column (Goheen and Engelhorn, 1984). Elution is carried out using a linear gradient from 100% A to 100% B (0.1 M sodium phosphate buffer, pH 7.0).

In the method of Szepesy and Rippel (1992), the gradient elution system was 2 M salt to zero salt in 0.02 M Tris/HCl buffer pH adjusted with 0.1 M HCl. Two other HIC columns, Synchropak-Propyl (Synchrom, Linden IN) and Spherogel CAA-HIC (polyether) (Beckman Instr. San Ramon, CA), were used in this method.

8.3.3 *Contact angle determination*

30 ml of a 1 or 2% protein solution are ultrafiltered on a 45 mm anisotropic cellulose acetate membrane without stirring, until the flow rate slows down considerably (van Oss *et al.*, 1981). After removal from the ultrafilter, the membrane with the built-up hydrated protein layer is air-dried. At various intervals 10 μl drops of isotonic saline are applied and contact angles are measured, using either a small telescope with cross-hairs attached to a goniometer, or a magnified photograph. The advancing contact angle should be utilized (Penn and Miller, 1980; van Oss, 1989). Initially, when the protein layer is still wet, contact angles close to 0° are obtained, but within 60 min a plateau value (θ) is reached. Along with known value of γ_{LV} (surface tension of saline in this case), γ_{PV} that is the surface tension of protein, can be calculated.

8.3.4 *SDS binding capacity*

Sodium dodecylsulfate (SDS) is added to 10 ml of a 0.1% protein solution to a final concentration of 0.07 mM (Kato *et al.*, 1984). After standing for 30 min, SDS–protein mixtures are dialysed against 25 volumes of 0.02 M phosphate buffer, pH 6.0, for 24 h. The SDS contents of inner dialysates are determined by the method of Epton (1948). 10 ml of $CHCl_3$ are added to 0.5 ml of the inner dialysates and mixed in a test tube. 2.5 ml of a 0.0024% methylene blue solution are then added to the $CHCl_3$ layer. After mixing, the mixture is centrifuged at 2500 rpm. The absorbance of the SDS–methylene blue mixture in the lower layer is measured at 655 nm. The SDS binding capacity is expressed as micrograms of SDS bound to 500 μg of protein.

8.3.5 *Triglyceride binding capacity*

10 μg of 1,6-diphenyl-1,3,5-hexatriene (DPH) are dissolved in 1 ml of heptane solution of corn oil (10 mg/ml) in a 22 × 92 mm flat-bottom test tube. The heptane is evaporated by nitrogen flushing (Tsutsui *et al.*, 1986). After adding 10 ml protein solution (0.02 to 0.12%), the contents of the test tube are blended using an UltraTurrax (Janke and Kunkel, Staufen, Germany) at 12 800 rpm for 5 min while cooling in ice water. After centrifuging the emulsion at 27 000 *g* for 30 min, the bottom layer is diluted 20-fold with 0.02 M phosphate buffer pH 7.4, and the RFI of the solution is measured (λ_{ex} = 366 nm, λ_{em} = 450 nm). RFI readings are standardized using DPH in heptane. If the protein precipitates during centrifugation, about 4.5 ml of the bottom layer solution are withdrawn and the precipitate is transferred to another tube. The bottom layer solution separated is then added to the precipitate to bring it up to the 4.0 ml mark. After homogenizing the content of the test tube as above for 1 min, 0.4 ml aliquot is mixed with 0.6 ml of the bottom layer solution and then diluted 20 times with the phosphate buffer.

8.3.6 *Hydrophobic partition*

8.3.6.1 Materials. Dextran T500 (mol. wt. 460 000) and T70 (mol. wt. 70 000) can be obtained from Pharmacia Fine Chemicals, Sweden. Poly(ethylene glycol) (PEG) as Carbowax 6000 can be obtained from Union Carbide. PEG palmitate is synthesized as described by Shanbhag and Johansson (1974) with approximately 60% substitution of the total hydroxyl groups in PEG.

8.3.6.2 Two-phase system. A system composed of 8% dextran T70, 8% PEG (including ligand derivative), 100 mM K_2SO_4 and 2 mM potassium phosphate buffer pH 7.1 is used (Shanbhag and Axelsson, 1975). The two-phase systems are prepared by mixing stock solutions of 20% dextran and 40% PEG with salt solutions and protein solutions at room temperature. After mixing by gentle inversion the systems are centrifuged at 1500 *g* for 5 min to speed up the separation. The composition of the two phases obtained are as follows: upper phase (13% PEG, 0.3% dextran and 86.7% water) and lower phase (0.9% PEG, 22% dextran and 77.1% water). The partition coefficient of salts is very close to unity.

8.3.6.3 Partition coefficient. The concentration of the partitioned protein in each phase is determined photometrically. An aliquot is withdrawn from each phase and diluted with water. The absorbance of the diluted phase is measured using as a blank an identically diluted solution of the corresponding phase from a system containing no protein. The absorbance is measured at 280 nm for proteins. Partition coefficient Δlog K is computed from:

$$\Delta \log K = \log [C_u/C_l]_{I} - \log [C_u/C_l]_{II} \quad (8.5)$$

where C_u and C_l are the concentrations of the partitioned protein in the upper and lower phases, respectively, for a two-phase system, I and II with and without the ligand derivative of PEG.

References

Absolom, D.R., van Oss, C.J., Zingg, W. and Neuman, A.W. (1981) Determination of surface tensions of proteins. II. Surface tension of serum albumin, altered at the protein–air interface. *Biochim. Biophys. Acta*, **670**, 74–8.

Ainsworth, S. and Flanagan, M.T. (1969) Effects that the environment exerts on the spectroscopic properties of certain dyes that are bound by bovine serum albumin. *Biochim. Biophys. Acta*, **194**, 213–21.

Albertsson, P.-A. (1986) *Partition of Cell Particles and Macromolecules*, 3rd edn, Wiley Interscience, New York.

Arteaga, G.E. (1994) *Assessment of Protein Surface Hydrophobicity by Spectroscopic Methods and its Relation to Emulsifying Properties of Proteins*. Ph.D. Thesis, University of British Columbia.

Barford, B.A., Kumosinski, T.F., Parris, N. and White, A.E. (1988) Salt-binding effects in hydrophobic-interaction chromatography. *J. Chromatogr.*, **458**, 57–66.

Bigelow, C.C. (1967) On the average hydrophobicity of proteins and the relation between it and protein structure. *J. Theor. Biol.*, **16**, 187–211.

Birdsall, B., King, R.W., Wheeler, M.R. *et al.* (1983) Correction for light absorption in fluorescence studies of protein–ligand interactions. *Anal. Biochem.*,**132**, 353–61.

Bonomi, F., Iametti, S., Pagliarini, E. and Peri, C. (1988) A spectrofluorimetric approach to the estimation of the surface hydrophobicity modifications in milk proteins upon thermal treatment. *Milchwissenschaft*, **43**, 281–5.

Borejdo, J. (1983) Mapping of hydrophobic sites on the surface of myosin and its fragments. *Biochemistry*, **22**, 1182–7.

Burley, S.K. and Petsko, G.A. (1985) Aromatic–aromatic interaction: a mechanism of protein structure stabilization. *Science*, **229**, 23–8.

Burley, S.K. and Petsko, G.A. (1988) Weakly polar interaction in proteins. *Adv. Protein Chem.*, **39**, 125–89.

Burstein, E.A., Vedenkina, N.S. and Ivkova, M.N. (1973) Fluorescence and the location of tryptophan residues in protein molecules. *Photochem. Photobiol.*, **18**, 263–79.

Cardamone, M. and Puri, N.K. (1992) Spectrofluorimetric assessment of the surface hydrophobicity of proteins. *Biochem. J.*, **282**, 589–93.

Casal, H.L., Kohler, U. and Mantsch, H.H. (1988) Structural and conformational changes of β-lactoglobulin B: an infrared spectroscopic study of the effect of pH and temperature. *Biochim. Biophys. Acta*, **957**, 11–20.

Charton, M. (1986) Quantitative description of side chain effects on binding to protein. *Int. J. Peptide Protein Res.*, **28**, 201–7.

Charton, M. (1990) The quantitation of protein bioactivity. Phage T4 lysozymes substituted at residue 86. *Collect. Czech. Chem. Commun.*, **55**, 273–81.

Cheung, H.C. and Morales, M.F. (1969) Studies of myosin conformation by fluorescent techniques. *Biochemistry*, **8**, 2177–82.

Chothia, C. (1984) Principles that determine the structure of proteins. *Annu. Rev. Biochem.*, **53**, 537–72.

Cid, H., Bunstein, M., Canales, M. and Gazitua, F. (1992) Hydrophobicity and structural classes in proteins. *Protein Eng.*, **5**, 373–5.

Damodaran, S. and Kinsella, J.E. (1982) Effects of ions on protein conformation and functionality, in *Food Protein Deterioration: Mechanisms and Functionality* (ed. J.P. Cherry), American Chemical Society, Washington, DC, pp. 327–57.

Diamond, A.D., Yu, T. and Hsu, J.T. (1990) Peptide and protein partitioning in aqueous two-phase systems, in *Protein Purification. From Molecular Mechanisms to Large-Scale Processes* (eds M.R. Ladisch, R.C. Willson, C.C. Painton and S.E. Builder), American Chemical Society, Washington, DC, Chapt. 4.

Edsall, J.T. and McKenzie, H.A. (1983) Water and protein. II. The location and dynamics of water in protein systems and its relation to their stability and properties. *Adv. Biophys.*, **16**, 53–183.

Eftink, M.R., Zajicek, J.L. and Ghiron, C.A. (1977) A hydrophobic quencher of protein fluorescence: 2,2,2-trichloroethanol. *Biochim. Biophys. Acta*, **491**, 473–81.

Epton, S.R. (1948) New method for the rapid titration analysis of sodium alkyl sulphates and related compounds. *Trans. Faraday Soc.*, **44**, 226–30.

Freifelder, D. (1976) *Physical Biochemistry. Application to Biochemistry and Molecular Biology*, W.H. Freeman & Co., San Francisco, pp. 410–21.

Fujita, T., Iwasa, I. and Hansch, C. (1964) A new substituent constant, π, derived from partition coefficients. *J. Amer. Chem. Soc.*, **86**, 5175–80.

Ganong, B.R. and Delmore, J.P. (1991) Phase separation temperature of Triton X-114 and Triton X-45: application to protein separation. *Anal. Biochem.*, **193**, 35–7.

Goheen, S.C. and Engelhorn, S.C. (1984) Hydrophobic interaction high-performance liquid chromatography of proteins. *J. Chromatogr.*, **317**, 55–65.

Griffin, M.C.A. and Roberts, G.C.K. (1985) A ^{1}H-n.m.r. study of casein micelles. *Biochem. J.*, **228**, 273–6.

Hammett, L.P. (1970) *Physical Organic Chemistry*, 2nd edn, McGraw-Hill, New York.

Hansch, C. (1978) Recent advances in biochemical QSAR, in *Correlation Analysis in Chemistry* (eds N.B. Chapman and J. Shorter), Plenum Press, New York, pp. 397–438.

Hansch, C. and Klein, T.E. (1991) Quantitative structure–activity relationships and molecular graphics in evaluation of enzyme–ligand interactions. *Methods Enzymol.*, **202**, 512–43.

Hansch, C. and Leo, A. (1979) *Substituent Constants for Correlation Analysis in Chemistry and Biology*, Wiley Interscience, New York.

Hayakawa, S. and Nakai, S. (1985) Relationships of hydrophobicity and net charge to the solubility of milk and soy proteins. *J. Food Sci.*, **50**, 486–91.

Hofstee, B.H. and Otillio, N.F. (1978) Modifying factors in hydrophobic protein binding by substituted agaroses. *J. Chromatogr.*, **161**, 153–63.

Holbrook, S.R., Muskal, S.M. and Kim, S.-H. (1990) Predicting surface exposure of amino acids from protein sequence. *Protein Eng.*, **3**, 659–65.

Ichikawa, T. and Terada, H. (1979) Estimation of state and amount of phenylalanine residues in proteins by second derivative spectrophotometry. *Biochim. Biophys. Acta*, **580**, 120–8.

Ichikawa, T. and Terada, H. (1981) Determination of phenylalanine, tryptophan and tyrosine in mixture of amino acids by second derivative spectrophotometry. *Chem. Pharm. Bulletin*, **29**, 438–44.

Kabasch, W. and Sander, C. (1983) Dictionary of protein secondary structure: pattern recognition of hydrogen-bonded and geometrical features. *Biopolymers*, **22**, 2577–637.

Kato, A. and Nakai, S. (1980) Hydrophobicity determined by fluorescence probe method and its correlation with surface properties of proteins. *Biochim. Biophys. Acta*, **624**, 13–20.

Kato, A., Matsuda, T., Matsudomi, N. and Kobayashi, K. (1984) Determination of protein hydrophobicity using a sodium dodecyl sulfate binding method. *J. Agric. Food Chem.*, **32**, 284–8.

Keshavarz, E. and Nakai, S. (1979) The relationship between hydrophobicity and interfacial tension of proteins. *Biochim. Biophys. Acta*, **576**, 269–79.

Klein, P., Jacquez, I.A. and Delis, C. (1986) Prediction of protein function by discriminant analysis. *Math. Biosci.*, **81**, 177–89.

Klotz, I.M. and Hunston, D.L. (1971) Properties of graphical representations of multiple classes of binding sites. *Biochemistry*, **10**, 3065–9.

Krystek, S.R., Jr. and Andersen, T.T. (1985) A program for hydropathy and antigenicity analysis of protein sequences. *Endocrinology*, **117**, 1118–24.

Kyte, J. and Doolittle, R.F. (1982) A simple method for displaying the hydropathic character of a protein. *J. Mol. Biol.*, **157**, 105–32.

Lau, S.Y.M., Taneja, A.K. and Hodges, R.S. (1984) Effects of high-performance liquid chromatographic solvents and hydrophobic matrices on the secondary and quaternary structure of a model protein. *J. Chromatogr.*, **317**, 129–40.

Lecomte, J.T.J., De Marco, A. and Linas, M. (1982) Analysis of the ^{1}H-NMR spectrum of crambin, a hydrophobic protein. *Biochim. Biophys. Acta*, **703**, 223–30.

Lesser, G.J. and Rose, G.D. (1990) Hydrophobicity of amino acid subgroups in proteins. *Proteins Struct. Function Genet.*, **8**, 6–13.

Li-Chan, E. (1991) Hydrophobicity in food protein systems, in *Encyclopedia of Food Science and Technology* (ed. Y.H. Hui), John Wiley & Sons, New York, pp. 1429–39.

Li-Chan, E. and Nakai, S. (1991) Raman spectroscopic study of thermally and/or dithiothreitol induced gelation of lysozyme. *J. Agric. Food Chem.*, **39**, 1238–45.

Lieske, B. and Konrad, G. (1994) A new approach to estimate surface hydrophobicity of proteins. *Milchwissenschaft*, **49**, 663–6.

Manavalan, P. and Ponnuswamy, P.K. (1978) Hydrophobic character of amino acid residues in globular proteins. *Nature*, **275**, 673–4.

Mant, C.T., Zhou, N.E. and Hodges, R.S. (1989) Correlation of protein retention times in reversed-phase chromatography with polypeptide chain length and hydrophobicity. *J. Chromatogr.*, **476**, 363–75.

Meirovitch, H., Rackowsky, S. and Scheraga, H.A. (1980) Empirical studies of hydrophobicity. I. Effect of protein size on the hydrophobic behavior of amino acids. *Macromolecules*, **13**, 1398–405.

Melander, W. and Horváth, C. (1977) Salt effects on hydrophobic interactions in precipitation and chromatography of proteins: an interpretation of the lyotropic series. *Arch. Biochem. Biophys.*, **183**, 200–15.

Melander, W.R., El Rassi, Z. and Horváth, C. (1989) Interplay of hydrophobic and electrostatic interactions in biopolymer chromatography. Effect of salts on the retention of proteins. *J. Chromatogr.*, **469**, 3–27.

Mohammadzade-K, A., Feeney, R.E. and Smith, L.M. (1969a) Hydrophobic binding of hydrocarbons by proteins. I. Relationship of hydrocarbon structure. *Biochim. Biophys. Acta*, **194**, 246–55.

Mohammadzadeh-K, A., Feeney, R.E. and Smith, L.M. (1969b) Hydrophobic binding of hydrocarbons by proteins. II. Relationship of protein structure. *Biochim. Biophys. Acta*, **194**, 256–64.

Mozhaev, V.V. and Martinek, K. (1984) Structure–stability relationships in proteins: new approaches to stabilizing enzymes. *Enzyme Microb. Technol.*, **6**, 50–9.

Mozo-Villaria, A., Morros, A. and Andreu, J.M. (1991) Thermal titrations in the structure of tubulin. *Eur. Biophys. J.*, **10**, 100–8.

Myers, C. (1988) Functional attributes of protein isolates, in *Characterization of Proteins* (ed. F. Felix), Humana Press, Clifton, NJ, pp. 491–549.

Nakai, S. and Li-Chan, E. (1988) *Hydrophobic Interactions in Food Systems*, CRC Press, Boca Raton, FL.

Nakai, S. and Li-Chan, E. (1993) Recent advances in structure and function of food proteins: QSAR approach. *CRC Rev. Food Sci. Nutr.*, **33**, 477–99.

Nakai, S., Li-Chan, E., Hirotsuka, M., Vazquez, M.C. and Arteaga, G.E. (1991) Quantitation of hydrophobicity for elucidating the structure–activity relationships of food proteins, in *Interactions of Food Protein* (eds N. Parris and R. Barford), American Chemical Society, Washington, DC, pp. 42–58.

Padros, E., Morros, A., Monosa, J. and Dunach, M. (1982) The state of tyrosine and phenylalanine residues in proteins analyzed by fourth-derivative spectrophotometry. *Eur. J. Biochem.*, **127**, 117–22.

Padros, E., Dunach, M., Morros, A., *et al.* (1984) Fourth-derivative spectrophotometry of proteins. *Trends Biochem. Sci.*, **9**, 508–10.

Penn, L.S. and Miller, B. (1980) Advancing, receding and 'equilibrium' contact angles. *J. Colloid Interfac. Sci.*, **77**, 574–6.

Penzer, G. (1972) 1-Anilinonaphthalene-8-sulfonate. The dependence of emission spectra on molecular formation studied by fluorescence and proton magnetic resonance. *Eur. J. Biochem.*, **25**, 218–28.

Pinaev, G., Tartakovsky, A., Shanbhag, V.P., *et al.* (1982) Hydrophobic surface properties of myosin in solution as studied by partition in aqueous two-phase systems: effects of ionic strength, pH and temperature. *Mol. Cell Biochem.*, **48**, 65–9.

Rekker, R.F. (1977) *The Hydrophobic Fragmental Constant*, Elsevier, New York.

Richards, F.M. (1979) Packing defects, cavities, volume fluctuations, and access to the interior of proteins, including some general comments on surface area and protein structure. *Carlsberg Res. Commun.*, **44**, 47–63.

Rose, G.D., Gierasch, L.M. and Smith, J.A. (1985) Turns in peptides and proteins. *Adv. Protein Chem.*, **37**, 1–109.

Segrest, J.P. and Feldmann, R.J. (1977) Amphipathic helixes and plasma lipoproteins: a computer study. *Biopolymers*, **16**, 2053–65.

Shanbhag, V.P. and Axelsson, C.-G. (1975) Hydrophobic interaction determined by partition in aqueous two-phase systems. Partition of proteins in systems containing fatty-acid esters of poly(ethylene glycol). *Eur. J. Biochem.*, **60**, 17–22.

Shanbhag, V.P. and Johansson, G. (1974) Specific extraction of human albumin by partition in aqueous biphasic systems containing poly(ethylene glycol) bound ligand. *Biochem. Biophys. Res. Commun.*, **61**, 1141–6.

Sklar, L.A., Hudson, B.S. and Simoni, R.D. (1975) Conjugated polyene fatty acids as membrane probes: preliminary characterization. *Proc. Nat. Acad. Sci.* (USA), **72**, 1649–53.

Sklar, L.A., Hudson, B.S. and Simoni, R.D. (1976) Conjugated polyene fatty acids as fluorescent membrane probes: model system studies. *J. Supramol. Structure*, **4**, 449–65.

Sklar, L.A., Hudson, B.S. and Simoni, R.D. (1977) Conjugated polyene fatty acids as fluorescent probes: binding to bovine serum albumin. *Biochemistry*, **16**, 5100–8.

Smith, L.M., Fantozzi, P. and Creveling, R.K. (1983) Study of triglyceride–protein interaction using a microemulsion–filtration method. *J. Amer. Oil Chem. Soc.*, **60**, 960–7.

Snyder, L.R. (1978) Classification of the solvent properties of common liquids. *J. Chromatogr. Sci.*, **16**, 223–34.

Stryer, L. (1965) The interaction of a naphthalene dye with apomyoglobin and apohemoglobin. A fluorescent probe of non-polar binding sites. *J. Mol. Biol.*, **13**, 482–95.

Stuper, A.J., Brugger, W.E. and Jurs, P.C. (eds) (1979) *Computer Assisted Studies of Chemical Structure and Biological Function*, John Wiley & Sons, New York.

Szepesy, L. and Rippel, G. (1992) Comparison and evaluation of HIC columns of different hydrophobicity. *Chromatographia*, **34**, 391–7.

Townsend, A.-A. and Nakai, S. (1983) Relationships between hydrophobicity and foaming characteristics of food proteins. *J. Food Sci.*, **48**, 588–94.

Tsutsui, T., Li-Chan, E. and Nakai, S. (1986) A simple fluorometric method for fat-binding capacity as an index of hydrophobicity of proteins. *J. Food Sci.*, **51**, 1268–72.

van Oss, C.J. (1989) Energetics of cell–cell and cell–biopolymer interactions. *Cell Biophys.*, **14**, 1–16.

van Oss, C.J. (1992) Hydrophilic and 'hydrophobic' interactions, in *Protein Interactions* (ed. H. Visser), VCH, New York, pp. 25–55.

van Oss, C.J., Absolom, D.R., Neumann, A.W. and Zingg, W. (1981) Determination of the surface tension of proteins. I. Surface tension of native proteins in aqueous media. *Biochim. Biophys. Acta*, **670**, 64–73.

Weber, G. and Young, L.B. (1964) Fragmentation of bovine serum albumin by pepsin. 1. The origin of the acid expansion of the albumin molecule. *J. Biol. Chem.*, **239**, 1415–23.

Wilson, K.J., Honneger, A., Stozel, R.P. and Hughes, G.J. (1981) The behavior of peptides on reverse-phase supports during high pressure liquid chromatography. *Biochem. J.*, **199**, 31–41.

Wolfenden, R., Andersson, L., Cullis, P.M. and Southgate, C.C.B. (1981) Affinities of amino acid side chains for solvent water. *Biochemistry*, **20**, 849–55.

Wu, S.-L., Figueroa, A. and Karger, B.L. (1986) Protein conformation effects in hydrophobic interaction chromatography. *J. Chromatogr.*, **371**, 3–27.

Yamagishi, T., Yamauchi, F. and Shibasaki, K. (1981) State of aromatic amino acid residues in soybean 11S globulin heated in the presence of *N*-ethylmaleimide by derivative spectrophotometry. *Agric. Biol. Chem.*, **45**, 459–67.

Yamagishi, Y., Ebina, F. and Yamauchi, F. (1982) Analysis of the state of aromatic amino acid residues in heated soybean 7S globulin by absorption derivative spectrophotometry and spectrofluorimetry. *Agric. Biol. Chem.*, **46**, 2441–8.

Index

Numbers in **bold** refer to figures and numbers in *italics* refer to tables

GPSR Compliance

The European Union's (EU) General Product Safety Regulation (GPSR) is a set of rules that requires consumer products to be safe and our obligations to ensure this.

If you have any concerns about our products, you can contact us on ProductSafety@springernature.com

In case Publisher is established outside the EU, the EU authorized representative is:

Springer Nature Customer Service Center GmbH
Europaplatz 3
69115 Heidelberg, Germany

Batch number: 10692380

Printed by Printforce, the Netherlands